Qualitative Research in Tourism

While qualitative approaches are beginning to be more commonly used and accepted in tourism, discussions of research methods have rarely moved beyond practical considerations. To date there has been limited attention given to the underlying philosophical and theoretical underpinnings that influence the research process. This book links the theory with research practice, to offer a more holistic account of how qualitative research can be used in tourism.

Qualitative Research in Tourism is the first book to focus solely upon this subject. It combines discussions of the philosophies underpinning qualitative research, with chapters written in a reflexive style that demonstrate the ways in which the techniques can be used. The book is based on a range of empirical tourism studies set in the context of theoretical discussion. It demonstrates the benefits of using a range of qualitative approaches to research tourism, and the text explores the ways in which a number of techniques, including participant observation, memory-work, biographical diaries and focus groups, have been adopted by researchers from a range of disciplinary backgrounds to undertake empirical research in tourism.

The book offers a range of case studies written by leading scholars from the United Kingdom, France, America, Australia, New Zealand and Indonesia. The book makes clear the ways in which these pieces of research have been informed by the authors' ontological, epistemological and methodological standpoint. *Qualitative Research in Tourism* will be indispensable to any final-year undergraduates, masters' and PhD students embarking on research in the field, and also academics with an interest in either tourism research or qualitative methodology.

Jenny Phillimore is a Lecturer and **Lisa Goodson** is a Research Fellow at the Centre for Urban and Regional Studies, The University of Birmingham.

Contemporary geographies of leisure, tourism and mobility

Series Editor: Michael Hall

Professor at the Department of Tourism, University of Otago, New Zealand

The aim of this series is to explore and communicate the intersections and relationships between leisure, tourism and human mobility within the social sciences.

It will incorporate both traditional and new perspectives on leisure and tourism from contemporary geography, e.g. notions of identity, representation and culture, while also providing for perspectives from cognate areas such as anthropology, cultural studies, gastronomy and food studies, marketing, policy studies and political economy, regional and urban planning, and sociology, within the development of an integrated field of leisure and tourism studies.

The series comprises two strands:

Contemporary Geographies of Leisure, Tourism and Mobility aims to address the needs of students and academics, and the titles will be published in hardback and paperback. Titles include:

The Moralisation of Tourism
Sun, sand . . . and saving the world?
Jim Butcher

The Ethics of Tourism Development
Mick Smith and Rosaleen Duffy

Tourism in the Caribbean
Trends, development and prospects
Edited by David Timothy Duval

Qualitative Research in Tourism
Ontologies, epistemologies and methodologies
Edited by Jenny Phillimore and Lisa Goodson

Routledge Studies in Contemporary Geographies of Leisure, Tourism and Mobility is a forum for innovative new research intended for research students and academics, and the titles will be available in hardback only. Titles include:

1. Living with Tourism
Negotiating identities in a Turkish village
Hazel Tucker

Qualitative Research in Tourism

Ontologies, epistemologies and methodologies

Edited by Jenny Phillimore and Lisa Goodson

Routledge
Taylor & Francis Group

LONDON AND NEW YORK

First published 2004
by Routledge
11 New Fetter Lane, London EC4P 4EE

Simultaneously published in the USA and Canada
by Routledge
29 West 35th Street, New York, NY 10001

Routledge is an imprint of the Taylor & Francis Group

Typeset in Times by Wearset Ltd, Boldon, Tyne and Wear
Printed and bound in Great Britain by TJ International, Padstow,
Cornwall

British Library Cataloguing in Publication Data
A catalogue record for this book is available from the British Library

Library of Congress Cataloging in Publication Data
Qualitative research in tourism : ontologies, epistemologies and
methodologies / edited by Jenny Phillimore and Lisa Goodson.
        p. cm. — (Contemporary geographies of leisure, tourism and
mobility)
Includes bibliographical references and index.
    1. Tourism—Research—Methodology. 2. Qualitative
research—Methodology. I. Phillimore, Jenny. II. Goodson, Lisa,
1972– III. Series: Routledge/contemporary geographies of leisure,
tourism, and mobility.

  G155.A1Q34 2004
  910'.72—dc22

                                                    2003018374

    ISBN 0-415-28086-9 (hbk)
    ISBN 0-415-28087-7 (pbk)

Contents

Illustrations

Contributors

Jill Belsky is Professor and Chair of the Sociology Department at the University of Montana. She received her PhD from Cornell University in 1991 in rural sociology. She teaches classes in environmental sociology, social theory, gender and global development. She has conducted research in Indonesia, the Philippines, Belize and the interior American West on the opportunities for, and constraints to, collaborative, community-based approaches in rural and environmental management, most recently focusing on rural ecotourism.

Stroma Cole studied social anthropology and environmental biology at Oxford Brookes University and continued her studies with a master's in social anthropology at the School of Oriental and African Studies, London University. Stroma has recently completed a PhD in the anthropology of tourism at the University of North London.

Following extensive travel of her own and on completion of her master's, Stroma started her own tour operating business in Indonesia. For seven years she led small groups all over the archipelago. After a period of consultancy for UNESCO and ADB she returned to the United Kingdom to take up her post in the Faculty of Leisure and Tourism at Buckinghamshire Chilterns University College, where she has been working for five years. Stroma's professional and research interests include tourism development in less developed countries, the anthropology of tourism, responsible tourism, and Web-based learning for tourism studies.

Alain Decrop is Assistant Professor at the University of Namur and Visiting Professor at the Catholic University of Louvain, Belgium. He holds masters' degrees in history and economics, and a PhD in business administration. His major research interests include consumer behaviour, qualitative interpretive methods, and tourism marketing. His works have been published in journals such as *Tourism Management*, the *Journal of Travel and Tourism Marketing* and *Tourism Analysis*, as well as in many books.

Heather Gibson is Assistant Professor in the Department of Recreation, Parks and Tourism and an Associate Director of the Center for Tourism Research and Development at the University of Florida. She teaches and researches in the sociology of leisure, tourism and sport. Some of her research projects over the years include change and stability in tourist role preference over the adult life course, the leisure-travel patterns of older adults, and a cross-cultural study of solo women travellers with Fiona Jordan at Cheltenham and Gloucester College of Higher Education, UK. Her current projects include an investigation of the meaning of leisure in the lives of retirement-aged women, the serious leisure associated with being a University of Florida football fan, and the sport tourism implications associated with US college sport. She has published in the *Annals of Tourism Research, Leisure Studies, Society and Leisure* and the *Journal of Vacation Marketing* among others.

Lisa Goodson is Research Fellow at the University of Birmingham, Centre for Urban and Regional Studies. Her PhD focused on community participation and sustainable development in urban tourism. Lisa specialises in qualitative research and has extensive experience working with a range of community groups on issues concerning social exclusion, regeneration, sustainable development and training. She has worked on a range of projects funded by organisations such as the United Nations, the DETR, the ODPM and the LSC. She also co-directs the postgraduate qualitative research methods programme at the university and runs day seminars for public service personnel on qualitative research and community consultation, and through the development of an innovative Peer Research Programme she, together with Jenny Phillimore, trains socially excluded community groups to become co-researchers.

Michael Hall is Professor and Head of the Department of Tourism at the School of Business, University of Otago in New Zealand. He is also co-editor of *Current Issues in Tourism* and chair of the International Geographical Union Study Group on Tourism, Leisure and Social Change. He has published widely on areas relating to tourism, heritage and environmental history. His current research interests include tourism as a component of economic restructuring, particularly in rural areas; contemporary mobility, second homes, wine and food tourism, and lifestyle.

Keith Hollinshead is Professor of Public Culture at the University of Luton in England. He is also Visiting Professor in Cultural Tourism at The University of Western Sydney, Australia. Originally a 'marketing' specialist, he has become an eclectic theorist drawing insight from anthropology, political science, cultural studies, human communications, and continental philosophy amongst a wide mix of domains.

Over the years, Professor Hollinshead has worked on the editorial board of eight international peer-review journals, and has served as Joint-Masthead Editor on three of them, namely *Tourism Analysis, Tourism, Culture and Communication* and *Current Issues in Tourism*. He also served as Vice President #2 for International Tourism for the International Sociological Association for 8 years, co-ordinating its pioneering *Paradigmatologie* in Finland in 1996, and its 'Worldscape 21' Colloquium in Belgium in 2001.

Barbara Humberstone is Professor of Sociology of Leisure at Buckinghamshire Chilterns University College. She teaches gender, difference and leisure, and outdoor education and adventure recreation. Her research interests include social and environmental equity/justice and leisure, (eco)feminist theories, cultural diversity and research methodologies. Barbara is a board member of the European Institute for Outdoor Adventure Education and Experiential Learning. She is a member of the editorial board of *Leisure Studies*, journal of the Leisure Studies Association.

She recently contributed to the development of a World Leisure Position Statement: Outdoor Leisure Education for the World Leisure Association in July 2001. She is a frequently invited keynote speaker to a number of national and international conferences, most recently Gender Transgressions and Contested Nature *Nature and Identity (programme for Cultural Research by the Research Council of Norway)*, Alta, Norway, 6–9 September 2001, and was Visiting Professor to the Department of Outdoor Education and Nature Tourism, La Trobe University, Bendigo, Australia, in spring 2001.

Guy Jobbins is a coastal ecologist specialising in the governance of human–ecosystem interactions. Guy has worked in a variety of roles in the United Kingdom and across the Middle East, most recently as a Research Fellow at University College London developing decision support tools for ecosystem management.

Fiona Jordan is Senior Lecturer in Leisure and Tourism Management in the School of Sport and Leisure at Cheltenham and Gloucester College of Higher Education, UK. She graduated from Manchester Metropolitan University with a degree in Hotel and Catering Management, has a master's degree in Leisure and Tourism Studies and is currently studying part-time for her PhD. Fiona has a general research interest in the relationship between women and tourism, both as producers and consumers. She has previously published in the areas of women in leisure and tourism management, and the experiences of women as solo tourists. Her work with Heather Gibson focuses on cross-cultural comparisons of the experiences of women travelling solo and her work with Cara Aitchison focuses on discourses of power and knowledge in leisure and tourism pedagogy and research. She is particularly interested in researching leisure, tourism and heritage as sites and processes of social, spatial and cultural inclusion and exclusion.

Adele Ladkin is Reader in Tourism at the International Centre for Tourism and Hospitality Research, School of Service Industries, Bournemouth University, UK. She is the Deputy Head of Research, and Head of MICE research unit within the School of Service Industries. Her research interests and publications are in the area of tourism education, career analysis and labour mobility in the tourism and hospitality industry, and the meetings, incentives, conventions and exhibitions industry. She is currently Assistant Editor in Chief for the *International Journal of Tourism Research*. She gained an MSc in Tourism Management and a PhD from the University of Surrey.

Dennison Nash is Emeritus Professor of Anthropology at the University of Connecticut, Storrs. He is the author of various writings, among which are *Anthropology of Tourism* (1996) and *A Community in Limbo* (1970); he also continues to teach a course in ethnographic fieldwork and is associated, as membership chairman, with Benmarl, the oldest surviving vineyard in the United States.

Jenny Phillimore completed her PhD, an ethnographic study of women's employment in rural tourism in the United Kingdom, in 1998, and since then has been Lecturer at the Centre for Urban and Regional Studies at the University of Birmingham. She co-directs the Post-Graduate Qualitative Research Programme at the university and works very closely with Lisa Goodson on a range of different initiatives. These include running seminars on qualitative research and inclusivity for public service professionals and undertaking a wide range of qualitative research projects largely concerned with involving excluded communities in regeneration initiatives and increasing the employability of hard-to-reach groups. Their work has included the development of an innovative Peer Research Programme which trains excluded individuals to become co-researchers. Beneficiaries of this programme have included offenders, homeless young people, and asylum seekers and refugees. She is currently leading a number of substantial research projects looking at the experiences of asylum seekers and refugees in the United Kingdom, wherein she has called upon a wide range of qualitative as well as quantitative research methods.

Jennie Small is Lecturer in the School of Leisure, Sport and Tourism at the University of Technology, Sydney, Australia. Her main teaching and research interest is tourist behaviour. She is the author of various published papers on memory-work and co-editor of a collection of memory-work papers that emanated from the Memory-work Research Conference, which she organised in Sydney in 2000. She is completing her doctoral studies investigating the positive and negative aspects of women's and girls' holiday experiences.

Margaret Byrne Swain is Director of the Center for Gender and Global Issues in the Women and Gender Studies Program at the University of California, Davis, and affiliated as an Adjunct Professor in Anthropology. Her dissertation research in the 1970s was on gender and tourism development issues among the San Blas Kuna of Panama. Continuing interest in minority indigenous peoples, missionaries and tourists informs her current work in south-west China and her forthcoming book on 'embodied cosmopolitanism'. She has edited several collections on gender in tourism (*Annals of Tourism Research* 22(2) 1995; with Janet Momsen, Cognizant Communication Corporation 2001), and authored numerous ethnographic and theoretical articles. Her work on gender and global environmental and population issues is both instructional and praxis-orientated, bridging academics and activism.

Karen Thomas is Senior Lecturer in Tourism Studies in the Department of Geography and Tourism, Canterbury Christ Church University College. She received her PhD from the University of Birmingham, which evaluates a tourist role scale technique as a foundation for the development of theory and practice of tourism marketing strategies. She has conducted consultancy research for both public- and private-sector organisations, most recently undertaking focus group research for Medway Council on the 'Optimum Branding of Medway and its Specific Attractions as a Total Visitor Destination: The Short Break Experience'. In her present post she specialises in lecturing in the areas of tourism marketing, tourism policy and development, and research methods, and is Programme Director for the MSc in Tourism and Environmental Management. Her key research interests are in destination marketing and branding, destination imagery, and tourism marketing communication strategies.

John Tribe is Head of Research in the Faculty of Leisure and Tourism at Buckinghamshire Chilterns University College. John graduated in economics from the University of London, and returned there to study for his MA and PhD. His MA was on business education, specialising in tourism. His PhD research was undertaken at the Centre for Higher Education Studies and his thesis was on the subject of tourism, knowledge and the curriculum. He joined the Faculty of Leisure and Tourism, BCUC, as Senior Lecturer in 1993, became Principal Lecturer in 1997 and was appointed Professor of Tourism and Head of Research in 2000. He is also Visiting Lecturer at the University of Surrey.

John has spent his career in education in both the USA and the

United Kingdom. His teaching has focused on economics, strategy and education aspects of leisure and tourism, and he has authored two textbooks in these areas, one of which is translated into Spanish. His research interests are focused on education and sustainability issues. He has published widely in these two areas in books and journal articles and has accepted invitations to give conference papers across Europe. He has directed two major European research projects: TOURFOR (Sustainable Tourism in Forests) and THEME (Tourism Curriculum Development in Moldova). He also writes occasional articles for *The Times* and the *Guardian* newspapers. John's professional interests include leisure and tourism economics; leisure and tourism strategy; higher education in leisure and tourism; and sustainability in leisure and tourism.

Foreword

This is a compelling book, to be read and thoughtfully considered by every serious researcher involved in tourism at the planning and management level.

The linguist Wick Miller once observed that no one has ever written an ethnography of academia – their goals, their methods and their behaviours. This unique volume might be the catalyst to prompt such a study. Dating to Aristotle, scholars in every discipline have diligently sought 'the truth' but their philosophies and products have varied according to the dictates of the individual field of study.

The global importance of tourism has generated the need for answers to problems such as economic development, social impact, stakeholder conflicts, environmental degradation and political control. These questions all seek 'the truth' but the orientation is different. The business world wants to know 'who, what, when and where' for that is their 'bottom line'. Their approach is essentially quantitative, and statistically oriented for forecasting. By contrast, researchers involved in heritage, habitat and history quest for 'why': what roles did or do these elements play in human society and its survival? To what degree have they changed, and how should they be interpreted now? What is appropriate authenticity?

As the editors point out, a resurgent interest in the qualitative methodology as it applies to the study of tourism surfaced a little over a decade ago but lacked a substantive base. Here, in this first-of-a-kind compendium, a body of recognised scholars have outlined diverse research techniques and illustrated them with case studies. Culture, as a set of human survival customs, may be a collective noun but the behaviours of individuals operating within their respective ethnic bonds can seldom be ranked on a scale of 1 to 10.

The phenomenal growth of tourism in the past five decades has dramatically changed global lifestyles to include tourism, and the impetus for still greater growth is rooted in globalisation and the expanding economies of Asia. We in the West have much to learn about these future 'new' tourists who will have discretionary income, leisure time, and even government sanctions (as a balance of trade factor) to visit Europe and the Americas in the next decade.

The great merits of this book are its new paths for studying tourism as a societal institution, as we once learned to study the family, the factory and world view as cultural components of a survival system. The case studies open wide opportunities for testing and theoretical validation. This is a *must read.*

Valene L. Smith
Chico, California

Acknowledgements

Some three years ago we decided that a book about qualitative research in tourism would make a useful contribution to the field as we had both had to rely heavily on qualitative methodology books from outside tourism when developing ideas for our PhDs. We put together a tentative proposal and sent it to Routledge for their consideration. Since that time the scope of the project has widened considerably. With the encouragement of Routledge and anonymous reviewers we approached a number of established tourism researchers around the world and asked them to make specific contributions exploring key social science research concepts in relation to the study of tourism. We are grateful for the enthusiasm with which the majority of them accepted the challenge offered them.

The successful completion of this project and its metamorphosis from an edited book of accounts of qualitative tourism research in the United Kingdom, to a collection of contributions considering the theoretical underpinnings of qualitative tourism research and how the theory is applied in practice in the field, is in no small part due to the efforts of a range of individuals. First and foremost, our contributors, to whom thanks are due for their thought-provoking contributions and passion for the project. Second to Routledge for being prepared to take on such a groundbreaking project and take a risk on two relatively unknown writers. Third to the reviewers, with whom we have not always agreed but who have provided us with extremely useful guidance on turning what was originally a disparate collection of papers to what we now hope is a coherent book. Finally, thanks are due to friends, family and colleagues, who have given us encouragement and support throughout the whole process.

Part I

Key ontological, epistemological and methodological issues in social science

1 Progress in qualitative research in tourism

Epistemology, ontology and methodology

Jenny Phillimore and Lisa Goodson

Aims of the chapter

- To provide an introductory discussion on qualitative research and how thinking about research and the way we view the social world has evolved in the social sciences.
- To review progress made in qualitative tourism research by considering the nature of research in the field, the range of approaches commonly adopted and areas for future development.
- To highlight the main themes and issues addressed in the remainder of the book.

Introduction

The term 'qualitative research' is something of an enigma. For some time, the issue of what exactly qualitative research is has been at the centre of a great deal of debate within social science, most notably in the disciplines of sociology and anthropology (see May 1993). More recently, these debates have gained greater prominence in the field of tourism research (Hollinshead 1996, 1999; Jamal and Hollinshead 2001; Riley 1996; Riley and Love 2000; Walle 1997; Dann 1996). Traditionally, qualitative research has been viewed somewhat simplistically as a set of different research methods that have certain features in common. In this respect, qualitative methods are employed to collect data about activities, events, occurrences and behaviours and to seek an understanding of actions, problems and processes in their social context. From this perspective, qualitative research is perceived as distinct from quantitative research as it does not produce quantified findings or have measurement and hypothesis-testing as an integral part of the research process. When qualitative research is thought of as a series of methods, it is often considered to be an approach to research that is 'other' to quantitative research because quantitative data are not collected (Bryman and Burgess 1994). As such, qualitative research has been prone to criticisms that it is a 'soft', 'non-scientific' and inferior approach to studying social life, and one that is often seen as useful only when accompanied by, or as a precursor to, quantitative techniques (Guba and Lincoln 1998: 196). This attitude to qualitative research was observed by Riley and Love (2000) in their review of the methodological approaches employed in the study of tourism. However, the labelling of qualitative research as a poor alternative to 'real', rigorous, 'scientific', quantitative studies has been questioned over the past 25 years in many social science disciplines. Those who viewed qualitative research merely as a set of methods have been accused of having an oversimplified view that fails to acknowledge the multiplicity of forms and functions of qualitative research (Silverman 2000). Indeed, over the past few decades, using qualitative approaches to study social life has been considered more acceptable within the mainstream, rather than being viewed as an adjunct to quantitative work. In fact, qualitative research has become increasingly valued as thinking about research developed and research began to be viewed as more of a process than an activity, with discussions about the appropriateness of method being superseded by concerns with methodology (Bryman and Burgess 1994). While this mind shift has been under way in the social sciences generally, it is fair to say that tourism scholars have generally been more hesitant in their adoption and acceptance of qualitative research, and more specifically in developing their understanding of the philosophical and theoretical process that underpins knowledge production and practices.

In recent times, textbooks concerned with the nature of social research are more likely to refer to qualitative research as a distinctive research strategy than simply a set of methods (Bryman 2001: 264). As a strategy, qualitative inquiry can generate theory out of research, should place emphasis on understanding the world from the perspective of its participants, and should view social life as being the result of interaction and interpretations. Rather than aping the study of natural phenomenon using a 'natural science' approach, which is the stated aim of quantitative research, qualitative studies are said to be 'located at the meeting place between art and science' (Sandelowski 1994: 312). This is not to say there is no place for quantification in research. In tourism research there is an ongoing need for statistical insights into aspects such as market and migration trends, income generation, and so forth. However, qualitative approaches offer a great deal of potential, much of which remains largely untapped, for helping us understand the human dimensions of society, which in tourism include its social and cultural implications. With qualitative approaches, the emphasis is placed upon studying things in their natural settings, interpreting phenomena in terms of the meanings people bring to them, humanising problems and gaining an 'emic', or insider's, perspective. Going further still to thinking of qualitative research as a different way of looking at social life from that of the 'etic', or outsider's approach, which formerly dominated social studies, Denzin and Lincoln

(1998: 5) have argued that qualitative research is a critique of the traditional positivistic approaches to researching. In this context it is not so much a distinct set of methods as a new way of approaching and undertaking research, an approach which seeks to highlight, and then remedy, the so-called deficiencies of 'natural science' methods which underpin quantitative research.

There are many different views about what qualitative research is. Is it a set of methods, a strategy, a critique or an approach? This book will consider some of the main arguments but does not seek to promote one particular approach or 'solution'. What this book does seek to demonstrate is that qualitative research is as much a way of conceptualising and approaching social inquiry as it is a way of doing research. It is argued that, to date, tourism research has, in the main, used qualitative research as a set of methods rather than as a set of thinking tools which enable researchers to consider different ways of approaching research and uncovering new ways of knowing. The main rationale behind the development of this book is to encourage tourism researchers to adopt a more sophisticated attitude to thinking about and using qualitative research. By viewing qualitative research as more than a simple set of methods, tourism researchers can begin to consider different ways of thinking about, and undertaking, research, and take a new look at the ways in which knowledge is produced in the field. This book introduces some of the debates that exist in social science research texts into discourses about tourism research. It then explores what might be gained from adopting different approaches to the more traditional natural science approaches.

This chapter provides an introductory context to the 17 contributions that follow. It includes a brief discussion about the ways in which thinking about research in the social sciences has developed and how these developments relate to debates about how we think about the social world more generally. It then utilises Denzin and Lincoln's (1998) five moments of qualitative research as a framework to introduce a number of key concepts and paradigmatic issues that have gained increasing recognition in social science research practice and discourse. The chapter then deliberates over the extent to which these paradigmatic issues have been incorporated into tourism research by exploring the research paradigms adopted, both explicitly and implicitly, in some of the main tourism journals. In so doing, it highlights dominant themes, ways of doing research and emerging ideas in tourism research. The chapter concludes by outlining the main issues that the contributors cover in the remainder of the book. Emphasis in Part I, is on issues around research paradigm and tourism. This is followed, in Part II, by a series of chapters which look in an open and reflexive way at tourism research processes and the ways in which qualitative research in tourism can be approached with a view to learning from these examples.

Thinking about researching the social world

Social science has been defined as 'the attempt to explain social phenomena within the limits of available evidence' (Lewins 1992: 41). The definition of what counts as 'evidence' is problematic, as is the question of how evidence is collected. Evidence has traditionally been understood as data collected through the counting of responses or observations. In circumstances where sociological concepts were not directly observable, they were instead grasped indirectly through the testing of hypotheses where ideas could be viewed as 'facts' if they were verifiable by certain observations. Society was seen as rule governed, with there being one 'true' reality, and research was based upon the idea of there being one universal knowledge which was created through the application of objective thinking and empirical research. Emphasis was placed on a reasoned, scientific approach to knowledge production, wherein any researcher operating under the same conditions would reach the same conclusion (Code 1993). However, attitudes about what constitutes valid and reliable knowledge, research and evidence have changed as social research has developed.

Five 'moments' of qualitative research

Denzin and Lincoln (1998: 13) divide the history of social research into five phases which they refer to as the 'moments of qualitative research'. Riley and Love (2000) have used this conceptual framework to explore paradigmatic influences and consequent use of qualitative methods within tourism research. They focused their evaluation upon articles published within four tourism journals: *Annals of Tourism Research* (ATR), *Journal of Travel Research* (JTR), *Tourism Management* (TM) and *Journal of Travel and Tourism Marketing* (JTTM). These have been published since 1973, 1970, 1980 and 1992 respectively.

Riley and Love's (2000) review concluded that the greatest concentration of qualitative tourism research was located in the first two moments, the 'traditional' and the 'modernist' phases. In order to aid consideration of qualitative research in tourism in this book, Riley and Love's (2000) commentary of tourism research was updated using a selection of tourism journals which were analysed from papers published from 1996 onwards, or from the journal's inception if publication commenced later than 1996 (see Table 1.1). This review enabled us to consider the most recent body of research to have emerged over the past seven years, review trends, and identify the use of new techniques and ways of approaching research in tourism. It also provided insight into the extent to which qualitative tourism research has developed in recent times. For ease of reference we refer to this analysis as the 'post-1996 review'.

When examining the articles for the post-1996 review, we considered it

Table 1.1 Tourism and leisure journals included in the post-1996 review

Journal	Years
Tourism Geographies	1999*–2003
Journal of Sustainable Tourism	1996– 2003
Tourism Management	1996–2003
International Journal of Tourism Research	1999*–2003
Progress in Tourism and Hospitality Research	1996–1998
Annals of Tourism Research	1996–2003
Leisure Studies	1996–2003

Note
*First volume.

important to view each contribution holistically, rather than adopt a mono-dimensional approach by looking solely at the research methods employed. This entailed viewing each piece of research from a number of different perspectives and asking a series of questions to take into account the complexity of various elements that shaped the nature of the final articles. These questions related to the traits which Denzin and Lincoln ascribed to each 'moment'. For example, in the first moment, the traditional period, the researcher's voice is heard as giving an expert and objective interpretation of the 'facts', while in other moments less authority is given to the researcher, and their voice is heard to consider a variety of possible interpretations mediated by interaction with the researched. The range of questions considered is outlined in Table 1.2.

The relationship between the research questions listed above, the traits and the five moments is described in Table 1.3.

First-moment research: the predominance of positivism in the traditional period

Denzin and Lincoln (1998) refer to the first moment in qualitative research as the 'traditional period'. Though emphasis in this time was upon quantitative research, not all research undertaken employed quantitative methods. Empirical research in this era was, however, exclusively informed by a positivistic, natural science approach in which the researcher was viewed an independent, objective expert whose judgement determined the validity of findings (Black 1993; Ryan 1979; Stern 1979). In this context came the development of evolutionary ethnography, a positivistic approach which employed qualitative methods as its main data collection tool. The discovery of the Other by Western explorers and missionaries was followed by the urge to record Other societies, as researchers/explorers collected data to support the assertion that all peoples were natives somewhere on an evolutionary chain with West Europeans at the top (Vidich and Lyman 1998: 48).

Table 1.2 Framework for the post-1996 review

	Questions
Research design	What methods were chosen and how creative were they? That is, to what extent are unconventional methods such as the use of images involved? How do the chosen tools fit with the overarching objective of the research? That is, are the methods fit for the purpose? Was the research governed by a deductive or inductive agenda? That is, at what stage were the question and framework for analysis formulated? Was the approach formulated/directed by specific rules?
Methodological reflexivity	Was there sufficient detail about the methods adopted and how they were applied to understand how the research was carried out? Can the reader follow the chain of inquiry? Does the research process appear transparent? Was the methodological approach considered from a critical perspective, with possible alternatives being evaluated? Was there any insight into the personal/intellectual biography of the author?
Analysis of findings	How was the data interpreted? Who/how many people interpreted the data? Did participants have opportunity to contribute to the process in any way? Were any feedback mechanisms put in place to enable participants to reflect on, or add to, the interpretation of the findings and conclusions?
Presentation of findings	How were the findings written up? What sort of language has the author used to represent their findings? Has space been made for the voices of the participants to be heard? To what extent were findings generalised? Where does the power and authority within the research process appear to lie?
Scale of the research	At what level (local, regional, national, international), was the research undertaken? What sorts of respondents were involved? What geographical area or groups are the findings said to represent? Are the findings offering a snapshot in time or conclusions that apply regardless of temporal moment?

This first stage in the development of qualitative research is attributed with the provision of 'objective colonising accounts of field experiences' (Denzin and Lincoln 1998: 15), whereby researchers co-operated with nineteenth-century imperial governments in the suppression and exploitation of natives (*sic*) (Vidich and Lyman 1998: 52). Researchers recorded through the use of observation, interview and analysis of documents how such 'primitive' societies operated. They issued their interpretation of what was going on as the truth about evolution and native peoples. The

Table 1.3 Relationship between the post-1996 review framework and Denzin and Lincoln's five moments of qualitative research

Moment	Traits
Traditional	• Positivist • Objective colonising accounts of field experience • Depersonalised accounts • Researcher seen as 'expert' – their judgement determines the validity of the findings • Research findings represented as fact
Modernist	• Move away from natural science notion that 'reality is out there' • Attempts to formalise qualitative research and maintain positivistic rigour • Interested in ways that people categorise the world and the meaning people place on events • Phenomenology, ethnomethodology, grounded theory
Blurred genres	• Indistinct boundaries between disciplines • Recognition of multiple approaches • Theories, techniques and approaches mixed and matched, e.g. more creative methods such as photography and advertising used with more conventional qualitative methods, such as participant observation, in-depth interviews and ethnography • Use of different theoretical models – feminist, ethics, semiotics
Crisis in representation	• Researchers knowledge challenged • Rigour and generalisability of social research questioned • Reflexivity, embodiment and personal biography of researchers critical • Recognition of multiple interpretations • Questions raised around issues of gender, class, race e.g. what kind of impact embodied characteristics have on the kinds of questions researchers ask and the ways they interpret data
Fifth moment	• End of grand narrative • Focus on specific delimitated local research – seen as a snapshot of a particular time • Theories are context specific • Authority of the researcher as 'objective expert' rejected • Researchers voice is one among many

tradition continued into the new century, and in the 1920s folk psychologists began to look inwards to their own societies, undertaking qualitative research often underpinned by a moralistic aim of uplifting the poor (Vidich and Lyman 1998; Flick 1998). First-moment studies in tourism are characterised by those which adopt a colonialist perspective, privilege the observer as 'knowing', 'museumify' cultures of the Other, and reports timeless and unchanging accounts of 'reality' (Riley and Love 2000). Examples cited by Riley and Love (2000) include Cohen's (1982) research in Thailand, Towner's (1985) study of the Grand Tour and Greenwood's (1977) work on the Arlarde in Fuenterribia.

The examination of post-1996 articles showed that research characteristic of the 'traditional period' still remains common. While the purpose of the review was not to quantify the numbers of studies falling into each moment – and indeed this would be difficult, given the issues raised above – it is fair to say that the majority of studies fall into this category. Notably, many ethnographic accounts still remained highly depersonalised, devoid of any reflexive accounts, and the researcher's voice as 'expert' dominates the text. Ryan and Crotts's (1997) work on the impact of tourism on Maori culture, Smith's (1998) ethnographic account on war and tourism, and Doorne *et al.*'s (2003) research on identity and tourism in China are examples of such research, which are based mainly on commentary and interpretation by the author(s) of experiences, events, processes and places. In these texts the 'authentic' voice of those researched remains largely invisible; research findings are written from an 'objective' perspective, although often in a story format rather than in a traditional natural science approach; and interpreted accounts are often represented as generalisable facts, while findings are offered as a single truth (see also Higham and Hinch 2002; Ladkin 2002; Preston-Whyte 2002; Steymeist 1996).

Post-colonial researchers in particular have challenged such approaches on the grounds that where an attempt has been made to explore tourism from the perspective of local people, their voice has been represented as a single unified view. It has been argued that through this simplistic viewing of the Other, without making visible the multiple voices of the hosts concerning their own culture, tourism research can serve to reinforce the sense of superiority of the researcher's dominant culture, and perpetuate, rather than break down, existing stereotypes (Wearing and Wearing 2001). A range of other tourism contributions were also considered to typify the 'traditional' period. Reasons for classifying research into this period varied, but generally related to the underlying positivist tendencies evident within the research design, process or presentation. These included:

- *Studies based on predetermined rigid research agendas.* These are most frequently used by those focusing on economics, marketing or management in survey and structured interviewing. In such research, the perspective and authority of the researcher(s) shape the entire nature of the study; there is no consideration of research participants' perspective in the research design since research questions and frameworks for analysis are determined well before the research commences. Examples include Jansen-Verbeke and van Rekom's (1996) consideration of museum visitors' motivations using qualitative techniques predetermined by the Kelly Grid method, and Enoch's (1996) use of brochures to compare the contents of tour packages across European countries, which were analysed according to pre-defined criteria without providing a discussion of the genesis of the criteria.

• *Studies that seek to quantify qualitative data.* Surveys and structured interviewing can be used to generate both quantitative and qualitative data. A common tendency in tourism studies of this nature has been to place the main emphasis on the quantitative dimension of the study and/or the tendency to quantify qualitative data to present a positivist interpretation of the findings (e.g. Andersen *et al.* 1997; Mitchell and Eagles 2001; Rodriguez 2002; Kozak 2002; Buhalis and Licata 2002; Weaver and Fennell 1997). These observations are hardly startling, given that throughout the 'traditional period' the mainstream approach to research, beyond ethnographic work, was quantitative.

• *Studies aimed at generating tourist typologies.* We placed a prominent body of work centred on the development of tourist typologies in the 'traditional' phase. These sought to conceptualise the tourist as 'wanderer'/'gazer'/'escaper', etc. (e.g. Cohen and Taylor 1976; Pearce 1984; Urry 1990; Rojek 1993, 1997) and hosts as 'positive'/'negative', 'active'/'passive', etc. (Butler 1975, 1980; cf. Doxey's (1975) 'Irridex' and Smith's (1977) seven-stage tourist typology). The presence of more recent typology studies suggests that this is still a prevalent area of tourism research (e.g. Selin 1999). In some cases the approach has evolved to take into account some of the developments emanating from the fourth and fifth moments; for example, Wickens (2002) in her work on Greece uses respondents' voices to support her arguments throughout the discussion and acknowledges ontological considerations. However, the underlying objective of these studies was to classify behaviour into a range of categories. This approach, we would argue, is underpinned by positivist modes of thinking which attempt to provide predictive, analytical and explanatory tools based on generalisations which are then applied to broad populations. Research aimed at generating these typologies may serve to strengthen or even construct stereotypes of the hosts, guests and/or the destination.

While such models acknowledge that individuals may hold a range of different perspectives, and recognise the potential existence of multiple realities, those perspectives are represented by frameworks that are rigid and often pre-set. Such frameworks cannot by their predestined nature account for the full range and diversity of tourism experiences. This homogenising of realities into categories fails to account for the significant range and diversity of individual experiences along dimensions such as age, gender, race, class, (able) bodiment, etc., or to consider the ways in which perceptions and emotions may change over time. As such, typology studies fail to provide insight into the complexities of tourism interactions at both the experiential and the emotional levels. Wearing and Wearing (2001: 151) argue that there is a need to move beyond simplistic typologies towards a more 'analytical flexible conceptualisation' that enables assumptions imbued within the research process – for example, the 'tourist gaze', the tourist 'destination', the marketing 'image' and the 'visit' – to be recognised.

- *Studies that place little or no emphasis on methodological issues.* A number of studies examined offered no description or evaluation of the methodological approach used – for example, Telfer and Wall's (1996) discussion of tourism and food production, Paradis's (2002) consideration of the political economy of urban places in New Mexico, and Scheyvens's (2000) exploration of women's empowerment through involvement in ecotourism, which draws on case studies from Nepal and Samoa, none of which contain any discussion of methodology. (See also Markwick 2000; Stymeist 1996; Ritchie and Ritchie 2002.) The lack of accountability and transparency about the research process presents the researcher as 'powerful/knowing', as the issues and justification surrounding 'fitness for purpose' are left unquestioned. This lack of methodological documentation makes it difficult for the reader to follow the chain of inquiry and leaves questions about how the research was constructed, how the findings were generated, analysed and interpreted, and how conclusions were drawn.

Modernist period

As the second, 'modernist' moment up to the 1970s began, there were attempts to formalise qualitative research as part of what Denzin and Lincoln (1998: 18) describe as 'the golden age of rigorous qualitative analysis'. Specific schools of thought evolved in the United States and Germany around the time of World War II which looked at the collection of data and the construction of knowledge in entirely different ways as compared with the quantitative approach. They have ultimately had a major impact on subsequent qualitative research. Phenomenology, symbolic interactionism and ethnomethodology are distinct approaches to thinking about and researching the social world. They focus on the ways in which people categorise the world by distinguishing certain phenomena, the meanings people place on events, the way people use and interpret symbols in order to communicate, and how social reality is reproduced through interactions. These schools of thought moved away from the notion that reality was 'out there' in the social world to consideration that there were multiple realities held by the inhabitants of the social world and that these could be reached by the use of qualitative techniques such as interviewing and conversational analysis. Attempts at formalising qualitative research were informed by post-positivism. These continued to seek a singular version of reality but recognised that flawed intellectual mechanisms made reality only imperfectly comprehensible. In that era,

Glaser and Strauss (1967) formulated grounded theory, setting down a distinct and rigorous procedure for creating theory using qualitative techniques that are still very much in use to this day. They, like ethnomethodologists and phenomenologists, had a profound impact on qualitative research, demonstrating, albeit in positivistic terms, that rigorous, rule-governed qualitative research was capable of theory generation.

In 'modernist' tourism research there is also evidence of a shift to post-positivist modes of thinking, although with clear evidence that researchers continue to be convinced of the fundamental importance of maintaining positivist rigour. Riley and Love note how the 'underclass', 'deviants' and social processes are frequently the focus of studies in the second moment, and cite Dann's (1992) work using travelogues and Gottlieb's (1982) investigation of tourism as inversionary behaviour as examples of second-moment tourism studies.

In contrast to Riley and Love (2000), we focused on the actual methodological approach adopted by the researcher in order to classify 'modernist-phase' studies. We looked specifically at the extent to which contributions were influenced by rule-bound methodologies which attempt to place a systematic procedure and structure on the research process. A number of examples identified in the post-1996 review sought to bring these positivistic rules into qualitative inductive research. An excellent example of this approach can be seen in Jutla's (2000) use of mapping and photography to explore people's images of Simla according to Lynch's legibility method, and also Miller's (2001) use of the Delphi technique to structure group communication about the development of indicators for sustainable tourism. In particular, the borrowing of grounded theory principles, used in other disciplines such as health care and marketing, could be seen to exemplify 'modernist' traits' in tourism. For example, Johnston (2001) uses a comparative technique to examine processes in resort development and reveal features of epistemological concern in relation to Butler's destination life cycle (1980). Burns and Sancho (2003) use an ethnographic approach to interview key stakeholders and use grounded theory principles to present oral data around six themes using direct quotations to allow 'authentic' voices to speak for themselves. Verbole's (2000) policy-orientated study on rural tourism in Slovenia adopts an 'actor perspective' using Strauss and Corbin's (1998) grounded theory procedures and technique to guide the research process (cf. Connell and Lowe 1997). Researchers are, however, rarely seen to adopt the full grounded theory technique developed by Glaser and Strauss (1967). These examples do, though, demonstrate how some researchers are leading the move from the traditional to modernist moments with research continuing to be underpinned by positivist ideologies but moving from a deductive towards an inductive approach.

A number of general methodological discussion pieces such as Decrop's (1999) and Davies's (2003) commentary on triangulation, and Morrison's

(2002) general reflection on hospitality research, which calls for an enhanced degree of 'research philosophy awareness' and the formulation of an internally (valid) conceptual framework for hospitality research, were considered to be underpinned by 'modernist'-phase principles, as they focus primarily on ways in which tourism research can be made more rigorous by incorporating certain rules and tests. These papers provide a useful contribution to the tourism literature as they help to develop thinking on methodological, epistemological and ontological issues.

Blurred genres

Denzin and Lincoln's third moment in the development of social research is termed 'blurred genres'. In the social sciences they refer to it as the period from the 'modernist' moment up to the mid-1980s. During this time, researchers began to choose from different theoretical models, and the boundaries between different disciplines became indistinct so that theories, techniques and approaches could be borrowed, mixed and matched according to the research task. The blurred genres stage marks a distinct difference from the 'modernist' and 'traditional' phases as researchers begin to move away from natural science and recognise multiple approaches embracing a more creative, artistic approach to research. Riley and Love (2000) reported few examples from the four journals of researchers in tourism operating within 'blurred genres'. Such articles are described as those that have employed more innovative approaches to data collation, utilising photographic and advertising media, and personal experiences, together with more traditional techniques for collecting qualitative data, such as participant observation, in-depth interviews and ethnography.

Since the mid-1990s there has been increased evidence that qualitative tourism researchers are beginning to embrace third-moment issues. For example, research beginning to engage with feminist debates, notably the special issue on gender and tourism in ATR (Swain 1995) and Kinnaird and Hall's (1994) book *Tourism: A Gender Analysis*, sparked interest in men's and women's differential experiences of tourism and has been followed by a wave of contributions focusing on gender issues, which have approached the study of tourism from a range of methodological, theoretical and thematic perspectives. These include studies on tourism marketing and imagery (Aitchison and Reeves 1998; Aitchison 1999; Pritchard and Morgan 2000a; Westwood *et al.* 2000), sexuality and holiday choices (Pritchard *et al.* 2000), employment (Jordan 1997) and planning (Marshall 2001; Scheyvens 2000). The dominant theoretical influences within much of this work are underpinned by thinking in feminist and cultural geography. Some of the debates emerging on issues such as power, the social construction of tourism imagery and landscapes and the construction of 'the self' in tourism take the reader into 'fourth- and fifth-moment' territory, and tackle a range of epistemological and ontological issues which have rarely been addressed by commentators in tourism (Rojek 1997; Jamal and Hollinshead 2001). These issues are discussed further in Chapter 2. It could be argued that the borrowing of different ideas, methods and theories simply reflects the highly interdisciplinary and multi-disciplinary nature of tourism (Botterill 2001). On the other hand, the resultant hybridisation of tourism research could be seen as a gauge of the way in which tourism researchers have become receptive to research of earlier moments while embracing emerging paradigms, subjects and techniques from other disciplines.

In this moment we also placed innovative experimentation with different methodological approachs such as semiotics. Echtner (1999) discusses the usefulness of the approach in uncovering the systems of signs and the 'deep structure' of meaning in tourism marketing; personal diaries have been used to explore young people's holiday behaviour (Carr 2002); and photographs of New Zealand landscape have been used to identify distinct visitor behaviour (Fairweather and Swaffield 2001; see also Pavlovich 2002; Jenkins 1999). In addition, there has been the use of multi-method approaches, whether mixing solely qualitative methods, such as Morgan's (2002) use of documents, interviews and observation to explore the symbolism associated with New Brighton's pier (cf. Maher *et al.* 2003; Ray and Ryder 2003), or both qualitative and quantitative methods, such as Marshall's (2001) combination of participant observation and interviews and quantitative methods to explore the different roles of men and women in relation to tourism on Grand Manan Island (cf. Davies 2003). These contributions all offer some indication that third-moment principles have started to be taken on board more widely. In this genre we see more attention given to methodological, epistemological and ontological issues, but the researcher's personal/intellectual biographies and participants' voices still remain hidden.

Crisis in representation

Denzin and Lincoln's fourth moment, 'the crisis in representation', is described as a profound rupture in thinking about research. Here, for the first time, the role of the researcher as all-knowing creator of knowledge was challenged, and the rigour of generalisation disputed. Questions were raised around issues of gender, class and race, such as what kinds of impact embodied characteristics have on the kinds of research questions researchers decide to ask, and the ways in which they interpret the answers to those questions. Critics began to argue that the personal biography of the researcher was critical in determining the way they approached research, and thus it was not possible, as had previously been argued, to simply replace one researcher with another and expect the same results, provided that the methods employed were unchanged. Instead of seeing research as a linear scientific process, researchers began to view all stages in the research as indistinct and overlapping. They acknowledge that there are multiple interpretations mediated by the personal biographies of researchers and their research subjects. Riley and Love (2000) found little evidence that tourism researchers have breached the territory of the fourth moment. They refer to only a single one such article (but fail to cite it) wherein the researcher adopts the use of the first person to write themselves 'into' the text, thereby considering the impact of their personal biography. They found no other articles that explored the fundamental role of the researcher in the interpretation of phenomena or construction of the text.

Our post-1996 review found that explicit examinations of researchers' embodied characteristics continue to be rare in tourism studies. Issues of embodiment and 'the self' have, however, started to gain attention, particularly the attention of feminist leisure writers who have turned to tourism research (Pritchard and Morgan 2000a,b; Fullagar 2002; Galani-Moutafi 2000; Wearing and Wearing 2001; Crouch 2000). Further, there is some evidence to suggest that a greater degree of reflexivity has started to influence more recent tourism works. For example, in Dyer *et al.*'s (2003) study of tourism impacts on the indigenous Australian Djabugay community, they use a critical interpretive methodology and discuss how two of the researchers could be described as 'Anglo-Australian descendants of colonisers' responsible for the legacy of the Djabugay people, and acknowledge the need to consider the colonial history of indigenous populations and how researchers' own historical identities may affect the research process. Further, they question the adequacy of the authors' interpretation and suggest that interpretation should be left open in order for readers to reach their own conclusions. Other examples include Ray and Ryder's (2003) research on travel needs and motivations of the mobility-disabled, in which they set the scene by explaining how the experiences of one of the authors of travelling with a disability had sparked their interest in this area of work. In Fullagar's (2002) contribution on narratives of travel and feminine subjectivity, she writes herself into the text, acknowledging the 'I', and as part of the overall method draws on excerpts from her own travel diaries to examine how different trajectories of desire structure the movement of feminine subjectivity (cf. Erb 2000).

Consideration of other aspects, particular those concerned with the validation of data interpretation, illuminated research characteristics closely aligned with the 'crisis in representation' genre. The existence of multiple interpretations and the questioning of the researchers' interpretation of data and privileging of the reader's own judgement of the text can be found, for example, in McGregor's (2000) examination of the relationship between guidebooks and tourists, and the importance placed on subjective accounts and individual interpretations of texts as socially constructed materials encoded by their authors (see also Ateljevic and Doorne 2002; Aitchison 2001). In these studies, researchers gave issues of interpretation greater resonance, with attempts to account for their own biases and subjectivities by opening up the analysis to multiple interpretations. In Bricker and Kerstetter's (2002) study of white-water destinations in the United States, a team of experts was assigned to verify and check the coding and thematic framework of their analysis. Others who have realised the value of multiple interpretations have circulated transcripts among peer researchers and/or research participants (Ap and Wong 2001). Another distinct feature of tourism studies that have entered the fourth moment is the purposeful use of personalised accounts and participants' voices to illustrate 'real', lived experience. While glimmers of this are scattered within research of the earlier moments, the distinction between those and fourth-moment studies is the prominence and priority placed on making space, both practically and theoretically, for voices to come alive and be heard in situations of some respect (Kayat 2002; Morgan 2002; Burns and Sancho 2003; Pritchard *et al.* 2000).

Fifth moment

The final stage in the development of qualitative research has been termed the fifth moment (Denzin and Lincoln 1998). This stage emanates from ideas about the pluralisation of social life which emerged during the 'blurred genres' stage. Rapid social change and the development of new social contexts make the traditional approaches to research – those based on the notion that data can be interpreted objectively and then generalised to become a fact which reveals some knowledge about a singular reality – redundant (Flick 1998). This stage sees the end of grand narratives aimed at explaining supposedly universal phenomena for everybody and focuses instead on specific, delimited, local research. Theories from this perspective are now read as being context specific, and the authority of the researcher as objective expert has been rejected. Examples of tourism research characteristic of the fifth moment had, until Riley and Love's (2000) evaluation, still to emerge. More recently, Jamal and Hollinshead have argued that in order to move towards more interpretive, qualitative tourism research, it is necessary to depart from more static, quantitative and positivist knowledge bases to more dynamic, experiential and reflexive approaches. Here there is recognition that social agents are central to the construction of knowledge and that the researcher's voice is one among many that influence the research process (2001: 67). Further, the researcher's standpoint, values and biases – that is, their cultural background, ethnicity, age, class, gender, sexuality, and so on – play a role in shaping the researcher's historical trajectory, and the way in which they interpret phenomena and construct texts. In Denzin's words (1997: 220), there are no 'stories out there waiting to be told and no certain truths waiting to be recorded'; there are only 'stories yet to be constructed'. Jamal and Hollinshead (2001) call for tourism to acknowledge these issues and for researchers to develop their approach to research accordingly.

The post-1996 review revealed a range of interesting discussion pieces that have engaged with issues and debates located within the fifth moment. Research located within an empirical context was less common. Elements of fifth-moment principles were evident in Ryan and Hall's (2001) book *Sex Tourism: Marginal People and Liminalities*. This book offers a more open and reflexive account than is the norm. In explaining the process used to develop the script, Ryan and Hall (2001) explain how it was initially their intention to examine issues of sex tourism from the 'classical positivistic stance of non-detachment', but how over time their intent and approach changed. From the approach finally pursued, it is possible to trace the intellectual biography and contribution each author makes to the book; they thus make themselves accountable for their own input, enabling the reader to make more of an informed judgement of the text. Further, they question the ways their own conclusions were reached and clearly state that it is 'not expected that all readers will share the same conclusions reached by the authors', thus acknowledging that the researchers' 'expert' opinion is something that is open to interpretation(s). Other aspects of the fifth moment which were identified were:

- Focus on the local in an attempt to give a 'snapshot' in time and place – for example, Bricker and Kersetter's (2002) exploration of the meanings attached by white-water recreationalists to the South Fork River and Kayat's (2002) consideration of residents' perceptions in Langkawi.
- Emphasis on the construction of identities from an individual perspective, such as the use of interviews by Doorne *et al.* (2003) to explore a range of arenas through which cultural identities are appropriated, constructed and traded through and around material objects of touristic exchange. Kayat's (2002) consideration of social exchange and power also falls into this area.

The challenges of applying Denzin and Lincoln's model to tourism research

There are certain difficulties with applying Denzin and Lincoln's (1998) model to examples of empirical tourism research. For instance, while Riley and Love (2000) examined journals from their launch through to 1996, the fact that the earliest published journal did not commence until the early 1970s demonstrates that tourism as a subject is relatively new, and consideration of research issues in tourism is immature relative to that in other social science fields. This means that we cannot expect Denzin and Lincoln's (1998) commentary on the temporal dimension of research trends in other social science disciplines to be directly mirrored by tourism research.

Additionally, undertaking this analysis highlighted the complexities of categorising research into categories. Though the analysis enabled developments within tourism research to be sketched over time, in practice the boundaries between different research moments were blurred. This partly comes from the artificial nature of the exercise undertaken. The expectation that researchers should be able to take hundreds of articles from almost as many individuals and break them down into five categories was unrealistic, and to some extent symptomatic of the oversimplification that occurs in research, a problem that is critiqued in some detail in this book. However, the exercise offered a means to an end, a snapshot of the developments under way in qualitative tourism research, and was suitably manageable for a chapter in an edited book. One of the problems of publishing research is that it has to be offered in manageable chunks. This, we suspect, and Michael Hall argues later in the book (Chapter 8), constrains many tourism writers from presenting their findings in more expressive and detailed ways that give justice to the complexities present in everyday life. Hence, to some extent it could be argued that the way in which research is presented is a product of the medium in which it is published as well as the dominant ideologies of academe (see Hall for further discussion). However, despite being flawed, the post-1996 review revealed a number of tendencies in tourism research that probably relate to its multidisciplinary nature and the relative infancy of the field. These were the selective way in which researchers moved between moments and the eclectic approach to developing a methodology.

Denzin and Lincoln (1998) argue that many researchers operate in the moment that best fits the researcher's needs in relation to the research problem and the research setting. The 1996 review showed that some tourism scholars dipped into and out of the different moments depending on the research task in hand. This point can be illustrated by drawing on a handful of the various contributions made by Ryan. He could be seen, within an eight-year period, to undertake work that moves from traditional research comprising an ethnographic account of Maori experiences of tourism narrated using the authoritative voice (Ryan and Crotts 1997), to using conversation analysis to explore the subjective experiences of over-55s in Majorca, arguably from a crisis of representation perspective (Ryan 1995), to the use of phenomenographic analysis to explore individual experiences via a fifth-moment approach (Ryan 2000), to a reflexive feminist perspective on sex tourism workers (Ryan and Hall 2001). Such moves appear to be undertaken by more established tourism researchers, and may relate to their confidence in pushing the boundaries of tourism research in an incremental way within the accepted realms of tourism publishing, an area with which they are very familiar. This selective approach to deciding to adopt a particular approach should be applauded, as it encourages experimentation and sets a precedent for other, less experienced academics.

A further strand of this 'selectivity' which is far more common is the eclectic way in which many qualitative tourism researchers adopt methods and approaches from the different, not necessarily neighbouring, moments, mixing and matching many of the traits outlined above. This eclecticism combines a cocktail of techniques, principles and ways of presenting findings (see Ritchie 1998; Weaver and Fennell 1997). Such hybridisation of research approaches appears to be increasingly common, and perhaps indicative of the interdisciplinary nature of tourism and the influence of research practices that have been imported from other, non-tourism-related disciplines. It could be argued that one of the strengths of tourism research is that it is not bound to fixed disciplinary boundaries with their associated methods and is therefore free to combine a range of approaches and even research paradigms to give a more fluid approach to research. The experimental nature of this type of research could be argued to increase the potential for discovery but may pose difficulties for less experienced researchers, who might struggle to position themselves from an epistemological perspective within the field. In addition, there are sometimes contradictions within a paper whose authors use experimental methods that very much rely on the perspectives on the individuals being researched, but then write up their findings in a positivistic 'expert' voice. Thus, given that some researchers operate selectively between moments and many more operate eclectically, it was a challenge to position all papers within Denzin and Lincoln's framework. There are some which are so eclectic that they cannot be positioned anywhere – for example, Ray and Ryder's (2003) study of 'ebilities' tourism, which uses a mixed methods approach typical of blurred genres; includes some discussion of personal biographies, which would be expected in the crisis of representation; but generalises findings using the 'expert' voice associated with the traditional period (see also Westwood *et al.* 2000; Gibson *et al.* 2003). For most examples we were able to identify the dominant approach within any study and positioned the paper accordingly.

This sweeping account of the development of thinking about and researching the social world, and indeed tourism research, has taken us a long way. It has demonstrated, albeit in rather general terms, the ways in which other social science disciplines have contested the dominance of positivism, and explored other ways of knowing. It has moved us from the point where, it could be argued, the researcher was seen as the all-powerful interpreter of a unitary social world and their commentary was accepted and venerated as fact, to a situation where the presence of multiple realities is more accepted and the role of the researcher is sometimes questioned. Researchers can choose, and tourism researchers often elect, to operate within any of the above phases in research. They can move within and between different approaches and advance new ways of thinking and researching which fall outside those developed so far. The change in thinking about the social world and how it should be researched has profound implications for qualitative researchers. Traditionally marginalised within the academy by positivists on the grounds that qualitative methods were not capable of reaching and verifying facts, qualitative researchers in the social sciences generally now find themselves in a position where some commentators argue that they are the best equipped for tackling the crisis in representation. Tourism researchers have yet to fully follow the lead offered by their counterparts outside the field but are beginning to engage tentatively with some of the issues. This book will encourage them to embrace those issues in a more wholehearted manner.

Structure of the text

The aim of this book is to explore the potential of qualitative research to aid the construction of tourism knowledge. First and foremost, it seeks to promote greater discussion and consideration in tourism research of the ontological, epistemological and methodological issues emerging from all five qualitative research 'moments'. Furthermore, in order for readers to learn from the approach others have taken, it also seeks to present some reflexive accounts of tourism research which have moved beyond the mere use of qualitative methods and adopted a qualitative approach. In order to address this task, the book is divided into two parts. The first looks at some of the key ontological, epistemological and methodological issues in social science and applies them directly to the study of tourism.

Part I continues in Chapter 2 as the editors build upon their discussion of developments in qualitative research generally with a consideration of the main critiques of quantitative research. They outline the four main inquiry paradigms that underpin knowledge production and demonstrate how it is the elements of the inquiry paradigm (ontology, epistemology and methodology) that determine how research is undertaken and interpreted. In this respect, they argue that these key philosophical issues are of greater importance in shaping any research study than are methods, which are merely tools applied according to the methodological approach adopted. Jenny Phillimore and Lisa Goodson then move on to discuss the main paradigms employed in tourism research and the hesitancy of tourism researchers in challenging the dominant inquiry paradigms. They argue that in order to advance the state of tourism research, researchers need either to embrace or to challenge emergent thinking.

Chapter 3 considers epistemological issues in tourism. Here John Tribe begins by evaluating the disciplinary status of tourism and arguing that tourism should be viewed as a field rather than a distinct discipline. He then seeks to map out the structure of tourism knowledge, and distinguishes between multidisciplinary, interdisciplinary and extradisciplinary approaches to tourism. Tribe moves on to explore different issues and how they impact on knowledge creation in tourism, including discussions on power, culture, hegemony, ideology and values. He finishes by highlighting the importance of epistemological issues to tourism researchers.

In Chapter 4, Keith Hollinshead begins his consideration of ontological issues in tourism research. He argues that ontological matters relate to important questions of seeing, experiencing, meaning, being and identity in tourism, and that the very nature of tourism – the interactions it embodies between so many different individuals and places – means that ontological issues are particularly significant. He outlines the reasons why ontological issues are so pertinent and argues that qualitative research, and particularly recent developments in approach, offer exceptional opportunities for tourism researchers to tackle ontological challenges relating to the multifarious interpretations of being, meaning and identity. In the following chapter, Hollinshead develops his argument further in a consideration of ten key ontological challenges in tourism. For each challenge he outlines a sample issue in tourism studies, the area of ontological difficulty it epitomises and the suggested strategic gain in insight that qualitative research could offer.

In Chapter 6, Margaret Byrne Swain provides an overview of issues of embodiment in qualitative research before giving examples of good practice of the discussion of embodiment in research generally and in tourism research in particular. She uses the issue of embodiment to raise questions about the power dynamics between the researcher and researched, and calls for greater reflexivity to make these issues more transparent in tourism research. In a related chapter, Chapter 7, Barbara Humberstone identifies, critiques and develops thinking that recognises the importance of Other, insider perspectives and the partial nature of knowledge. She examines the constructs of standpoint and 'post'-standpoint research and the connections between standpoint and ecological perspectives, and in particular the links between standpoint research and research into ecotourism. Developing this example, she concludes by considering the usefulness of standpoint research for tourism research in general.

Chapter 8 builds upon the calls made by contributors in the previous chapters for greater reflexivity in tourism research. Here Michael Hall provides a wry account of the various issues which constrain tourism researchers, the methodologies they develop and indeed the issues they research. He calls for greater accountability and openness about the ways in which tourism research is commissioned, sponsored, judged and published, as well as for greater openness about how researchers' personal biography and ideological stance impact on their research.

Alain Decrop focuses upon the issue of trustworthiness in Chapter 9. He argues that positivist critiques of qualitative research and the questioning of its reliability have led to the need for alternative measures of reliability to be introduced. He sets out the basic criteria of trustworthiness and uses examples from tourism research to show how trustworthiness can be adopted in the field. Emphasis is placed on triangulation as one of the most effective techniques for achieving trustworthiness. Decrop suggests that this and other approaches to increasing trustworthiness have different applications depending on the research moment and inquiry paradigm within which the researcher operates.

In Chapter 10, Dennison Nash concludes Part I (except for a short linking chapter by the editors) by outlining concerns that much of the focus of anthropologists in tourism has been the nature of the relationship between hosts and guests. He argues that a more comprehensive picture could be constructed if researchers were also to look at tourism generation in Western metropolitan centres. Furthermore, he suggests that anthropological research in tourism could be greatly improved by being theoretically informed, empirically grounded and self-reflexive.

The second part of the book is very much aimed at providing examples of how qualitative research, using a range of techniques, can be applied in different ways in the field. This part of the book focuses on making the research process more transparent so that the reader can gain an understanding of how research works in practice and how it is influenced by the research paradigm of the individual researcher(s). It is hoped that readers will be able to gain from the experiences of these contributors, who have been charged with the task of describing the research process 'warts and all' so that others get the chance to learn from their experiences, good and bad.

Part II begins with Karen Thomas's discussion in Chapter 12 of the research process as a journey. Thomas considers some of the wide range of influences which impacted upon her research into tourist motivations in the United Kingdom. She looks at the use of focus groups in a multi-method study underpinned by a positivist inquiry paradigm and explores how her philosophical stance evolved over time as decisions were made throughout the research process. In Chapter 13, Fiona Jordan and Heather Gibson continue the theme of research as a constantly evolving process, but this time from an interpretive perspective. Their research concerned a cross-cultural comparison of women solo travellers from the United States and the United Kingdom. They consider how their approach to recruiting interviewees and the development of research questions evolved over time in relation both to events occurring in the field and to their joint working. They also explore the challenges of working collaboratively across continents and the issues they needed to consider when disseminating their findings.

In Chapter 14, Adele Ladkin discusses the use of life history analysis, its historical development and the strengths and weaknesses of the method. She looks specifically at how the technique was applied in an exploration of the hospitality labour market underpinned by a positivist inquiry paradigm, and considers wider applications for the method in tourism. In Chapter 15, Jennie Small explores the use of memory-work in the study of women's experiences of tourism in New Zealand. While this method could be considered to come out of the life history stable, Small's approach to research is very different to that of Ladkin. She explores the way in which an interpretive approach underpins use of the technique to recruit the researched as co-researchers in a bid to reduce power differentials that was ultimately compromised by the constraints of the doctoral process.

The final three chapters look at ethnographic studies in developing countries and the different relationships that the researchers developed with their research participants. In Chapter 16, Jill Belsky explores issues surrounding the politics of ecotourism in Belize using participant observation and in-depth interviewing. She discusses the opportunities and challenges that the use of these methods afforded the project, the benefits of working with a team of researchers from different cultural backgrounds and the ways in which participatory research could benefit both the researchers and the researched. In contrast, in Chapter 17, Stroma Cole considers her experiences as a lone researcher undertaking a range of different projects in Indonesia. This chapter focuses on the importance of longitudinal research in tourism settings and the impact that building lasting personal relationships has on both the research product and the researcher as an individual. Cole outlines how she adapted techniques such as focus groups from their application in Western settings to much less conventional usage. The book concludes with a quirky contribution in Chapter 18 from Guy Jobbins. While this chapter also focuses on relationships between the researcher and his Moroccan and Tunisian respondents, it begins from a much more positivistic stance, with the researcher questioning the reliability of data collected and eventually realising that misinterpretations came from cultural misunderstandings on the part of the researcher rather than, as he anticipated, the researched.

In order to give some continuity to the book and to aid learning throughout the text, each chapter has some features that need a little explanation. All chapters end with a bullet-pointed summary that sets out the key issues which the reader might care to consider when embarking on tourism research – whether this be at the stage of thinking about problems or that of developing a methodology. In addition, each chapter offers a number of think-point questions. These give the reader the opportunity to reflect over the text and to consider in greater detail some of the key issues that have been discussed.

Questions

1 Denzin and Lincoln (1998) discuss five 'moments' in relation to the evolution and development of qualitative research. What are these five moments called and what are the key characteristics of each?

2 Select a qualitative tourism research journal and examine the methodological approach taken. Decide which of Denzin and Lincoln's (1998) five moments it is best aligned to. Outline:

   • your reasons for arriving at that decision;
   • any difficulties you encountered in making your decision;
   • how useful Denzin and Lincoln's (1998) framework is in analysing methodological approaches.

3 The review of tourism articles in this chapter highlights the fact that

the 'traditional' moment has been the predominant approach influencing qualitative tourism research to date. What are the key factors that may have contributed to this trend?

References

Aitchison, C. (1999) 'Heritage and nationalism: gender and the performance of power'. In D. Crouch (ed.) *Leisure/Tourism Geographies: Practices and Geographical Knowledge*, London: Routledge.

Aitchison, C. (2001) 'Theorising other discourses of tourism, gender and culture: can the subaltern speak (in tourism)?', *Tourist Studies*, 1 (2): 133–147.

Aitchison, C. and Reeves, C. (1998) 'Gendered (bed) spaces: the cultural and commerce of women only tourism'. In C. Aitchison and F. Jordan (eds) *Gender, Space and Identity: Leisure, Culture and Commerce*, Brighton: Leisure Studies Association.

Andersen, V., Prentice, R. and Guerin, S. (1997) 'Imagery of Denmark among visitors to Danish fine arts exhibitions in Scotland', *Tourism Management*, 18 (7): 453–464.

Ap, J. and Wong, K. (2001) 'Case study on tour guiding: professionalism, issues and problems', *Tourism Management*, 22 (5): 551–563.

Ateljevic, I. and Doorne, S. (2002) 'Representing New Zealand: tourism imagery and ideology', *Annals of Tourism Research*, 29 (3): 648–667.

Black, R. (1993) *Evaluating Social Research*, London: Sage.

Botterill, D. (2001) 'The epistemology of a set of tourism studies', *Leisure Studies*, 20: 199–214.

Bricker, K. and Kerstetter, D. (2002) 'An interpretation of special place meanings whitewater recreationists attach to the South Fork of the American river', *Tourism Geographies*, 4: 396–425.

Bryman, A. (2001) *Social Research Methods*, Oxford: Oxford University Press.

Bryman, A. and Burgess, R. (eds) (1994) *Analysing Qualitative Data*, London: Routledge.

Buhalis, D. and Licata, C. (2002) 'The future eTourism intermediaries', *Tourism Management*, 23: 207–220.

Burns, P. and Sancho, M. (2003) 'Local perceptions of tourism planning: the case of Cuéllar, Spain', *Tourism Management*, 24: 331–339.

Butler, R (1975) 'Tourism as an agent of social change', in F. Helleiner (ed.) *Tourism as a Factor in National and Regional Development*, Occasional Paper no. 4. Department of Geography, Trent University, Ontario.

Butler, R. (1980) 'The concept of a tourism area cycle of evolution: implications for management of resources', *Canadian Geographer*, 24: 5–12.

Carr, N. (2002) 'Going with the flow: an assessment of the relationship between young people's leisure and holiday behavior', *Tourism Geographies*, 4: 115–134.

Code, L. (1993) 'Taking subjectivity into account'. In L. Alcoff and E. Potter (eds) *Feminist Epistemologies*, London: Routledge.

Cohen, E. (1982) 'Marginal paradises: bungalow tourism on the islands of southern Thailand', *Annals of Tourism Research*, 9: 189–228.

Cohen, S. and Taylor, L. (1976) *Escape Attempts*, Harmondsworth: Penguin.

Connell, J. and Lowe, A. (1997) 'Generating grounded theory from qualitative data: the application of inductive methods in tourism and hospitality management research', *Progress in Tourism and Hospitality Research*, 3: 165–173.

Crouch, D. (2000) 'Places around us: embodied lay geographies in leisure and tourism', *Leisure Studies*, 19: 63–76.

Dann, G. (1992) 'Travelogues and management of unfamiliarity', *Journal of Travel Research*, 30: 59–63.

Dann, G. (1996) *The Language of Tourism: A Sociolinguistic Perspective*, Oxford: CAB International.

Davies, B. (2003) 'The role of quantitative and qualitative research in industrial studies of tourism', *International Journal of Tourism Research*, 5: 97–111.

Decrop, A. (1999) 'Triangulation in qualitative tourism research', *Tourism Management*, 20: 157–161.

Denzin, N. K. (1997) *Interpretive Ethnography*, London: Sage.

Denzin, N. K. and Lincoln, Y. S. (1998), *The Landscape of Qualitative Research: Theories and Issues*, Thousand Oaks, CA: Sage.

Doorne, S., Ateljevic, I. and Bai, Z. (2003) 'Representing identities through tourism: encounters of ethnic minorities in Dali, Yunnan Province, People's Republic of China', *International Journal of Tourism Research*, 5: 1–11.

Doxey, G. (1975) 'A causation theory of visitor–resident irritants, methodology and research inferences'. In *The Impact of Tourism, Sixth Annual Conference, Proceedings of the Travel Research Association*, San Diego.

Dyer, P., Aberdeen, L. and Schuler, S. (2003) 'Tourism impacts on an Australian indigenous community: a Djabugay case study', *Tourism Management*, 24: 83–95.

Echtner, C. (1999) 'The semiotic paradigm: implications for tourism research', *Tourism Management*, 20 (1): 47–57.

Enoch, Y. (1996) 'Contents of tour packages: a cross-cultural comparison', *Annals of Tourism Research*, 23 (3): 599–616.

Erb, M. (2000) 'Understanding tourists: interpretations from Indonesia', *Annals of Tourism Research*, 27 (3): 709–736.

Fairweather, J. and Swaffield, S. (2001) 'Visitor experiences of Kaikoura, New Zealand: an interpretative study using photographs of landscapes and Q method', *Tourism Management*, 22: 219–228.

Flick, U. (1998) *An Introduction to Qualitative Research*, London: Sage.

Fullagar, S. (2002) 'Narratives of travel: desire and the movement of feminine subjectivity', *Leisure Studies*, 21: 57–74.

Galani-Moutafi, V. (2000) 'The self and the other traveller, ethnographer, tourist', *Annals of Tourism Research*, 27 (1): 203–224.

Gibson, H., Willming, C. and Holdnak, A. (2003) 'Small-scale event sport tourism: fans as tourists', *Tourism Management*, 24: 181–190.

Glaser, B. and Strauss, A. (1967) *The Discovery of Grounded Theory*, Chicago, IL: Aldine.

Gottlieb, A. (1982) 'Americans' vacations', *Annals of Tourism Research*, 9: 165–187.

Greenwood, D. (1978) 'Culture by the pound: an anthropological perspective on tourism as cultural commoditization'. In V. Smith (ed.) *Hosts and Guests*, Oxford: Blackwell.

Guba, E. and Lincoln, Y. S. (1998) 'Competing paradigms in qualitative research'. In N. K. Denzin and Y. S. Lincoln (eds) *The Landscape of Qualitative Research: Theories and Issues*, Thousand Oaks, CA: Sage.

Higham, J. and Hinch, T. (2002) 'Tourism, sport and seasons: the challenges and potential of overcoming seasonality in the sport and tourism sector', *Tourism Management*, 23: 175–185.

Hollinshead, K. (1996) 'The tourism researcher as *bricoleur*: the new wealth and diversity in qualitative inquiry', *Tourism Analysis*, 1 (1): 67–74.

Hollinshead, K. (1999) 'Surveillance of the worlds of tourism: Foucault and the Eye-of-Power', *Tourism Management*, 20: 7–23.

Jamal, T. and Hollinshead, K. (2001) 'Tourism and the forbidden zone: the under-served power of qualitative inquiry', *Tourism Management*, 22: 63–82.

Jansen-Verbeke, M. and van Rekom, J. (1996) 'Scanning museum visitors: urban tourism marketing', *Annals of Tourism Research*, 23 (2): 364–375.

Jenkins, O. (1999) 'Understanding and measuring tourist destination images', *International Journal of Tourism Research*, 1 (1): 1–15.

Johnston, C. (2001) 'Shoring the foundations of the destination life cycle model, part 1: Ontological and epistemological considerations', *Tourism Geographies*, 3: 2–28.

Jordan, F. (1997) 'An occupational hazard? Sex segregation in tourism employment', *Tourism Management*, 18 (8): 525–534.

Jutla, R. (2000) 'Visual image of the city: tourists' versus residents' perception of Simla, a hill station in northern India', *Tourism Geographies*, 2: 404–420.

Kayat, K. (2002) 'Power, social exchanges and tourism in Langkawi: rethinking resident perceptions', *International Journal of Tourism Research*, 4: 171–191.

Kinnaird, V. and Hall, D. (1994) *Tourism: A Gender Analysis*, Chichester: Wiley.

Kozak, M. (2002) 'Comparative analysis of tourist motivations by nationality and destinations', *Tourism Management*, 23: 221–232.

Ladkin, A. (2002) 'Career analysis: a case study of hotel general managers in Australia', *Tourism Management*, 23 (4): 379–388.

Lewins, F. (1992) *Social Science Methodology: A Brief but Critical Introduction*, Melbourne: Macmillan Education Australia.

McGregor, A. (2000) 'Dynamic texts and tourist gaze: death, bones and buffalo', *Annals of Tourism Research*, 27 (1): 27–50.

Maher, P., Steel, G. and McIntosh, A. (2003) 'Examining the experiences of tourists in the Antarctic', *International Journal of Tourism Research*, 5: 29–67.

Markwick, M. (2002) 'Golf tourism development: stakeholders, differing discourses and alternative agendas: the case of Malta', *Tourism Management*, 21 (5): 515–524.

Marshall, J. (2001) 'Women and strangers: issues of marginalization in seasonal tourism', *Tourism Geographies*, 3: 165–186.

May, T. (1993) *Social Research: Issues, Methods and Process*, Buckingham: Open University Press.

Miller, G. (2001) 'The development of indicators for sustainable tourism: results of a Delphi survey of tourism researchers', *Tourism Management*, 22: 351–362.

Mitchell, R. and Eagles, P. (2001) 'An integrative approach to tourism: lessons from the Andes of Peru', *Journal of Sustainable Tourism*, 9 (1): 4–28.

Morgan, D. (2002) 'A new pier for New Brighton: resurrecting a community symbol', *Tourism Geographies*, 4: 426–439.

Morrison, A. (2002) 'Hospitality research: a pause for reflection', *International Journal of Tourism Research*, 4: 161–169.

Paradis, T. (2002) 'The political economy of theme development in small urban places: the case of Roswell, New Mexico', *Tourism Geographies*, 4: 22–43.

Pavlovich, K. (2003) 'The evolution and transformation of a tourism destination network: the Waitomo Caves, New Zealand', *Tourism Management*, 24: 203–216.

Pearce, P. (1984) 'Tourist–guide interaction', *Annals of Tourism Research*, 11: 12–46.

Preston-Whyte, R. (2002) 'Constructions of surfing space at Durban, South Africa', *Tourism Geographies*, 4: 307–328.

Pritchard, A. and Morgan, N. (2000a) 'Privileging the male gaze: gendered tourism landscapes', *Annals of Tourism Research*, 27 (4): 884–905.

Pritchard, A. and Morgan, N. (2000b) 'Constructing tourism landscapes: gender, sexuality and space', *Tourism Geographies*, 2 (2): 115–139.

Pritchard, A., Morgan, N., Sedgley, D., Khan, E. and Jenkins, A. (2000) 'Sexuality and holiday choices: conversations with gay and lesbian tourists', *Leisure Studies*, 19: 267–282.

Ray, N. and Ryder, M. (2003) '"Ebilities" tourism: an exploratory discussion of the travel needs and motivations of the mobility-disabled', *Tourism Management*, 24: 57–72.

Riley, R. (1996) 'Revealing socially constructed knowledge through quasi structured interviews and grounded theory analysis', *Journal of Travel and Tourism Marketing*, 5: 21–40.

Riley, R. and Love, L. (2000) 'The state of qualitative tourism research', *Annals of Tourism Research*, 27 (1): 164–187.

Ritchie, B. (1998) 'Managing the human presence in ecologically sensitive tourism destinations: insights from the Banff–Bow Valley study', *Journal of Sustainable Tourism*, 6 (4): 293–313.

Ritchie, R. and Ritchie, B. (2002) 'A framework for an industry supported destination marketing information system', *Tourism Management*, 23: 439–454.

Rodriguez, A. (2002) 'Determining factors in entry choice for international expansion: the case of the Spanish hotel industry', *Tourism Management*, 23: 597–607.

Rojek, C. (1993) *Ways of Escape: Modern Transformations in Leisure and Travel*, London: Macmillan.

Rojek, C. (1997) 'Indexing, dragging and the social construction of tourist sights'. In C. Rojek and J. Urry (eds) *Touring Cultures: Transformations of Travel and Theory*, London: Routledge.

Ryan, A. (1979) *The Philosophy of the Social Sciences*, London: Macmillan.

Ryan, C. (1995) 'Learning about tourists from conversations: the over 55s in Majorca', *Tourism Management*, 16 (3): 207–215.

Ryan, C. (2000) 'Tourist experiences, phenomenographic analysis, post-positivism and neural network software', *International Journal of Tourism Research*, 2 (2): 119–131.

Ryan, C. and Crotts, J. (1997) 'Carving and tourism: a Maori perspective', *Annals of Tourism Research*, 25 (4): 898–918.

Ryan, C. and Hall, M. (2001) *Sex Tourism: Marginal People and Liminalities*, London: Routledge.

Sandelowski, M. (1994) 'Notes on transcription', *Research in Nursing and Health*, 17: 311–314.

Scheyvens, R. (2000) 'Promoting women's empowerment through involvement in ecotourism: experiences from the Third World', *Journal of Sustainable Tourism*, 8 (3): 232–249.

Selin, S. (1999) 'Developing a typology of sustainable tourism partnerships', *Journal of Sustainable Tourism*, 7 (3 and 4): 260–273.

Silverman, D. (2000) *Doing Qualitative Research: A Practical Handbook*, London: Sage.

Smith, V. (1977) *Hosts and Guests: The Anthropology of Tourism*, Philadelphia, PA: University of Pennsylvania Press.

Smith, V. (1998) 'War and tourism: an American ethnography', *Annals of Tourism Research*, 25 (1): 202–227.

Stern, P. (1979) *'Evaluating Social Science Research*, Milton Keynes: Open University Press.

Strauss, A. and Corbin, J. (1998) *Basics of Qualitative Research: Techniques for Developing Grounded Theory*, 2nd edn, Thousand Oaks, CA: Sage.

Stymeist, D. (1996) 'Transformation of Vilavilairevo in tourism', *Annals of Tourism Research*, 23 (1): 1–18.

Swain, M. (1995) 'Gender in tourism', *Annals of Tourism Research*, 22 (2): 247–267.

Telfer, D. and Wall, G. (1996) 'Linkages between tourism and food production', *Annals of Tourism Research*, 23 (3): 635–653.

Towner, J. (1985) 'The grand tour: a key phase in the history of tourism', *Annals of Tourism Research*, 12: 297–333.

Urry, J. (1990) *The Tourist Gaze*, London: Sage.

Verbole, A. (2000) 'Actors, discourses and interfaces of rural tourism development at the local community level in Slovenia: social and political dimensions of the rural tourism development process', *Journal of Sustainable Tourism*, 8 (6): 479–490.

Vidich, A. J. and Lyman, S. M. (1994) 'Qualitative methods: their history in sociology and anthropology'. In N. K. Danzin and Y. S. Lincoln (eds) *Handbook of Qualitative Research*, Thousand Oaks, CA: Sage.

Wall, G. (1996) 'Perspectives on tourism in selected Balinese villages', *Annals of Tourism Research*, 23 (1): 123–137.

Walle, A. (1997) 'Quantitative verses qualitative tourism research', *Annals of Tourism Research*, 24 (3): 524–536.

Wearing, S. and Wearing, B. (2001) 'Conceptualizing the selves of tourism'. *Leisure Studies*, 20: 143–159.

Weaver, B. and Fennell, D. (1997) 'The vacation farm sector in Saskatchewan: a profile of operations', *Tourism Management*, 18 (6): 357–365.

Westwood, S., Pritchard, A. and Morgan, N. (2000) 'Gender-blind marketing: businesswomen's perceptions of airline services', *Tourism Management*, 21 (1): 353–362.

Wickens, E. (2002) 'The sacred and the profane: a tourist typology', *Annals of Tourism Research*, 29 (3): 834–851.

2 The inquiry paradigm in qualitative tourism research

Lisa Goodson and Jenny Phillimore

Aims of the chapter

- To examine debates concerning quantitative verses qualitative research *methods* and highlight why greater consideration of *methodology* is pertinent to the further advancement of qualitative tourism research.
- To discuss issues relating to the concept of *inquiry paradigm* and how this is inextricably linked to wider ontological, epistemological and methodological issues which underpin the research process.
- To highlight some of the emergent epistemological and methodological debates in tourism research.

Introduction

In Chapter 1, consideration of Denzin and Lincoln's (1998) framework on the evolution and development of qualitative research provided some indication of the range of issues yet to be explored by tourism researchers. Further, the review of tourism journals discussed in that chapter has served to illustrated that, to date, qualitative tourism studies have been predominantly located within Denzin and Lincoln's (1998) first two 'moments' ('traditional' and 'modernist'), with some evidence of third-moment studies ('blurred genre') emerging and discussions of fourth- ('crisis in representation') and fifth-moment issues beginning to filter through to influence research practices.

Although tourism researchers have started to question the shortcomings of positivism and quantification on the grounds that they are not fully equipped to explore questions of meaning and understanding (e.g. Hollinshead 1996; Riley 1996; Walle 1997; Dann 1996), they have yet to expose themselves to the wider range of qualitative approaches which will enable them to begin to tackle issues around the authority of interpretation and access the multiple realities associated with lived experience. The recognition and incorporation of a variety of qualitative methods has become more commonplace in tourism, along with the mixing of both quantitative and qualitative methods. However, although epistemological and ontological issues are beginning to be more visible at a theoretical level, the influence of the full range of research paradigms is yet to emerge fully in practice. Over time, the quest for more in-depth insights should, one hopes, fuel exploration of more innovative research practices and engagement with more sophisticated and contemporary debates surrounding epistemology, ontology and methodology which have gained momentum in other disciplines and form the focus of the latter part of this chapter.

This chapter considers some of the debates surrounding the primacy of quantitative *vis-à-vis* qualitative research and focuses attention on the main critiques of quantitative approaches with a view to uncovering some of the reasons why qualitative approaches have been gaining momentum in the social sciences. It then seeks to highlight the growing importance of issues concerning methodology and how these have superseded rudimentary considerations of methods. The concept of the inquiry paradigm is then developed in a discussion which focuses on the four main research paradigms, the relationship between paradigm and research approach, the increasing importance of relationships between researcher and researched, and how reflexive exploration of inquiry paradigms can benefit research. The chapter concludes by considering some of the key issues emerging from recent fourth- and fifth-moment contributions in tourism.

Moving from quantitative to qualitative

The quantitative approach to research has traditionally been informed by a deterministic outlook, with the focus being on producing a hypothesis, indicating how it will be tested, testing it, and then verifying or modifying the hypothesis on the basis of the research findings (Robson 1993). Accordingly, quantitative researchers have often sought to abstract the phenomenon that is being studied from the rest of the social world and to fix meaning within what might be described as a contextual vacuum. Generalisation of findings to the entire social world was seen as the main aim, with the intention of extrapolating results to include entire social groups if, for example, there was statistical evidence to suggest that on the balance of probabilities, this was possible (Guba and Lincoln 1998). This approach assumed that the relationship between researcher and researched was unidirectional. In addition, it was argued that while the researcher sought information from or about the researched, the process of seeking that information had little impact on the behaviour of the individual being researched. This often meant that little account of any impact was made in the interpretation of the data because it was assumed that carefully designed methods would completely overcome demand characteristics.

For reasons of brevity, we have offered almost a caricature of quantitative research based upon the criticisms made of its main premises. As Chapter 1 has suggested, these criticisms have emerged with changing attitudes about the nature of knowledge. The view emanating from the 'blurred genres' phase (Denzin and Lincoln 1998) suggested that empirical fact could only be described as contemporary knowledge, given rapid changes in the social world and the impossibility of knowledge being able to remain static in this mercurial environment (Hamilton 1994; Flick 1998). More specifically, some of the premises on which the production of quantitative knowledge was based have been challenged in the light of the crisis in representation as researchers began to view the research product as an account of the research process even when it employed so-called objective statistics (Farran 1990). Guba and Lincoln (1998: 196) discuss in some depth the main criticisms of quantification, raising concerns, echoed elsewhere (e.g. Maynard and Purvis 1994; Warren 1988; Bryman and Burgess 1994; Ritchie and Spencer 1994; Maynard 1994; Flick 1998; May 1993), about the appropriateness of generalising data from one context, for example a laboratory or sample, and extending the generalisation across the entire social world. Feminists have been at the forefront of these criticisms, pointing out that much quantitative research extrapolated results from research into male behaviour across the entire population (Alcoff and Potter 1993; Stanley and Wise 1993; Harding 1993; Nash 1994; Smith 1988). Other researchers have demonstrated that findings were also based on research into white Europeans (Bhavnami 1993; Fine 1998; hooks 1989, 1990; Ladner 1971; Phoenix 1994). Naive quantification paid little attention to gender, ethnicity, age, class or able-bodiedness while being overly concerned with theorising across the entire social world. The differences which exist within and between all communities were given little attention despite, it is argued, their being critical to the understanding of social life.

In addition to problems with what was being researched and what was achievable using the quantitative approach, there were criticisms about quantitative researchers' contribution to the development of social science knowledge. Quantitative researchers were said to gloss over methodological issues that were critical to knowledge production. Little attention was paid to understanding the creative process of knowledge production in order to share vital experience with other researchers. Where methodologies were written, they focused entirely on an objective account of methods used. Furthermore, most emphasis was placed on predicting what might happen in particular circumstances rather than seeking explanation and understanding of processes which determined behaviour. Smith (1988) describes one of the main problems as the production of bifurcated consciousness: a situation where the subjects of research do not recognise themselves in the research findings. Put simply, the production of grand, generalised theory, no matter how statistically meaningful, made little sense to individuals within the population researched. Other criticisms include the likelihood that one set of facts will support a whole range of different theories, throwing doubt on the scientific argument that there is one 'real' truth. Also discounted was the possibility that research was a one-way process. Indeed, it was argued that it should be viewed as an interactive process whereby the researcher, merely through being present to observe or question, impacts upon the responses or behaviour of the researched. Following on from this, the perspective the researcher develops about the researched then impacts on the way in which they collect and interpret findings. Finally, the whole notion of objective, value-free research has been challenged (Guba and Lincoln 1998) with the argument that every researcher brings something different to a study: different attitudes, values, perspectives, ideologies, etc., all of which impact upon the research from its inception to its dissemination. Perhaps the inevitable and unavoidable impact of the individual upon their research would be viewed as more acceptable if people from across the whole of society were involved in research. However, critics point out that knowledge production is a reflection of existing power relations (Warren 1988). It is often only those individuals with power who are in a position to be able to undertake research and to have their findings supported by the establishment. They are generally white Western, middle-class males (Code 1993; Alcoff and Potter 1993; Stanley and Wise 1990, 1993; Kincheloe and McLaren 1993).

Severe though these criticisms might be, their intent was not to undermine the production of knowledge but to highlight shortcomings in existing approaches so that they can be overcome and ultimately the production of knowledge made more rigorous. While these criticisms have been made about quantification, the aim is not to dispense with the use of quantitative methods. The intention is to resolve problems of naive knowledge production: that which does not take account of the interactive, value-laden nature of the research process. Critics were calling for a change in the way knowledge was created so that its production could become more accountable. Accountability has become more important as we begin to move from seeing social science research as a scientific activity to viewing it as a knowledge-production process that is socially situated. So where do qualitative research methods sit within this process? Quantitative and qualitative methods of all kinds are embedded within the research process, but while they are critical to the process, they are merely one aspect of it (Flick 1998). Other aspects often include the biography and politics of the researcher, the nature of the research sponsor, the purpose of the research, the interpretation of data, and the presentation and dissemination of findings (Denzin and Lincoln 1998). Qualitative researchers have often raised the importance of these facets of the research process, and initiated discussions about their role in research. Although during the 'modernist' periods it was possible (and indeed still is) to locate arguments that called for the privileging of qualitative over quantitative, their main aim is to improve the quality of research. This can be achieved by choosing a range of methods to increase research potential (Maynard and Purvis 1994) and, most importantly, by understanding and explicating the role of the inquiry paradigm in the research process.

The role of the inquiry paradigm

Denzin and Lincoln (1998) make a cogent argument for viewing the researcher as *bricoleur* – an individual who pieces together sets of practices to make a solution to a puzzle. Here they view research as a messy, non-linear process, with the researcher being innovative and creative in seeking out the different pieces of the puzzle until they reach a point when they are able to present as complete a picture as possible. There is no one set of methods that can bring total insight, the concept of objectivity is rejected, and consequently there is no perfect outcome – no 'right' answer to research questions posed. The messy research process is highly subjective not through choice but because that is the nature of social research. The aim of the researcher is to take account of subjectivity, of their ethics, values and politics, and use a range of appropriate interconnected interpretive methods to maximise understanding of the research problem. The researcher's actions are underpinned by a basic set of beliefs that define their worldview. This basic set of beliefs is known as a paradigm. Inquiry paradigms define for researchers what falls within and outside legitimate inquiry (Guba and Lincoln 1998). A paradigm is a human construction which 'represents simply the most informed and sophisticated view that its proponents have been able to devise' (p. 202). There are three main elements to an inquiry paradigm: ontology, epistemology and methodology. Ontology is the study of being, and raises questions about the nature of reality while referring to the claims or assumptions that a particular approach to social inquiry makes about the nature of social reality (Denzin and Lincoln 1998). Epistemology is the theory of knowledge, and is interested in the origins and nature of knowing and the construction of knowledge, and the claims and assumptions that are made about what the nature of knowledge is (Longino 1990; Dalmiya and Alcoff 1993). Methodology is the study of how we collect knowledge about the world. Put simply, knowledge production relies heavily upon the ontology of the researcher – their definition of reality. Their epistemology – what they count as knowledge – depends on what they want knowledge about, while the kind of knowledge that they seek determines their methodology (Jones 1993). Within this methods are merely tools which take on meaning according to the methodology within which they are employed (Silverman 2000).

The researcher can identify their inquiry paradigm by answering three interconnected questions (Guba and Lincoln 1998: 201):

- the ontological question – what is the form and nature of reality, and what can be known about reality?
- the epistemological question – what is the nature of the relationship between the researcher and what can be known?
- the methodological question – how can the researcher find out what she/he believes can be known?

There are four major paradigms which structure research: positivist, post-positivist, critical and interpretive. Each provides flexible guidelines that connect theory and method and help to determine the structure and shape of any inquiry. Guba and Lincoln give an example of a positivist paradigm where the researcher believes only in the existence of the 'real' (observable) world. Thus all issues that relate to moral or aesthetic matters are excluded from inquiry. Given that a 'real' reality is assumed, the relationship between the researcher and reality can only be one of objective detachment or value freedom to determine how things really work. An objective inquirer would wish to control possible confounding factors and thus would seek methods that enable manipulation of variables.

As we have established in the above discussion, the positivist and post-positivist paradigms provide the context within which many researchers, particularly tourism researchers, operate. These paradigms tend to be associated with a quantitative approach, and are certainly associated with a particular view on the production of knowledge, namely that researchers are value free and neutral and can be substituted for one another without having an impact on findings. Indeed, they argue that researchers must be able to transcend subjectivity and disconnect knowledge from everyday life if there is to be any knowledge worthy of analysis (Kincheloe and McLaren 1994; Code 1993). These beliefs are disputed by those thinkers influenced by the critical or interpretive paradigms. Feminists in particular have argued that the ideal objectivity portrayed by positivists is actually a generalisation from the subjectivity of a small, select group of people, 'albeit a group that has power, security, prestige', that believes 'it can generalise its experiences and normative ideals across the whole social order' (Code 1993: 22). Instead they argue that values, politics and knowledge are interconnected rather than hierarchical, and thus there is a need to explore how knowers' values and politics impact upon the ways in which they undertake research and create knowledge (Alcoff and Potter 1993). In order to create a stronger, more defensible knowledge, researchers need to try to become aware 'of the ideological imperatives and epistemological presuppositions that inform their research as well as their own subjective, intersubjective and normative reference claims' (Kincheloe and McLaren 1994: 265).

Researchers influenced by interpretivist inquiry paradigms turn the conventional positivistic approach to knowing upon its head. Rather than arguing that only the qualified researcher is capable of knowledge production, they consider that the complex social world can be understood only from the point of view of those who operate within it. Thus, research is undertaken in a collaborative fashion, with the researcher and the researched viewed as partners in the production of knowledge and the interaction between them being a key site for both research and understanding (Schwandt 1998). Argument and discussion are central to this approach to knowledge production, while it is accompanied by an acceptance of academic and intellectual pluralism. All findings are the product of an interaction between researcher and researched, and viewed as fundamentally value mediated (Holland and Ramazanoglu 1994; May 1993; Guba and Lincoln 1998). The critical roles of both values and context in knowledge production mean that these two aspects of the research process have to be explored in some depth. This means undertaking research in a reflexive way whereby ethical, political and epistemological dimensions of research are explored as an integral part of producing knowledge (Marcus 1998). It is also important to ensure that the indexicality – that is, the contextual position – of knowledge is acknowledged through exploring how claims for knowledge relate to a particular temporal, geographical or social moment. From this perspective, only through openly reflexive interpretation can validity be claimed for any research, regardless of whether it is quantitative or qualitative (Holland and Ramazanoglu 1994).

We have argued that the traditional, positivistic approaches to research have produced forms of knowledge that some researchers consider questionable. This was partly because, in the first instance, they focused on phenomena abstracted from their social context and sought to explore what was going on without seeking explanations as to why, for example, people behave in a particular way. Second, the belief that there is one 'true' reality was considered to be flawed, as social life develops in a pluralistic fashion. Third, the failure of traditional research to take account of the means of its own production meant that critical parts of the research process – ethics, values and context – were not considered when conclusions were being drawn. These criticisms have, in the main, been accepted by qualitative researchers, largely because they are influenced by interpretive inquiry paradigms. In these quarters the need to undertake, present and disseminate research in a reflexive way has been accepted. The rationale behind this is that we need to explore exactly how knowledge was produced in order to ascertain the validity of the claims being made. Unless we can take account of the researcher's inquiry paradigm and how it influences the choices they make throughout the research process, then we are unable to explore how the values associated with their worldview may have impacted on judgements about issues ranging from selection of research topic to deciding what conclusions to reach. Furthermore, as Denzin and Lincoln (1998) have demonstrated in their discussion of the five moments in qualitative research (see Chapter 1 for further discussion), the ways in which we undertake research are constantly evolving. Research needs to be presented in a transparent way in order that we learn from it as well as judge it.

Understanding what can be offered by doing research in different ways is the first step along the long journey towards greater acceptance and legitimisation of non-'traditional' techniques. Clearly, such an understanding will emerge only if researchers enable others to view the various decisions made throughout the research process, the justification and evaluation of decisions taken in the light of their influence and ontological and epistemological standpoints. Through this opening up of the research process it is hoped to encourage others to develop other innovative approaches and advance the state of qualitative research in tourism. Although tourism research has, throughout its development, employed qualitative methods, it has yet to embrace, with any conviction, a qualitative approach to research. Despite the prominence of anthropological and sociological influences in qualitative tourism research, there would appear to be few who have pushed the paradigmatic boundaries. While it is clear that qualitative methods have become more widely used and, arguably, more accepted as a legitimate approach to research, it would appear that many researchers are still operating within the boundaries of a limited range of epistemological, ontological and methodological frameworks.

Research and discourse relating to tourism remain multidisciplinary, with a range of foci, approaches and styles that are not always congruent with one another. On the one hand, this may give the field dynamism and potentially offer new opportunities for cross-fertilisation of ideas, practices and processes. However, on the other hand there appears to have occurred a fragmentation of the field, with, for example, those interested in a practitioner and economic perspective holding competing and different research objectives as compared with those interested in social research and theory generation. Riley and Love (2000) noted that far fewer qualitative articles were published in journals with a practitioner emphasis (*Tourism Management, Journal of Travel and Tourism Marketing, Journal of Travel Research*), while *Annals of Tourism Research*, with its underlying academic bent, offered more extensive use of qualitative techniques. However, even here there was limited evidence of researchers challenging the dominant inquiry paradigms. The post-1996 review, discussed in Chapter 1, provided some evidence of a move towards constructivist and interpretive paradigms but with the focus being on discussion pieces about how such a move might be facilitated rather than on research embracing these new approaches.

A number of issues can be identified which shed some light on the possible reasons why positivist approaches have continued to dominate research in tourism. It is widely acknowledged that the interpretive paradigm has generally lagged behind its positivist predecessors in other research fields. What is more, it has been argued that tourism is less methodologically and theoretically advanced than other fields in the social sciences. The relative newness of the field, its late engagement with methodological issues and its tendency to embrace the more conventional approaches in social science may help explain the limited use of interpretive paradigms in the field.

The dominance of qualitative research in the first ('traditional') and second ('modernist') moments may be due to the reliance on familiar and more commonly documented qualitative methods well established by earlier anthropological and sociological influences. The transfer of these perspectives into the field of tourism underpinned the early seminal qualitative works in *Hosts and Guests* (Smith 1977), and to some extent have become markers for the style of subsequent work in tourism. For example, the 'gatekeepers' – the editors of tourism journals and organisers of forums – may still view works in the positivist, 'scientific' style, and written in the third person, as the expected protocol (Aitchison 2000; and see Hall in Chapter 8). Among the 'gatekeepers' of research, qualitative approaches remain less familiar. As a result, qualitative practices and processes tend to be viewed with caution or even suspicion and doubts cast over the reliability and validity of such work (Riley and Love 2000).

It is essential that those tourism researchers engaging with interpretive paradigms and qualitative methods and methodologies clearly justify their choice of approach and make visible their data collection and analysis procedures. Such an approach, or at the very least increased transparency, is essential to enable the audience to judge the quality of the research, the justifications for the approach taken and the rigour inherent in the research process. While sociologists and anthropologists themselves have progressed towards such approaches and practices to embrace stages 3, 4 and 5 of Denzin and Lincoln's (1998) framework, few of their tourism peers have yet succeeded in moving much beyond their earlier conceptual legacy. However, while the findings from the post-1996 review support Riley and Love's (2000) assertion that tourism research still appears to be dominated by the first and second moments, it did illuminate more examples of research that has emerged with elements characteristic of the latter moments: 'crisis in representation' and 'fifth moment'. There was also further evidence to suggest that some of the key debates central to the fourth and fifth moment have formed the focus of a number of discussions, with key considerations including the social construction of tourism space, place, reality and knowledge; the conceptualisation of 'self' in tourism; and issues of subjectivity, embodiment and the nature of power relations.

These issues were considered worthy of further discussion in this chapter for a number of reasons. First, they form some of the key themes running through this book, as they have received scant attention in tourism discourse to date despite being central to developments in social science research more generally. Second they are considered to offer some potential for the further development of fourth- and fifth-moment studies in tourism and the advancement of research in the field generally. It would seem that there is potential for tourism scholars to draw on examples of good practice from critical and interpretive paradigms, which would offer the opportunity for tourism researchers to engage with a wider variety of research issues than has historically been the case. Throughout this book it is argued that tourism researchers need to engage with emergent thinking, whether it be to embrace or to challenge it, in order to further advance the state of qualitative tourism research.

Subjectivities

The endeavour to produce scientific, neutral, non-personal accounts has often led to the purposeful separation of self and personal narrative from 'objective' observations (Galani-Moutafi 2000) in first-, second- and third-moment tourism research. This is illustrated in much of the tourism literature, whereby categories of self and Other are conceptualised as distinct opposites. In the past few years a growing number of writers have begun to problematise the subject/object dichotomy that has often formed the basis of tourism research frameworks (Wearing and Wearing 1996, 2001; Pritchard and Morgan 2000a,b). It is argued that the resultant outcome of tourism research based on such foundations has led to the objectification of cultures, societies, geographies and people (Galani-Moutafi 2000; Wearing and Wearing 1996). From such a perspective it is contended that tourists, hosts and researchers appear disembodied in much of the mainstream tourism research, and there has been little real attempt to understand individual experiences of tourism. It is suggested that placing emphasis on how the relationship with others involved in the research process is played out and translated into authority through analysis and writing of tourism text can help to reduce the problem of researcher subjectivity (Galani-Moutafi 2000). In much of the tourism-related literature the concept of embodiment, in terms of tourists, hosts (Wearing and Wearing 2001) *and*, as is argued here, researchers, has largely been ignored. Consequently, it is important for tourism researchers to consider how the research process can become more embodied and move away from a simplistic, de-personalised dichotomy. This issue is discussed in considerable depth by Swain in Chapter 6 of this book.

Social construction of tourism realities

Recent contributions have generated increasing recognition of tourism spaces as socio-cultural constructions rather than physical locations (Pritchard and Morgan 2000b; Aitchison 1999; Aitchison *et al.* 2001; Crouch 2000; Rojek 1997). Given that tourism spaces are not physically but socially constructed, it is important to consider how the meanings relating to those spaces are constructed, deconstructed and reconstructed over time. Tourism is a complex phenomenon based on interrelations and interactions, but the tendency in tourism research has been to focus on the tangible, and arguably the 'objective' and readily measurable interrelationships and interdependencies between people and places, frequently from an eco-nomics marketing and/or management perspective. A more person-focused approach which takes account of the individual's subjective experiences and perceptions and the roles these play in constructing the tourist, or indeed host, experience has so far received scant attention.

It is argued that in order to move towards fifth-moment considerations, the theorisation and study of tourism need to focus not only on the desti-nation, divorced from human subjectivity, but also on the subjectivity and the socio-historical and socio-cultural antecedents of such subjectivity. This is not to remove studies from the wider sociological context and vari-ables that impact upon individuals' daily lives, but to say that research should also focus upon the individual's role in the active construction and reconstruction of reality through interactions with others and the mean-ings they attach to various aspects of tourism.

While adopting such an approach would suggest more qualitative methods of data collection, at least until sufficient data have been collected to inform the design of more quantitative research tools, there must also be an aware-ness of the researcher's position and practice within the research process. Qualitative approaches would appear to facilitate the collection of a greater variety of responses and enable the researcher to tap into the impacts on the self (Denzin and Lincoln 1994), but we also need to consider the role of the researcher in this dynamic. These issues are considered further in Humber-stone's contribution on standpoints in Chapter 7 of this book.

Self-awareness/reflexivity

Some of the critics of tourism research have called for a greater level of self-awareness and self-reflexivity within research agendas. It is suggested that the individual identities that researchers bring with them into the field should not be hidden or masked by labels such as ethnographer, anthro-pologist, scholar, and so on. Greater consciousness about the way in which tourism images and texts about Others are related to one's own identity and culture are needed in order to advance qualitative research in tourism (Galani-Moutafi 2000). As Wearing and Wearing suggest, the 'self' is a political construct and there are many subjectivities, many 'I's' (Wearing and Wearing 2001). This notion of the 'I's' to which they refer could, in the research process, be extended to all actors involved in the tourism experience. Researchers as well as the researched construct their own mul-tiple versions of reality. Consequently, we need to consider the ways in which tourism researchers are active in the construction of interpretations of these experiences into text.

Reflexivity is one of the ways in which these issues can be explored. Galani-Moutafi (2000) refers to reflexivity as 'the conscious use of the self as a resource for making sense of others', which requires researchers to acknowledge and question their own culture and identity in order to provide some insight into their understanding of themselves in the context of their interactions with others. Furthermore, it is argued that the tourism research experience should be acknowledged as being integral to the interpretive process, as this can be a factor in the establishment of identity (Wearing and Wearing 2001; Cohen and Taylor 1976). A good example of the self-reflexive process is described by Dubisch (1995 cited in Galani-Moutafi 2000). The issue of reflexivity is one of the key themes underpinning Part I of the book and forms the focus of the contribution by Michael Hall in Chapter 8.

Conclusion

The post-1996 review of journals showed that research located in the 'traditional period' continues to dominate the study of tourism. However, a major advance since Riley and Love's review is that there is now a growing body of work that has started to breach the frontiers of the 'crisis of representation' and fifth moments of Denzin and Lincoln's (1998) framework of qualitative research. The horizons of tourism research are widening, and even papers that are rooted in the traditional era are beginning to show signs of dabbling with 'blurred genre' or 'crisis in representation' ideas such as multi-methods and employing at least some examples of the voice of the researched. Indeed, selectivity and eclecticism are the order of the day, with researchers moving within and between moments as they experiment with new techniques and seek new ways of writing their research. We would argue that this portrays a somewhat optimistic future for tourism research. Finally, while theoretical and methodological boundaries are beginning to be stretched, it should be stressed that there is still a great deal of unexplored territory to venture into and still plenty of scope for mainstream tourism scholars to incorporate into their work some of the new ways of thinking about research which have begun tentatively to infiltrate tourism discourse.

In this chapter we have highlighted some of the critiques of tourism research and its over-reliance on positivist modes of thinking and ways of doing research (Botterill 2001). It has been argued that this has led to the production of knowledge which is partial (Wearing and Wearing 2001). Certainly this book argues that one of the reasons that tourism research has lagged behind research in the social sciences more generally is that it has failed to take advantage of the full range of research approaches which have become more accepted in other research fields and it has been slow to address many of the epistemological and ontological issues that have been debated in wider social science disciplines.

Issues of epistemology and ontology have rarely been at the forefront of tourism discourse, but there are signs that in recent years such concerns have started to be addressed by a small but growing number of researchers. The latter half of the past decade in particular has seen significant developments. The post-1996 review has demonstrated that tourism research is developing in both an interdisciplinary and a multidisciplinary manner (Botterill 2001). Researchers are beginning to consider important issues concerning the nature of research and knowledge, the role of researchers, the ways in which research questions are determined and the ways in which data are interpreted. Following on from these deliberations there is some evidence that tourism researchers are starting to adopt a broader range of approaches. This chapter has suggested that consideration of issues such as subjectivity, identity and reflexivity, which are commonplace in other disciplines, will help to open the doors for future researchers to enter the largely unexplored territory of fourth- and fifth-moment research, thereby broadening their horizons while improving the quality of the research they produce (Denzin and Lincoln 1998).

The chapters of this book have been drawn together to provide insights into a range of different ways of doing qualitative tourism research. These contributions aim to go some way to fill a number of gaps in tourism research, through addressing a broad range of theoretical and practical issues, many of which have still yet to be fully developed in tourism research. It is hoped that this compilation of contributions will offer alternative ways of thinking about research, and spark the imagination and creativity of those about to embark on their own research journey.

Questions

1   The quantitative versus qualitative debate has a long history in social science research. Outline the pros and cons of each approach in relation to tourism research. Consider what can be gained from each and the main shortcomings of each approach.

2   This chapter has outlined four main research paradigms. What are they called? What are the underlying principles? That is, what are the different epistemological and ontological worldviews underpinning each perspective?

3   Think about a tourism-related issue that intrigues you. Consider the following:

   • What do we already know about it? From what perspective do we know about it?
   • What else do we need to know? What would be interesting to know?

   Design a research proposal to address these issues. Outline and justify your methodology by considering:

   • What is your epistemological and ontological standpoint? That is, how do you view the social world and what sort of knowledge can be gained from this perspective?

- What paradigm best fits your standpoint?
- What methods are best suited to your approach and why?

4   This chapter has outlined a number of methodological challenges facing tourism researchers. What are they? Devise a checklist of good practice points to help translate these challenges into research practice.

References

Aitchison, C. (1999) 'Heritage and nationalism: gender and the performance of power'. In D. Crouch (ed.) *Leisure/Tourism Geographies: Practices and Geographical Knowledge*. London: Routledge.

Aitchison, C. (2000) 'Poststructural feminist theories of representing Others: a response to the "crisis" in leisure studies discourse', *Leisure Studies*, 19: 127–144.

Aitchison, C., MacLeod, N. and Shaw, S. (2001) *Leisure and Tourism Landscapes: Social and Cultural Geographies*. London: Routledge.

Alcoff, L. and Potter, E. (1993) 'Introduction: when feminisms intersect epistemology'. In L. Alcoff and E. Potter (eds) *Feminist Epistemologies*. London: Routledge.

Bhavnami, K. (1993) 'Talking racism and the editing of women's studies'. In D. Richardson and V. Robinson (eds) *Introducing Women's Studies*. London: Macmillan.

Botterill, D. (2001) 'The epistemology of a set of tourism studies', *Leisure Studies*, 20: 199–214.

Bryman, A. and Burgess, R. (eds) (1994) *Analysing Qualitative Data*. London: Routledge.

Code, L. (1993) 'Taking subjectivity into account'. In L. Alcoff and E. Potter (eds) *Feminist Epistemologies*. London: Routledge.

Cohen, S. and Taylor, L. (1976) *Escape Attempts*. Harmondsworth: Penguin.

Crouch, D. (2000) 'Places around us: embodied lay geographies in leisure and tourism', *Leisure Studies*, 19: 63–76.

Dalmiya, V. and Alcoff, L. (1993) 'Are old wives' tales justified?' In L. Alcoff and E. Potter (eds) *Feminist Epistemologies*. London: Routledge.

Dann, G. (1996) *The Language of Tourism: A Sociolinguistic Perspective*. Oxford: CAB International.

Denzin, N. K. and Lincoln, Y. S. (eds) (1994) *Handbook of Qualitative Research*. Thousand Oaks, CA: Sage.

Denzin, N. K. and Lincoln, Y. S. (1998) *The Landscape of Qualitative Research: Theories and Issues*. Thousand Oaks, CA: Sage.

Dubisch, J. (1995) *In a Different Place: Pilgrimage, Gender and Politics at a Greek Island Shrine*. Princeton, NJ: Princeton University Press.

Farran, D. (1990). ' "Seeking Susan": producing statistical information on young people's leisure'. In L. Stanley (ed.) *Feminist Praxis: Research Theory, and Epistemology in Feminist Sociology*. New York, NY: Routledge.

Fine, M. (1998) 'Working the hyphens: reinventing self and other in qualitative research'. In N. K. Denzin and Y. S. Lincoln (eds) *The Landscape of Qualitative Research: Theories and Issues*. Thousand Oaks, CA: Sage.

Flick, U. (1998) *An Introduction to Qualitative Research*, London: Sage.

Galani-Moutafi, V. (2000) 'The self and the other traveler, ethnographer, tourist', *Annals of Tourism Research,* 27 (1): 203–224.

Guba, E. and Lincoln, Y. S. (1998) 'Competing paradigms in qualitative research'. In N. K. Denzin and Y. S. Lincoln (eds) *The Landscape of Qualitative Research: Theories and Issues.* Thousand Oaks, CA: Sage.

Hamilton, D. (1994) 'Traditions, preferences and postures in applied qualitative research'. In N. K. Denzin and Y. S. Lincoln (eds) *Handbook of Qualitative Research.* Thousand Oaks, CA: Sage.

Harding, S. (1993) 'Rethinking standpoint epistemology: what is "strong objectivity"?' In L. Alcoff and E. Potter (eds) *Feminist Epistemologies.* London: Routledge.

Holland, J. and Ramazanoglu, C. (1994) 'Coming to conclusions: power and interpretation in researchers of young women's sexuality'. In M. Maynard and J. Purvis (eds) *Researching Women's Lives from a Feminist Perspective.* London: Taylor & Francis.

Hollinshead, K. (1996) 'The tourism researcher as *bricoleur*: the new wealth and diversity in qualitative inquiry', *Tourism Analysis,* 1 (1): 67–74.

hooks, b. (1989) *Feminist Theory: From Margin to Centre.* Boston, MA: South End Press.

hooks, b. (1990) *Yearning: Race, Gender and Cultural Politics.* Boston, MA: South End Press.

Jones, P. (1993) *Studying Society: Sociological Theories and Research Practices.* London: Collins Educational.

Kincheloe, J. L. and McLaren, P. L. (1998) 'Rethinking critical theory and qualitative research'. In N. K. Denzin and Y. S. Lincoln (eds) *Handbook of Qualitative Research.* Thousand Oaks, CA: Sage.

Ladner, J. (1971) *Tomorrow's Tomorrow.* New York, NY: Doubleday.

Longino, H. (1990) *Science as Social Knowledge: Values and Objectivity in Scientific Inquiry.* Princeton, NJ: Princeton University Press.

Marcus, G. (1998) *Critical Anthropology Now: Unexpected Contexts, Shifting Constituencies, New Agendas.* Santa Fe, NM: School of American Research Press.

May, T. (1993) *Social Research: Issues, Methods and Process.* Buckingham: Open University Press.

Maynard, M. (1994) 'Methods, practice and epistemology: the debate about feminism and research'. In M. Maynard and J. Purvis (eds) *Researching Women's Lives from a Feminist Perspective.* London: Taylor & Francis.

Maynard, M. and Purvis, J. (1994) 'Doing feminist research'. In M. Maynard and J. Purvis (eds) *Researching Women's Lives from a Feminist Perspective.* London: Taylor & Francis.

Nash, K. (1994) 'The feminist production of knowledge: is deconstruction a practice for women?', *Feminist Review,* 47: 65–77.

Phoenix, A. (1994) 'Practising feminist research: the intersection of gender and "race" in the research process'. In M. Maynard and J. Purvis (eds) *Researching Women's Lives from a Feminist Perspective.* London: Taylor & Francis.

Pritchard, A. and Morgan, N. (2000a) 'Privileging the male gaze: gendered tourism landscapes', *Annals of Tourism Research,* 27 (4): 884–905.

Pritchard, A. and Morgan, N. (2000b) 'Constructing tourism landscapes: gender, sexuality and space', *Tourism Geographies,* 2 (2): 115–139.

Riley, R. W. (1996) 'Revealing socially constructed knowledge through quasi-structured interviews and grounded theory analysis', *Journal of Travel and Tourism Marketing*, 5: 21–40.

Riley, R. W. and Love, L. L. (2000) 'The state of qualitative tourism research', *Annals of Tourism Research*, 27 (1): 164–187.

Ritchie, J. and Spencer, L. (1994) 'Qualitative data analysis for applied policy research'. In A. Bryman and R. Burgess (eds) *Analysing Qualitative Data*. London: Routledge.

Robson, C. (1993) *Real World Research: A Resource for Social Scientists and Practitioner-Researchers*. Oxford: Blackwell.

Rojek, C. (1997) 'Indexing, dragging and the social construction of tourist sights'. In C. Rojek and J. Urry (eds) *Touring Cultures: Transformations of Travel and Theory*. London: Routledge.

Schwandt, T. (1998) 'Constructivist, interpretivist approaches to human enquiry'. In N. K. Denzin and Y. S. Lincoln (eds) *Handbook of Qualitative Research*. Thousand Oaks, CA: Sage.

Silverman, D. (2000) *Doing Qualitative Research: A Practical Handbook*. London: Sage.

Smith, V. (1977) *Hosts and Guests: The Anthropology of Tourism*. Oxford: Blackwell.

Smith, V. (1998) 'War and tourism: an American ethnography', *Annals of Tourism Research*, 25 (1): 202–227.

Stanley, L. and Wise, S. (1990) 'Method, methodology and epistemology in feminist research processes'. In L. Stanley (ed.) *Feminist Praxis: Research Theory, Theory and Epistemology in Feminist Sociology*. London: Routledge.

Stanley, L. and Wise, S. (1993) *Breaking Out Again: Feminist Ontology and Epistemology*. London: Routledge.

Walle, A. (1997) 'Quantitative versus qualitative tourism research', *Annals of Tourism Research*, 24 (3): 524–536.

Warren, C. (1988) *Gender Issues in Field Research*. Qualitative Research Methods series 9. London: Sage.

Wearing, B. and Wearing, S. (1996) 'Refocusing the tourist experience: the "Flaneur" and the "Choraster",' *Leisure Studies*, 15: 229–244.

Wearing, S. and Wearing, B. (2001) 'Conceptualizing the selves of tourism', *Leisure Studies*, 20: 143–159.

3   Knowing about tourism

Epistemological issues

John Tribe

Aims of the chapter

- To evaluate the disciplinary status of tourism.
- To map the structure of tourism knowledge.
- To distinguish between multidisciplinary, interdisciplinary and extra-disciplinary approaches to tourism.
- To demonstrate the operation of culture, power, hegemony, ideology and values in knowledge creation in tourism.
- To highlight the importance of epistemological issues to tourism researchers.

Introduction

Researchers generally come to tourism after a period of intense acquisition of knowledge in higher education. It is of course possible to do tourism research by the application of knowledge and research techniques in an unreflexive way. Indeed, Franklin and Crang (2001: 6) criticise 'a tendency for studies [by tourism researchers] to follow a template', and Botterill (2001: 199) suggests that 'the assumptions that underlie social science research in tourism are seldom made explicit'. This chapter holds up epistemological issues in tourism research for scrutiny. Knowing about how and what we know in tourism is an epistemological question, epistemology being that branch of philosophy which studies knowledge. Indeed, epistemology explores the theory of knowledge, and its essential concerns are the meaning of the term 'knowledge', the limits and scope of knowledge and what constitutes a valid claim to know something. The first part of this chapter analyses the character of tourism knowledge. Based largely on the work of Tribe (1997), it provides an initial mapping of the epistemological territory of tourism and a review of the status of tourism as a discipline and the creation of tourism knowledge through multidisciplinarity, interdisciplinarity and extradisciplinarity (Gibbons *et al.* 1994).

The second part of the chapter problematises the concept of tourism knowledge and considers how knowledge can distort understanding. The sociology of knowledge provides the initial foray into the territory of tourism knowledge as a social and political construct. The chapter then explores the implications of the work of contemporary theorists on the development of tourism knowledge. These include Kuhn's (1970) concept of *paradigms* and Becher's (1989) notion of *academic tribes*. The idea of knowledge-constitutive interests (Habermas 1978) is used to alert us to the fact that there is no such thing as interest-free knowledge. The relationship between knowledge and power is further explored using Foucault's (1971) work on discourse. In particular, the concept of discourse helps us to understand the social and political processes that produce meaning and how certain discourses are legitimised and others marginalised. The chapter concludes by summarising the key epistemological issues and the consequences of these for tourism research.

Why tourism is not a discipline

Is tourism studies a discipline? (Ryan 1997). Though Goeldner (1988) described tourism as a discipline and Hoerner (2000) considers the development of tourism science (or 'tourismology'), the view of Cooper *et al.* (1998: 3) is that 'While tourism rightly constitutes a domain of study, at the moment it lacks the level of theoretical underpinning which would allow it to become a discipline.'

Hirst's (1965, 1974) work can serve as a useful framework for the evaluation of tourism studies as a discipline. Hirst explained the meaning of a form of knowledge, or discipline, as being 'a distinct way in which our experience becomes structured round the use of accepted public symbols' (1974: 44). Hirst proposed that forms of knowledge, or disciplines, are distinct, and explains their distinctness in four ways. First, each form has a network of interrelated concepts particular to that form of knowledge. Second, these concepts form a distinctive network, which give the form its distinctive logical structure. Third, each form has expressions or statements which are in some way testable against experience using criteria which are particular to that form. Fourth, disciplines are irreducible, which means that it is not possible to reduce these forms of knowledge any further; in other words, these are the basic building blocks.

Testing against these criteria demonstrates that tourism studies cannot be regarded as a discipline. First, while tourism studies contains a number of concepts (e.g. tourism motivation), these are rarely particular to tourism studies. Generally, they have started life elsewhere and been stretched or contextualised to give them a tourism dimension. Second, tourism concepts do not form a distinctive network. They need to be understood generally within the logical structure of their provider discipline. They do not form a cohesive theoretical framework, so that tourism studies, of itself, does not provide a distinctive, structured way of analysing the world, as does, say, sociology.

Third, tourism studies does not have expressions or statements which are testable against experience using criteria that are particular to tourism studies, but rather utilises those criteria which are found in its contributory disciplines. Neither does tourism pass the test of irreducibility. Taking, for example, 'tourism satisfaction' as a typical tourism problem, we find it is indeed reducible. The term 'satisfaction', for example, may be approached as a philosophical question when we probe the aspect of 'satisfaction with what'. Satisfaction may contain psychological elements when we ask how satisfaction is perceived by the subject. In fact, the substantive concept to be investigated within the concept of 'tourism satisfaction' is not tourism but 'satisfaction'. Tourism merely provides the context.

Thus, on the basis of Hirst's tests, tourism is found not to be a discipline. Its main shortcomings in this respect are, first, a lack of internal theoretical or conceptual unity, and second, a ready reliance on contributory disciplines. Toulmin's (1972) and Donald's (1986) epistemological tests for a discipline are similar to those of Hirst. However, King and Brownell (1966) also include other criteria such as the existence of a community, a network of communications, a tradition and a particular set of values and beliefs.

To what extent are these additional criteria met by tourism studies? First, considering the community aspects of tourism studies, Cooper *et al.* (1994: 54) assert that 'tourism has its own, albeit small academic community'. But there are very few faculties or departments of tourism. Also, tourism academics often identify themselves within a disciplinary or functional community before placing themselves within a tourism community. Second, tourism has developed a network of communications which includes professional associations, conferences, books and journals. But perusal of the burgeoning range of journal titles demonstrates some basic fault lines in tourism. On the one hand, there are those which are primarily about the business of tourism (e.g. *Tourism Management, Journal of Travel and Tourism Marketing, International Journal of Hospitality Management*). Then there are those which have a more open agenda (e.g. *Annals of Tourism Research, Journal of Tourism Studies* and *Travel and Tourism Analyst*). A third category is discipline-specific journals (e.g. *Tourism Economics* and *Tourism Geographies*).

Graburn and Jafari (1991: 1) traced scholarship in tourism and reflected that 'most studies have taken place since 1970 and 50 per cent of them since 1980'. Thus, tourism studies has not established anything that could be called a tradition that might impose its own unity. Neither is there a shared set of values among tourism scholars. Cotgrove investigated distinct and different 'value camps' in Business Studies (1983) and categorised its academics as, first, those who worked within the dominant social paradigm, and second, those who worked within an alternative environmental paradigm. Similarly, in tourism those working as insiders and outsiders to the business of tourism represent two distinct camps.

Tourism: two fields studied through multidisciplinarity, interdisciplinarity and extradisciplinarity

In response to Tribe's (1997) argument that tourism is not a discipline, Leiper (2000) (and previously Leiper 1981) made a case for tourism as an emerging discipline, an argument that was subsequently rebutted by Tribe (2000). It is Tribe's (1997) contention that tourism is a field rather than a discipline. Here, Henkel's (1988: 185) distinction between disciplines, which 'are held together by distinctive constellations of theories, concepts and methods', and fields, which 'draw upon all sorts of knowledge that may illuminate them', provides a useful starting point to develop this argument.

Writers such as Gunn (1987) classify tourism as a field. He lists the main disciplines contributing to tourism as marketing, geography, anthropology, behaviour, business, human ecology, history, political science, planning and design, and futurism. Additionally, Jafari and Ritchie (1981) presented a model of tourism studies as a field in their tourism wheel model, with its inner circle of boxes representing tourism courses and the outer ring of boxes denoting disciplines or departments. However, the mixing of disciplines and departments can cause confusion, and the model could be made clearer by putting together the objects of study on the inner ring and the disciplinary approaches (e.g. geography (Crouch 1999) and economics (Tribe 1995)) on the outer ring. In much of the literature, tourism has been viewed as a single entity called tourism studies, but this approach does not adequately reflect the tensions within tourism studies. Rather, there seem to be (at least) two fields of study discernible. One field is readily identifiable as tourism business studies. The identity of this is borrowed from the increasingly mature field of business studies but in a tourism context. It thus includes the marketing of tourism, tourism corporate strategy, and the management of tourism.

The other field of tourism studies does not have such an obvious title, because it is little more than just the rest of tourism studies, or non-business tourism studies. It lacks any unifying framework other than the link with tourism. It includes areas such as environmental impacts, tourism perceptions and social impacts. This tourism field is labelled as TF2, while TF1 denotes tourism business studies. Therefore:

$$\text{The field of tourism (TF)} = \text{TF1} + \text{TF2}$$

Echtner and Jamal (1997) refer to these as the 'impacts-externalities' camp (TF1) and the 'business development' camp (TF2). It should be noted that there is some overlap between the fields TF1 and TF2.

Figure 3.1 is used to demonstrate developments and knowledge creation in the fields of tourism. The outer circle represents the disciplinary tools of analysis (using discipline 'n' to denote other key disciplines). The

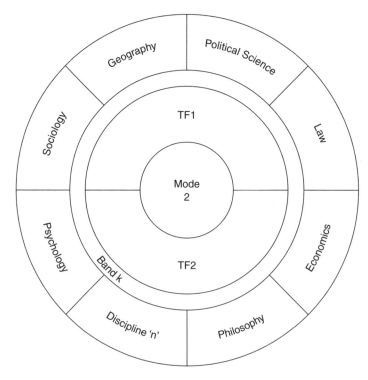

Figure 3.1 The creation of tourist knowledge. Outer circle = disciplines and subdisciplines; middle circle = fields of tourism; inner circle = world of tourism. TF1 = business interdisciplinarity; TF2 = non-business-related tourism.

middle circle (TF1 + TF2 = TF) represents the fields of the study of tourism. Notice that between the outer circle and the circle TF, which represents the field of tourism, is an area denoted band k. Band k represents an area where tourism knowledge is created. Several activities can be seen to be taking place in band k. First, at a simple level, it represents the interface between the disciplines and the fields of tourism. Thus, where economics enters a field of tourism, the theory of the tourism multiplier is born. This is little more than the application of an existing theory to a new field. Tourism knowledge here is multidisciplinary (Boyer 1997). Epistemologically speaking, each discipline provides the methodology to justify knowledge claims. Jafari and Aaser's (1988) study of US tourism doctoral dissertations (from 1951 to 1987) found that the disciplines of economics, anthropology, geography and recreation were paramount. However, it is also possible for band k to represent a place where disciplines interact with one another and the field of tourism. For example, the concept of carrying capacity emerges from a combination of disciplines including sociology, economics and biology. This combining of disciplines to create new insights into the external world of tourism represents an interdisciplinary approach (Przeclawski 1993). Interdisciplinarity generates an epistemology 'characterised by the explicit formulation of a uniform, discipline-transcending terminology or a common methodology' (Jantsch 1972).

Gibbons *et al.* (1994) refer to the mode of knowledge production that includes multidisciplinarity and interdisciplinarity as mode 1. Mode 1 knowledge is 'generated within a disciplinary, primarily cognitive context' (1994: 1). It is knowledge which has been primarily generated and nurtured within institutions of higher education. But Gibbons *et al.* have identified a new form of knowledge production which they label mode 2:

> The new mode operates within a context of application in that problems are not set within a disciplinary framework.... It is not being institutionalised primarily within university structures ... [and] makes use of a wider range of criteria in judging quality control.
>
> (1994: vii)

The majority of mode 2 tourism knowledge production takes place in the upper part of the centre circle in Figure 3.1 and relates to the TF1 area of the world of tourism. This is because the main sites of mode 2 knowledge production include industry, government, think-tanks, interest groups, research institutes and consultancies. It includes developments and applications such as smart hotel rooms, yield management systems and computerised reservations developments – developed in the industry for the industry.

Gibbons *et al.* (1994: 168) explain mode 2 knowledge in terms of transdisciplinarity – that is 'knowledge which emerges from a particular *context of application* with its own distinct theoretical structures, research methods and modes of practice but which may not be locatable on the prevailing disciplinary map'. However, it is proposed to use the term 'extradisciplinarity' to describe mode 2 knowledge production. This is because the term 'transdisciplinarity' (meaning 'across the disciplines') is easily confused with interdisciplinarity. Mode 2 knowledge is, however, being produced outside the disciplinary framework, hence the term 'extradisciplinarity' is seen as being more appropriate.

The important points to note about mode 2 knowledge production are, first, that it occurs outside higher education, the traditional centre for knowledge production. Second, it is developing its own epistemology. Disciplinary-based methodology and peer review are the hallmarks of quality control for mode 1 knowledge. Mode 2 knowledge, however, judges success by its ability to solve a particular problem, its cost-effectiveness, its ability to establish competitive advantage – that is, its effectiveness in the real world. Its results are often highly contextualised for a specific project.

Problematising tourism knowledge

Whereas the previous section provided a descriptive analysis of its territory of tourism knowledge, this section problematises tourism knowledge.
Drawing on the sociology of knowledge, it focuses particularly on culture,
power, hegemony, ideology and value freedom. Sociologists suggest that
knowledge is not independent of, but rather is conditioned by, the particular culture or society in which it is produced. Mannheim (1960) argued
that different groups in society experience the world in different ways and
have competing claims as to what is knowledge. In this view, knowledge
loses its claim to being an objective account of the world, and there is no
universal epistemology. An example of this is found in the work of Whittaker (1999), who investigates the issue of indigenous tourism and specifically considers the challenge of reclaiming knowledge, culture and
intellectual property in Australia. Whittaker investigates the moral claims
of indigenous people (Aboriginal Australians), and the political and social
implications that these have for the tourist industry and for knowledge
industries. One of his findings is that the depiction of Aborigines in tourist
literature is exploitative and racist, based upon and perpetuating histori-
cised images of Aboriginal groups. In other words, here tourism knowledge about indigenous people is generated within a Western (developed)
culture with a history of colonialism. The knowledge which is offered
about them is different from the knowledge they would offer about themselves. It is generated from different traditions, and different knowledge
traditions would result in (possibly radically) different projections and
understandings of the world or its inhabitants. Similarly, Hollinshead
(1999a) argues that postmodern 'nationalist' tourism acts as a coding
machine serving the interests of certain privileged groups. Perhaps one
way out of this cultural bind is offered by Ryan (2000), who argues that
phenomenography is a potent method of tourism research because of its
concern with individuals' understanding. Ryan also suggests that the use of
phenomenographic analysis is consistent with the epistemology of modes
of research which extend beyond positivism.

Following in the tradition of the sociology of knowledge and noting that
'facts do not speak for themselves', Kuhn (1970) showed that science operates in a historical rather than a transcendental realm. He emphasised the
importance of research communities and the paradigms they work within
for the construction of knowledge. For Kuhn, a paradigm represents
'accepted examples of actual scientific practice ... from which spring
particular coherent traditions of scientific research' (1970: 10). Kuhn looks
at knowledge creation not from the level of societal culture but at the level
of the culture of disciplines. Indeed, he noted that 'research is a strenuous
and devoted attempt to force nature into the conceptual boxes supplied by
professional education' (1970: 5), and that those things 'that will not fit
into the box are often not seen at all' (1970: 24). This alerts us to the pos-

sible blind spots not seen by those working within accepted paradigms. The point about paradigms is that they define the boundaries of accepted methods and knowledge for disciplines. It can be argued that tourism studies are pre-paradigmatic largely because of the lack of agreement about common rules and the schisms between those operating in TF1 and TF2. However, it could be argued that each field of tourism constitutes in itself a loose paradigm gradually establishing rules and norms. Dann (1997: 474) reports on the paradigm debate in tourism hosted by the International Sociological Association (Jyväskylä, Finland, July 1996), summarising its discussions as 'a willingness to progress beyond the theorizing of the 70s, a realization that there was no single truth underlying tourism inquiry, . . . a disposition to go beyond "prostration before methodology" [and] . . . an appreciation of the sociolinguistic underpinnings of tourism'.

Becher's (1989) investigation of 'academic tribes' focused on disciplinary cultures – that is, the rules, norms and acceptable behaviour patterns upheld by disciplinary communities – and these cultural influences are evident in tourism. We may consider the 'elders' of the various tourism tribes as those holding distinguished chairs, those on the panels of research grant award bodies, those on editorial boards and acting as referees of key journals, those on the executives of learned and professional associations, and (for the United Kingdom) those assessing Research Assessment Exercise (RAE) submissions. The key point here is that these 'elders' act as gatekeepers in the dissemination of tourism knowledge (Seaton 1996; McKercher 2002; see also Hall's discussion on gatekeepers in Chapter 8 of this book) and the steering of research agendas, and exert an important but unseen power in knowledge production and legitimation. Of course, tourism researchers also sit within and are sometimes judged by the wider academic community. In this context, Tribe (2003b) explains how tourism research in the United Kingdom was marginalised by those in positions of power (particularly those elders in traditional disciplines and 'old' universities) who set up the structures and systems in the 2001 RAE. Tribe has termed this unfortunate result 'the RAE-ification' of tourism research.

Following Becher, Aitchison's (1996) specific study into the fields of leisure and tourism identifies the key mechanisms by which patriarchal power and control are exerted. These are research and consultancy, publications, professional associations, educational management and teaching. Aitchison (1996) shows that these mechanisms are crucial, first, to the construction of knowledge and pedagogy within leisure and tourism studies, and second, to the communication, legitimation and reproduction of such knowledge. She thus offers an understanding of the underlying structures which shape leisure and tourism research and its patriarchal nature.

Aitchison follows this up (2001) by a gender analysis of authors in international refereed journals in leisure and tourism studies, where she finds a ratio of male to female authors of four to one. Additionally, she finds that practices that promote a more inclusive approach towards gender representation have not been adopted by any of the leisure and tourism journals audited. She argues that 'the codification of knowledge' is a product of both structural and cultural power, and advocates a combination of material and discursive analysis to examine the socio-cultural nexus of knowledge production, legitimisation and reproduction.

Foucault (1980) foregrounded the importance of discourse, disciplines, power and surveillance. He describes discourses as ' "regimes of truth" giving as examples medicine, psychiatry, and other forms of disciplinary knowledge' (Usher and Edwards 1994: 85). Foucault notes further that 'each society has its regime of truth, its "general politics" of truth: that is the type of discourse which it accepts and makes function as true' (1980: 131). He claims that discourses of knowledge have important power dimensions and that they are both expressions of power relations and embodiments of power. Discursive formations perform an including and excluding function because they provide the rules for what counts as knowledge and what does not, and who speaks with authority and who does not. On Foucault's view, therefore, a discourse can perform a repressive role furthering the interests of some groups while suppressing the interests of others. So, for example, it is possible to identify managerialism as a key discourse in the TF1 field of tourism. Ball (1990: 156) drawing on Foucauldian analysis, notes that 'Management is a professional, professionalizing discourse which allows its speakers to lay exclusive claims to certain sorts of expertise ... that casts others, subordinates, as objects of those procedures.' In other words, as a discourse it lays claim to being *the* way of doing things (i.e. it excludes other ways), and often carries the implicit assumption that it is 'the one best method' (Ball 1990: 156). Management discourse supports a number of dichotomies (the us/the other; the manager/the managed; the expert/the amateur; the developed/the underdeveloped; the successful/the failed), all of which are encouraged by an insider, technical specialist language and all of which are apparent in managerialist approaches to tourism. Echtner (1999) extends this analysis to tourism marketing. She uses a semiotic approach and uncovers a system of signs and the 'deep structure' of meaning which points to the existence of an established discourse.

Discourses also 'systematically form the objects of which they speak ... [they] are not about objects: they constitute them' (Foucault 1974: 49). For example, Foucault demonstrated how 'madness' is not understood as an ahistoric, universal, unchanging original state but rather is constituted by a current discourse of madness. This argument can be applied to many aspects of tourism. For example, Hughes (1995) exposes the discourse of sustainable tourism, showing that the dominant approach to sustainable tourism is technical, rational and scientific and that this has eclipsed the emergence of an ethical response to sustainability. In other words, 'sustainable tourism' is formed through our ways of analysing it. Cheong and Miller (2000) also use Foucault to emphasise the productive effects of power demonstrated in the formation of tourism knowledge.

Hollinshead (1999b) offers a critique of the power of surveillance (Foucault's *le regard*) in tourism, discussing how the Foucauldian eye-of-power acts not only through the organisations and agencies of travel and tourism but also in travel and tourism research. The power of surveillance is shown by Hollinshead (1999b) as an authoritative mix of normalising discourse and universalising praxis which routinely privileges particular understandings of heritage, society and the world in and through tourism (see Hollinshead's discussion in Chapter 4, on the ontology of tourism). Hollinshead argues that tourists and those who work in tourism can be seen as *Homo docilis*, participating in the regulation of the world and in the mastery of its social, cultural, natural and geographical environments, but also regulating and constraining themselves. By drawing attention to this, Hollinshead (1999b) seeks to make those involved in tourism (including researchers) more Other-regarded (and also self-aware) in terms of the governing suppositions and presuppositions they work within.

Another dimension to knowledge and power is offered by Lyotard (1984: 3) in *The Postmodern Condition*, where he develops the concept of performativity. The argument is that as science becomes more complex, it requires ever more technologically complex proofs. The importance of this is that 'an equation between wealth, efficiency and truth is thus established' (Lyotard 1984: 45). In other words, knowledge which is useful to the economy will tend to be favoured because science demands complex proofs which cost money, and thus: 'The production of proof ... thus falls under control of another language game, in which the goal is no longer truth, but performativity – that is the best possible input/output equation' (Lyotard 1984: 46). One consequence of Lyotard's analysis is that the business of tourism part of the field of tourism exerts a strong pull on knowledge production and that much tourism knowledge is generated for profitability. Epistemology is led by functionalism, and the aim of knowledge production becomes not an impartial uncovering of truth but a search for truths which are useful in terms of marketability and efficiency.

Tourism knowledge is generated using a variety of research methods (Dann *et al.* 1988), and Walle (1997) offers a comparison of qualitative and quantitative methods in tourism research. But a deeper insight is found in Habermas's (1978) theory of knowledge-constitutive interests, where he demonstrates that the pursuit of knowledge is never interest free but rather that human inquiry is motivated by one of three interests. First, the technical interest seeks control and management; second, the practical interest seeks understanding; and third, the emancipatory interest seeks freedom from falsehood and emancipation from oppression. Each of these interests is served by a different methodological paradigm. Scientific positivism serves the technical; interpretive methods seek understanding;

and critical theory seeks emancipation. Tribe (2001) considers the effects of these different research paradigms on curriculum development in tourism and notes that the use of interpretive methods in tourism enables meaning to be understood in terms of the actors in tourism. This goes some way to addressing the problem of unequal power relations between the researcher and the researched, common in scientific-positivist methods. But interpretive methods do not necessarily escape the effects of ideology and hegemony.

We must turn to critical theory to foreground these problems of ideology and hegemony in research: 'The job of critical theory is initially to identify which particular ideological influences are at work. Ideology critique then asks whose interests are being served by a particular ideology' (Tribe 2001: 446). Ideology may be considered as a regime of beliefs that directs the thought and practices of those who inhabit it. Hegemony, as developed by Gramsci (1971), offers an account of how particular ideas dominate thinking in society: 'It refers to an organised assemblage of meanings and practices, the central, effective and dominant system of meanings, values and actions which are *lived*' (Apple 1990: 5; emphasis in original).

What is of particular significance in hegemony is how it 'saturates our very consciousness' (Apple 1990: 5) so that it becomes the taken-for-granted way of thinking and doing. Tribe (2003a) reviews a tourism curriculum development project that he led in Moldova and, intriguingly, declares the project to be both a success and failure. The project is deemed successful since it achieved the objectives agreed with its EU funders (i.e. within a managerialist discourse), but in terms of failure Tribe notes that:

> far from being a neutral exercise in curriculum development, the project is saturated with 'taken for granted' meanings and values. This in turn undermines the notion of meaningful 'bottom up' participation and contribution. Rather the project is open to criticisms that its implicit values and meanings – its ideological baggage – facilitate the process where a developing country – Moldova – may be subsumed into Western hegemony and neo-colonialism.

This reinforces the point made in Kobasic's (1996) study, which noted that the dissemination of tourism knowledge is mainly one-way – from developed countries of a dominant language to others – and underlines the influences of neo-colonialism and imperialism in knowledge production.

Conclusion

There are several important implications for researchers in tourism that result from the above analysis. First, while there are four main approaches, tourism studies tends to be crystallising around the business interdisciplinary approach. This is because the field of tourism business studies has some coherence and structure, and a framework of theories and concepts (borrowed from the field of business studies). Interdisciplinary and multi-disciplinary knowledge that is created around the other field of tourism – TF2 – has no framework upon which to crystallise. Here, the major gravitational pull upon the atomised knowledge emanates from the disciplines themselves, so it still makes sense here to talk of the anthropology of tourism, the sociology of tourism, etc. Performativity reinforces the importance of the business of tourism. However, there are two caveats here. Although the business discourse exerts a particular power over tourism, Foucault observed that 'where there is power there is resistance' (1980: 95), and the multiple discourses that run through tourism are significant in providing sites for resistance. Also, Dann and Phillips (2001) have reported a recent trend whereby tourism research is moving away from pure quantification and towards a more qualitative approach.

Second, the external world of tourism which is actually distilled into tourism studies depends crucially on what we have gone looking for and how we have gone about looking for it. Hughes (1992) argued that epistemologically, tourism has existed in different forms and that our contemporary understanding is but another social construction. Furthermore, he points out (after Foucault) that we are actually creating rather than discovering the phenomenon called tourism. Research into tourism studies turns out to be not an objective, value-free search for tourism knowledge since the epistemological characteristics of the approaches of different fields and the rules of tourism discourses perform a selector role. For example, each discipline provides us with a particular pair of disciplinary spectacles. These spectacles illuminate certain parts of the territory of tourism. Thus, the economist may see tourism in terms of its resources while the anthropologist may wish to explore host–guest relations. Boyer (1997) found that tourism has always fallen under the influence of various disciplines which have attempted to control either some of or all its aspects, and Jamal and Hollinshead (2001), in exploring the constructive power of tourism, discuss the suitability of tourism's analytical tools to, for example, do justice to marginalised voices.

Third, it is not just disciplines that perform a selector role in determining how the external world is to be conceptualised, categorised, described and predicted. Disciplines can work within particular discursive formations (Foucault) which privilege some groups over others. Disciplines also have their distinctive cultures, and work within overarching societal cultures. What is significant about different cultures is that they uphold different value systems which are inevitably absorbed by those working within them (albeit generally at a tacit, implicit level). Since researchers themselves have biographies and are culturally located, they will have formulated particular interest standpoints, and thus the idea of value-free research cannot be upheld (see Chapter 7 for a further discussion on reflexivity and standpoints).

Fourth, the values held by those operating from different approaches to tourism may be quite different, making communication between the two fields difficult. An example here is the difficulties in communication that exist between those operating in the business of tourism and those operating from an environmental tourism approach. The different camps speak a different technical language, use different techniques, legitimate knowledge and truth in different ways, and frame problems differently, causing a lack of inter-subjectivity. This may result in a condition that Lyotard (1988: xi) termed a differend: 'a case of conflict between at least two parties that cannot be equitably resolved for lack of a rule of judgement applicable to both arguments'.

Fifth, the academic world has tended to overlook mode 2 production of knowledge. This is because it is not communicated in academic journals and is not validated in higher education. Cooper *et al.* (1994: 126) seem to dismiss mode 2 knowledge, observing that:

> the big problem with applied research is that it usually fails to add anything substantial or significant to the body of knowledge. . . . This is because the problem is too company- or sector-specific and relatively limited in its scope, i.e. it is usually concrete and operationally-orientated rather than abstract or conceptual in its nature ... and therefore, frequently does not progress the body of knowledge.

There is a danger here of a potential schism between mode 1 and mode 2 production, although Jenkins (1999) and Ayala (2000) make suggestions about how industry and academia can work together.

Sixth, much of the analysis of tourism knowledge points to hidden values, tacit understandings, taken-for-granted assumptions and therefore unintended consequences in the outcomes and results of research.

Seventh, tourism knowledge exerts a subtle power to define, to objectify, to foreground some issues, to privilege some groups, to leave some issues untouched, to exclude other groups.

Above all, to return to the opening remarks of this chapter, when we approach research in tourism we are armed with the acquisition of specialist knowledge that seems to equip researchers to pursue a disinterested quest for truth. But as Botterill (2001) counsels, we need to explore the assumptions underlying our research practices in tourism, for knowledge applied unreflexively can lead us to truths which are blind, partial, value-laden and overbearing.

Summary

This chapter analyses epistemological issues in tourism research. An initial mapping of the epistemological territory of the subject reveals tourism to be a field of studies rather than a discipline, and one in which tourism knowledge is created through multidisciplinary, interdisciplinary and extradisciplinary approaches. The existence of two distinct groups of researchers is noted: those interested in business issues and those interested in non-business issues. The second part of the chapter uses the sociology of knowledge to problematise tourism knowledge and reveal it as a social and political construct. The concept of *paradigms*, the notion of *academic tribes* and the idea of knowledge-constitutive interests demonstrated that there is no such thing as interest-free knowledge. Additionally, the concept of discourse revealed the social and political forces at work in the production of tourism knowledge and, through its operation, how some understandings of tourism are legitimised and others marginalised. Finally, the chapter underlined the importance of reflexivity in research.

Questions

1 To what extent is tourism a discipline and why does this debate matter?
2 What factors inhibit the search for objective truths in tourism?
3 Examine the implicit hidden values in a named journal article and analyse the consequences of this.
4 Explain the significance of each of the following to tourism research:

- 'the constructive power of tourism';
- discourse;
- mode 2 tourism knowledge;
- patriarchal power and control.

References

Aitchison, C. (1996) 'Patriarchal paradigms and the politics of pedagogy: a framework for a feminist analysis of leisure and tourism studies', *World Leisure and Recreation*, 38 (4): 38–40.

Aitchison, C. (2001) 'Gender and leisure research: the "codification of knowledge"', *Leisure Sciences*, 23 (1): 1–19.

Apple, M. (1990) *Ideology and the Curriculum*. London: Routledge and Kegan Paul.

Ayala, H. (2000) 'Surprising partners: hotel firms and scientists working together to enhance tourism', *Cornell Hotel and Restaurant Administration Quarterly*, 41 (3): 42–57.

Ball, S. (1990) 'Management as moral technology: a Luddite analysis'. In S. Ball (ed.) *Foucault and Education: Disciplines and Knowledge*. London: Routledge.

Becher, T. (1989) *Academic Tribes and Territories*. Buckingham: Open University Press.

Botterill, D. (2001) 'The epistemology of a set of tourism studies', *Leisure Studies*, 20 (3): 199–214.

Boyer, M. (1997) 'Tourism, a specific epistemology', *Loisir et Société*, 20 (2): 455–477.

Cheong SoMin and Miller, M. (2000) 'Power and tourism: a Foucauldian observation', *Annals of Tourism Research*, 27 (2): 371–390.

Cooper, C., Shepherd, R. and Westlake, J. (1994) *Tourism and Hospitality Education*. Guildford: University of Surrey.

Cooper, C., Fletcher, D., Gilbert, R., Shepherd, R. and Wanhill, S. (1998) *Tourism: Principles and Practices*, 2nd edn. London: Pitman.

Cotgrove, S. (1983) 'Risk, value conflict and political legitimacy'. In D. Groves (ed.) *The Hidden Curriculum in Business Studies*. Chichester: Higher Education Foundation.

Crouch, D. (1999) *Leisure/Tourism Geographies: Practices and Geographical Knowledge*. London: Routledge.

Dann, G. (1997) 'Paradigms in tourism research', *Annals of Tourism Research*, 24: 472–474.

Dann, G. and Phillips, J. (2001) 'Qualitative tourism research in the late twentieth century and beyond'. In B. Faulkner, G. Moscardo and E. Laws (eds) *Tourism in the Twenty-First Century: Reflections on Experience*. London: Continuum.

Dann, G., Nash, D. and Pearce, P. (1988) 'Methodology in Tourism Research', *Annals of Tourism Research*, 15 (1): 1–28.

Donald, J. (1986) 'Knowledge and the university curriculum', *Higher Education*, 15: 267–282.

Echtner, C. (1999) 'The semiotic paradigm: implications for tourism research', *Tourism Management*, 20 (1): 47–57.

Echtner, C. and Jamal, T. (1997) 'The disciplinary dilemma of tourism studies', *Annals of Tourism Research*, 21 (4): 868–883.

Foucault, M. (1971) *L'Ordre du discours*. Paris: Gallimard.

Foucault, M. (1974) *The Archeology of Knowledge*. London: Tavistock.

Foucault, M. (1980) *Power/Knowledge: Selected Interviews and Other Writings, 1972–77*. Brighton: Harvester Press.

Franklin, A. and Crang, M. (2001) 'The trouble with tourism and travel theory', *Tourist Studies*, 1 (1): 5–22.

Gibbons, M., Limoges, C., Nowotny, H., Schwartzman, S., Scott, P. and Trow, M. (1994) *The New Production of Knowledge*. London: Sage.

Goeldner, C. (1988) 'The evaluation of tourism as an industry and a discipline', International Conference for Tourism Educators, mimeo. Guildford: University of Surrey.

Graburn, N. and Jafari, J. (1991) 'Tourism social science', *Annals of Tourism Research*, 18 (1): 1–9.

Gramsci, A. (1971) *Selections from the Prison Notebooks*, trans. Q. Hoare and G. Smith. New York, NY: International Publishers.

Gunn, C. (1987) 'A perspective in the purpose and nature of tourism research methods'. In J. Ritchie and C. Goeldner (eds) *Travel, Tourism and Hospitality Research*. Chichester: Wiley.

Habermas, J. (1978) *Knowledge and Human Interests*, London: Heinemann.

Henkel, M. (1998) 'Responsiveness of the subjects in our study: a theoretical perspective'. In C. Boyes, J. Brennan, M. Henkel, J. Kirkland, M. Kogan and P. Youl (eds) *Higher Education and the Preparation for Work*. London: Jessica Kingsley.

Hirst, P. (1965) 'Liberal education and the nature of knowledge'. In R. Archambault (ed.) *Philosophical Analysis and Education*. Henley: Routledge and Kegan Paul.

Hirst, P. (1974) *Knowledge and the Curriculum*. London: Routledge and Kegan Paul.

Hoerner, J. M. (2000) 'The recognition of tourist science', *Espaces* (Paris), 173: 18–20.

Hollinshead, K. (1999a) 'Tourism as public culture: Horne's ideological comment-ary on the legerdemain of tourism', *International Journal of Tourism Research*, 1 (4): 267–292.

Hollinshead, K. (1999b) 'Surveillance of the worlds of tourism: Foucault and the eye-of-power', *Tourism Management*, 20 (1): 7–23.

Hughes, G. (1992) 'Changing approaches to domestic tourism', *Tourism Manage-ment*, 13 (1): 85–90.

Hughes, G. (1995) 'The cultural construction of sustainable tourism', *Tourism Management*, 16 (1): 49–59.

Jafari, J. and Aaser, D. (1988) 'Tourism as the subject of doctoral dissertations', *Annals of Tourism Research*, 15 (3): 407–429.

Jafari, J. and Ritchie, J. (1981) 'Towards a framework for tourism education', *Annals of Tourism Research*, 8: 13–33.

Jamal, T. and Hollinshead, K. (2001) 'Tourism and the forbidden zone: the under-served power of qualitative inquiry', *Tourism Management*, 22 (1): 63–82.

Jantsch, E. (1972) *Technological Planning and Social Futures*. London: Cassell.

Jenkins, C. (1999) 'Tourism academics and tourism practitioners: bridging the great divide'. In D. Pearce and R. Butler (eds) *Contemporary Issues in Tourism Development*. London: Routledge.

King, A. and Brownell, J. (1966) *The Curriculum and the Disciplines of Knowledge*. New York, NY: Wiley.

Kobasic, A. (1996) 'Level and dissemination of academic findings about tourism', *Turizam*, 44 (7–8): 169–181.

Kuhn, T. S. (1970) *The Structure of Scientific Revolutions*. Chicago, IL: University of Chicago Press.

Leiper, N. (1981) 'Towards a cohesive curriculum in tourism: the case for a distinct discipline', *Annals of Tourism Research*, 8 (1): 69–83.

Leiper, N. (2000) 'An emerging discipline', *Annals of Tourism Research*, 27 (3): 805–809.

Lyotard, J. (1984) *The Postmodern Condition: A Report on Knowledge*, trans. G. Bennington and B. Massumi. Manchester: Manchester University Press.

Lyotard, J. (1988) *The Differend: Phases in Dispute*, trans. G. van den Abbeele. Minneapolis, MN: University of Minneapolis Press.

McKercher, B. (2002) 'The privileges and responsibilities of being a referee', *Annals of Tourism Research*, 29 (3): 856–859.

Mannheim, K. (1960) *Ideology and Utopia: An Introduction to the Sociology of Knowledge*. London: Routledge and Kegan Paul.

Przeclawski, K. (1993) 'Tourism as the subject of interdisciplinary research'. In D. Pearce and R. Butler (eds) *Tourism Research: Critiques and Challenges*. London: Routledge.

Ryan, C. (1997) 'Tourism: a mature discipline?', *Pacific Tourism Review*, 1 (1): 3–5.

Ryan, C. (2000) 'Tourist experiences, phenomenographic analysis, post-positivism and neural network software', *International Journal of Tourism Research*, 2 (2): 119–131.

Seaton, A. (1996) 'Blowing the whistle on tourism referees', *Tourism Management*, 17: 397–399.

Toulmin, S. (1972) *Human Understanding*, vol. 1. Oxford: Clarendon Press.

Tribe, J. (1995) 'Tourism economics: life after death?', *Tourism Economics*, 1 (4): 329–340.

Tribe, J. (1997) 'The indiscipline of tourism', *Annals of Tourism Research*, 24 (3): 638–657.

Tribe, J. (2000) 'Undisciplined and unsubstantiated', *Annals of Tourism Research*, 27 (3): 809–813.

Tribe, J. (2001) 'Research paradigms and the tourism curriculum', *Journal of Travel Research*, 39: 442–448.

Tribe, J. (2003a) 'Curriculum development and conflict: a case study of Moldova', *Journal of Teaching in Travel and Tourism*, 3 (1): 25–46.

Tribe, J. (2003b) 'The RAE-ification of tourism research in the UK', *International Journal of Tourism Research*, 5: 225–234.

Usher, R. and Edwards, R. (1994) *Postmodernism and Education*. London: Routledge.

Walle, A. (1997) 'Quantitative versus qualitative research in tourism', *Annals of Tourism Research*, 24 (3): 524–536.

Whittaker, E. (1999) 'Indigenous tourism: reclaiming knowledge, culture and intellectual property in Australia'. In M. Robinson and P. Boniface (eds) *Tourism and Cultural Conflicts*, Wallingford: CABI Publishing.

4 A primer in ontological craft

The creative capture of people and places through qualitative research

Keith Hollinshead

Aims of the chapter

- To register the fact that the choice of qualitative research approach depends on the adoption of lines of inquiry which have the richest sustained ontological and epistemological fit with the problem area being investigated.
- To clarify that, ontologically, qualitative researchers must work to come to understand 'the real cultural world' of the individuals being investigated.
- To spell out that recent methodological advances have considerably added to the pool styles of interpretation which are available for researchers to harness ontologically alongside (or alternative to) so-called 'natural' hard science techniques of investigation.
- To explain that in tourism studies the multi-sited character of tourism and the international nature of the subject mean that ontological issues tend to be particularly significant because of the range of interests with which tourism developments have routine interface.
- To clarify that investigation of the ontologies of being, meaning and identity in the contemporary age is frequently a messy matter of infinite interpretive possibility, where interpretations need to be placed within the specific populations or insider-groups being studied.

Introduction: the necessity for emergent, engaged and dynamic non-linear understandings

This chapter assesses the platform considerations of being, meaning and identity – that is, the ontological assumptions which inevitably precede the conduct of each and every research study. In focusing upon those sorts of ontological questions of perceived reality and human experience that underpin qualitative inquiry, the chapter declares that the investigator's understanding about ontology is fundamentally a matter of sociological or anthropological awareness. This awareness constitutes the degree to which the researcher has competency in envisioning 'the real cultural world' of the local groups and other involved populations which are concerned in the given study locale. This circumspection and rigour is particularly required in tourism studies because researchers engaged in the field customarily have to consider the impacts or the influences of tourism upon a wide variety of different regional, ethnic, religious, subcultural and other 'special-interest' sanctioning groups. It is also deemed to be essential because tourism is customarily a quite magnetic force these days in the very demaking and very remaking of held individual and societal 'realities'. The chapter therefore argues that all who probe tourism settings must seriously be tuned into such matters of reflective ontological concern.

In this and the following chapter, it is suggested that the importance of ontological issues in an international cross-cultural field such as tourism studies renders the choice of qualitative inquiry method something other than the mere deployment of technique. The author argues that the choice of qualitative research instrument ought to be seen not simply as a methods-level matter of technical accuracy, but as a critical skill of applied philosophical awareness and applied critical literacy. Thus, in studies of international tourism in particular, methods-level decision-taking should be taken after primary methodological-level decisions have been taken, in which ontological concerns of 'being' are assessed in tandem with epistemological concerns about 'knowing'. Hence, in this first of the two chapters on ontology, decision-taking is seen to be an important paradigmatic matter: the qualitative researcher must take pains to consider a host of empowerments, anxieties and limitations which have become associated with qualitative research since the 1980s. In this light, this first chapter on ontological concerns addresses the relevance of approaches within post-positivistic, critical theory and constructivist thought at the paradigmatic level. It also weighs up the need for considerable creative craft on the part of the researcher(s) to carefully and reflexively select reality-aware, context-appropriate qualitative techniques.

Tourism and human knowing: the ontological dynamics of being and becoming

In recent years, tourism has come to be recognised as a very important player in the development and management of local and global places (Lanfant 1995; Meethan 2001). In Kirshenblatt-Gimblett's (1998) judgement, tourism plays a prodigious but underestimated role not only in the making of the world's culture, but also in its re-manufacture and de-manufacture. By Rothman's (1998) assessment, tourism is increasingly the lead vehicle by which the vital relationships of people to precious places are recognised, ascribed and scripted. By Fjellman's (1992) account, tourism is an industry with immense performative power, which corporate and other routinely collaborative agencies pointedly target in order to control what is known about the history, geography and nature of the world. By Buck's (1993) appraisal, tourism is increasingly a paramount means through which certain regions/states/nations, often working in cahoots with transnational corporations, can transform themselves into new paradisiacal lands, or can otherwise reinvent themselves as rare idyllic, bounteous or undisturbed 'treasures'. Given the sheer scale and scope of all these ways in which tourism helps lead the re-representation of populations and their places, it has been found to have dynamic authority in the way the world's locales are seen, experienced and known. It has been found to be a most significant people-maker, place-maker and past-maker as various groups use its articulations to identify or culturally invent themselves (Jamal and Hollinshead 2001: 63–65).

While a number of researchers have come to conclusions about the rising role tourism plays in our local representation of self and society (Selwyn 1996), tourism researchers have been neither theoretically nor methodologically advanced in the ways in which they have probed the various ontological identifications which exist between 'population' and 'place', and the various interconnectivities which lie between 'culture' and 'power' (Cohen 1984; Crick 1989; Featherstone 1995). For instance, we learn from Meethan (2001: 161) that the majority of management and research activities in tourism are routinely carried out *uncritically*. We learn that while many of the lead destinations of tourism are almost customarily settings of deep-seated importance to local populations, these complex symbolic values are not routinely appreciated by those who work in tourism studies. We learn that the field is still one 'where researchers tend to approach [the phenomenon of tourism] from within the specific boundaries of the main discipline in which they have been trained' (Echtner and Jamal 1997: 868). In the managerialist institutions which stand tall almost everywhere where tourism studies practitioners learn their trade, there is little genuine cross-disciplinary coverage of the tacit, the subjective, the discursive or the interpretive – that is, the stuff of qualitative research (Hollinshead 2002). We learn from Lidchi (1997: 153–154) that in tourism studies we are all relative novices in terms of our capacities to trace ways in which human meaning is created through the everyday systems of representation deployed at our various sites of cultural exhibition. In Lidchi's view, we are still relative probationers in terms of the accumulated experience we have so far gained in interpreting the agency of the active discourse of tourism. Then, we learn from Hall (1994) that tourism studies researchers have invariably concentrated upon the *prescriptive* and the *economic* worth of tourism, leaving the *descriptive* and *political* importance of the field relatively uncovered. And we learn from Horne (1992) that the field of tourism studies has generally been much too simplistic in the manner in which it has framed the scenes, the sites and the storylines it purveys.

Given the weight of these unfolding judgements on the current commonplace limitations of tourism studies, one may consonantly conclude that tourism studies is not yet in rude 'qualitative' health. The ontological logic its practitioners and researchers embrace is not advanced. Indeed, it is Kirshenblatt-Gimblett's (1992) view that there is a poverty of understanding about the ontological *hereness* of places (i.e. in this context, how the sites and settings are selected and projected in tourism). And likewise, it is her judgement that the field is not accomplished in its collective ability to gauge the ontological *madeness* of places (i.e. in this context, how the very special senses of 'being' highlighted at such places of attraction are locally mediated by interest groups). Furthermore, in Hall's (1994) view, tourism studies is a domain where supposedly value-free approaches to management and to research predominate. It is a field littered with standardised 'how-to-do' operational styles of research which are inclined to be weak in their political dimensionality and in their ontological rigour. And thereby, to Hollinshead (1998: 71) the field of tourism studies is one which is not yet rich in its capability to engage the fresh ways in which particular regions and destination-locations are strategically re-imagining themselves. In his view, tourism studies researchers are not consummately skilful in the capacity to monitor or map the imaginatively conceived realms which particular populations want their precious localities to become.

Outlooks on being, meaning, and becoming in and through tourism

This frame-making 'focus' for this first of two chapters on ontology has stressed the fact that although tourism is an immense international business and transformative inter-societal cultural phenomenon, the field of tourism studies does not appear to be advanced in its use of critical qualitative research approaches. Yet tourism studies has an enhanced need to develop a much more varied mix of qualitative research lines of thinking and techniques of investigation which are more discerning in their individual and collective capacity to probe the encountered ontologies of tourism. It should be realised that in many 'new encounters' in tourism today, there is an increasing likelihood that populations will clash over vital, group-affirming, ontological matters of being, becoming and meaning. Many of these potential gaps in ontological understanding will starkly lie between 'host' and 'guest' populations. Many others will be more nuanced, and will lie either between different sections within a host population, or between different sectors within the visiting travel market.

Background to the new profusion of qualitative research approaches: the continuum for 'naturalistic' techniques to 'progressive' techniques

Given that it is the purpose of this chapter to provide a basic introduction to the ontological considerations involved in the selection of particular qualitative research approaches in general, it is important that an early effort is made to explain how (after Holliday 2002: ix) the *progressivist* ontology common in qualitative research usually differs from the *naturalist* ontology which is ubiquitous in quantitative research. Thus, the current chapter seeks to reveal how the way investigators think about communal matters of seeing, cultural matters of experiencing, group matters of knowing, and individual/institutional matters of identifying is generally quite different in the day-by-day ontology of qualitative research from how it is within quantitative research. A key message of this chapter is therefore that researchers in tourism studies – indeed, in any field – must be very careful about the ways they go about investigating existential, aspirational and experiential issues – that is, the very ontological matters of being, becoming and meaning.

Clearly, qualitative research is no perfectly seeded terrain for the conduct of humanistic or equity-orientated research. Researchers in all disciplines must learn to appreciate not only the scale of illusions of objectivity which come hand in glove with many *quantitative* approaches to research, predicated upon the naturalistic worldviews which underpin them, but the like dangers that can also crop up in the *qualitative* domain, too. Although they may have been neatly dressed up in the rhetoric of liberalism or in the garb of progressivism, many qualitative approaches to research turn out to be poorly handled, and end up just as objectivist, presumptive or pre-calculative as the most presupposing of positivist quantitative methods. All too often, qualitative investigators unquestioningly adopt a pre-formulated, generalised or etic orientation to their subject of study that has been inclined to platform the majority of quantitative approaches to understanding (see Phillimore and Goodson's review of tourism research in Chapter 1). Too many researchers in all fields tend axiomatically to work on qualitative issues or tools with ultra-orthodox quantitative mindsets (Holliday 2002). Too many researchers in all fields tend to regard all research 'problems' as the same – that is, something which can be 'solved' via the fast deployment of takeaway quantitative approaches or 'take-home' statistical measures of frequency. Too many researchers in all fields have been taught to be, and have nurtured themselves into becoming and remaining, what Sorokin (cited in Converse 1987) would call 'quantophrenic' ontologists and 'quantophrenic' epistemologists. Too few researchers in all fields are prepared to indulge in what must admittedly be recognised as the longer reflective and necessarily deeper reflexive effort that the logic of qualitative methodologies is inclined to demand. Thus, in oh-so-many fields, the problem-fit and the interpretive-potency of qualitative approaches both tend to lie acutely under-serviced. In the light of this routine and ubiquitous preference for McDonaldised methodologies of calibration, this current chapter seeks to encourage researchers in tourism studies towards more situationally sympathetic and more contextually pertinent thinking about the issues of *being, seeing, experiencing, knowing* and *becoming* which they wish to explore. It seeks to stress that some problems are complex in their interpretability.

Robust and sustained forms of investigation which can be thickly descriptivist (see Denzin 1989: 11–12, 87–102, 144) rather than merely offering superficial depiction can offer benefits to tourism research exploring the complexity of some research problems. For those who work in tourism, it is important to raise levels of understanding about the options which are available to researchers to help them delve into the ways in which, in tourism locales or travel encounters, reality is understood. It is necessary to situate those qualitative approaches to being and to experience within the everyday social, cultural and lifespace practices of the different populations which are of concern to researchers within those particular contexts.

To this end, tourism studies investigators are encouraged towards the adoption of *verstehen* (after Weber 1978) styles of research that are differentially empathetic to the lived experience or to the worldview realities that are 'real' to distinct institutions or to particular interest groups (Schwandt 1994: 118). Such highly interpretive *emic* approaches to research generally permit the researcher to, for instance, scrutinise the experience of the found tourism promotion or the meaning of the given tourism development in terms of the operating local worldviews of the various different 'populations' involved. Such *verstehen* styles are generally intended to equip the researcher to be able to immerse themself within the multiple realities or the vogue mindsets that operate, for instance, within:

- a given tourism development body/the tourism planning organisation;
- its agent publics;
- its consumer publics;
- its supplier publics;
- its employee publics;
- its support publics;
- the competitor publics which are active in or have impact upon the given issue;
- the other *within-society* special-interest publics which are likewise active/involved;
- the other *beyond-society* special-interest publics which are active/involved;

- the *within-society* 'general public' which is active/involved;
- the *beyond-society* 'general public' which is active/involved.

(This mix of 'publics' has been adapted from an introductory list of publics provided in Howard and Crompton (1980: 321–324).) For a basic introduction to the application of a number of *verstehen* styles of approach in tourism studies, see Jennings (2001: 158–161).

The worth of qualitative styles of inquiry lies predominantly in their diverse and emergent capacities to capture the contesting worldviews that such publics have on any matter as it pertains to their held realities. Fundamentally, a qualitative researcher who is up to date in terms of what Denzin and Lincoln (1994: 11) would call his or her *fifth-moment* thinking (and, thereby, his or her critically aware, action-oriented, reflexive sense of engagement) would tend to work from the assumption that no single objectivity unites all of the involved publics which surround the tourism development or compose the particular travel encounter. It is therefore an essential ontological requirement on the part of the informed 'fifth-moment' tourism studies researcher to map the contours of the worldviews of each significant population as they pertain to the context in question. Put simply, it is to adopt a 'stranger position' within that involved population in order *to make the familiar strange* (Holliday 2002: 13). Thus, the key function of the qualitative researcher is to communicate to the reader a sense of context-revealing strangeness about the subject being analysed. It is to interpret the found human actions and aspirations of that context variably or differentially *vis-à-vis* the different in-group vantage points which are found to be significant at the given tourism site, in the given travel trade setting, or in the otherwise-defined human-societal encounter. See Chapters 1 and 2 of this book for further consideration of these issues.

In order to ontologically capture and, importantly, to thickly describe the different worldviews which are substantively relevant at or on a particular tourism site, the engaged tourism studies researcher may draw upon an ever-expanding lexicon of qualitative research techniques. While the range of these techniques is comprehensively surveyed at some depth of analysis by Denzin and Lincoln (1994), Holliday (2002) offers a helpful rudimentary study of qualitative research thinking. Table 4.1 has been adapted from Holliday to help readers comprehend the width of scope and diversities of fit that qualitative research techniques have.

Holliday's attempt to distinguish *naturalistic* frames of reference from *progressivist* frames of reference should be regarded not as a definitive indexing of available qualitative approaches, but merely as a suggestive and illustrative mapping of the new options in latter-day research. Indeed, many researchers will want to take issue with some of Holliday's categorisations and placements. For instance, if it were listed by Holliday, would feminist ethnography be progressivist (after 'feminism', in Table 4.1) or naturalistic (after 'ethnography', in Table 4.1)? And is that contentious

Table 4.1 Strategies and methods in qualitative research: Holliday's continuum from naturalistic to progressive outlooks

Paradigms and perspectives	Strategies of inquiry[a]	Methods
Largest		*Smallest*
Naturalist qualitative outlooks: post-positivism, realism	**Case study:** study of a specific 'bounded system', e.g. person or institution (Stake 1994: 236)	Structured interviewing
(a) Reality is quite plain to see.	**Ethnography:** explores 'the nature of a specific social phenomenon'; unstructured data, 'a small number of cases'; 'interpretation of the meanings and functions'; 'participant observation' (Atkinson and Hammersley 1994: 248).	Observational techniques
(b) Deeper social reality needs qualitative enquiry.		Interpreting documents
(c) The probable truth is supported by extensive, substantiated record of real settings.	**Ethnomethodology:** investigates people's 'practical everyday procedures . . . for creating, sustaining, and managing a sense of objective reality' involving 'rules, values, motives' (Holstein and Gubrium 1994: 264)	Content analysis
(d) Researchers must not interfere with real settings.		Semiotic analysis
	Phenomenology: 'focus on the way life-world – that is the experiential world every person takes for granted – is produced and experienced by members'; 'bracketing the life-world, that is, setting aside one's taken-for-granted orientation to it' (Holstein and Gubrium 1994: 263)	Unstructured interviewing
Progressive qualitative outlooks: critical theory, constructivism, postmodernism, feminism		
(a) Reality and science are socially constructed.	**Grounded theory:** theory that is grounded in data systematically gathered and analysed; 'continuous interplay between analysis and data collection' (Strauss and Corbin 1994: 273)	
(b) Researchers are part of research settings.	**Participatory action research:** 'emphasises the political aspects of knowledge production' (Reason 1994: 328); 'involves the individual practitioner in continually reflecting on his or her own behaviour-in-action' so that 'other members of the community do the same' (Reason 1994: 328, 331)	
(c) Investigation must be in reflexive, self-critical, creative dialogue.		
(d) Aim to problematise, reveal hidden realities, initiate discussions.		

Source: Adapted from Holliday (2002: 18).

Note

a All the strategies listed are usable with any of the items in the other two columns.

subject, grounded theory, faithfully progressivist? Denzin and Lincoln (1994: 205) do not think so, famously stating that grounded theory has yet to feel 'the direct influence' of feminist and postmodern arguments which have freshened up the ontology and epistemology of social science and which have beneficially oxygenated other forms of qualitative research. Overall, Holliday's (2002) book provides a very useful general pathway through the maze of qualitative approaches which have recently emerged in social science. Those who are troubled by some of Holliday's philosophical categorisations and operational placements, and those who want a more specific treatment of qualitative approaches within tourism studies *per se*, to thereby serve as a companion to their Phillimore and Goodson, are advised to inspect the recent work of Jennings (2001).

The profusion of choice in qualitative inquiry is characterised by a welter of tensions, contradictions and hesitations about which methods are safe, and appropriate to conduct (Denzin and Lincoln 1994: 15–16, 576–580). Ontologically, many of these tensions, contradictions and hesitations revolve around the degree to which ideological positions are seen to be embedded within particular qualitative research methods. Epistemologically, these problematics revolve around debates as to whether any single qualitative research method can afford the researcher a privileged place in gaining authoritative knowledge about the sites and settings they explore (Denzin and Lincoln 1994: 578, 581–582). In terms of both ontology and epistemology, these difficulties revolve around debates as to whether any single, emergent, fifth-moment qualitative research method can ever comprehensively capture the emic worldviews of the particular public being inspected (Harr 1980: 135–137). This ontological and epistemological dilemma is articulated by Vidich and Lyman (1994: 26) as 'how is it [ever] possible to understand the other when the other's values are not one's own?'.

But back to Table 4.1, and Holliday's distinction between 'naturalistic' and 'progressivist' styles of qualitative research. As Holliday argues, it is not easy for social science researchers in many institutions to engage in progressivist forms of qualitative inquiry. Their new freedoms of approach and open liberalities of outlook are often foreign to the traditional ways in which their host institutions think, act and value (Slife and Williams 1995: 6), an issue which Hall considers in some detail in Chapter 8. Since naturalistic styles of research are much more similar to those styles of positivistic inquiry which, we have noted, have generally ruled the roost in social science investigation, they tend to be much easier for most researchers to grasp. This is notably the case for those whose own working regime is highly quantitative or is otherwise predicated upon conventional scientific notions of universalism, validity and reliability (Chadwick *et al.* 1984: 6–10, 42–49; Nachmias and Nachmias 1987: 3–28). While naturalistic researchers tend to gravitate towards *like-it-is* forms of understanding which seek to capture the assumed single reality of a specific physical or geographical

place, progressivist forms of research tend to take no such supreme univer-
salisms for granted. Consonantly, progressivist qualitative researchers tend
to inspect not only how the people within that specific place construct the
world, but how different groups of those people *differentially construct the
world*, and, more especially, how they each differentially construct the
world in various settings. And throughout the conduct of progressivist
approaches to research, it is recognised that the effort faithfully to capture
those held ontologies is highly demanding of self-vigilance on the part of
the investigator. They must be perpetually alive to the influence or impact
of their own standpoints as they mature in sensibility or sympathy with
those contexts (Holliday 2002: 20–21). (Humberstone gives greater attention to standpoint research in Chapter 7.) Accordingly, the researcher
themself tends to become the required sensitive research instrument under
the presumptions of progressivist styles of research. Under progressivist
approaches, the researcher is engaged in a struggle to gauge the pertinence
of the various contingencies which flow and mingle around them
(Gubrium and Holstein 1997: vi). Hence, qualitative researchers often talk
of the need to conduct 'human instrument' assessments as opposed to (or
perhaps sometimes alongside!) the calibrated instrument forms of investigation which have tended to have been privileged under orthodox versions
of 'scientific method' (Chadwick *et al.* 1984: 6–14; Nachmias and Nachmias
1987: 301–307).

While quantitative research techniques tend to be applied uniformly,
and are inclined to be 'relatively inflexible' in terms of their capacity to
respond differentially to ontological considerations, qualitative studies
tend to be much more idiosyncratic, vexatious and time-consuming. Under
quantitative approaches, the measures or sources of validity are assumed
to be knowable, hence, under most positivist and neo-positivist lines of
inquiry, it is frequently assumed that the subject matter of quantitative
study may be observed similarly and measured similarly here, there and
everywhere, without much regard to contextual influences (Slife and
Williams 1995: 177). Under qualitative forms of research, such matters of
ontological concern are much more difficult to corral. The qualitative
researcher has to map out the profile of competing measures of reality. As
has been stated above, the contemporary or fifth-moment qualitative
investigator is inclined to operate under the view that a number of alternative worldviews inevitably hold true among different groups and within
different settings. Thus, while there may always indeed be significant differences between pure-form 'realist' and 'positivist' quantitative
researchers (Slife and Williams 1995: 176), broadly speaking the *quantitative* researcher often needs only establish the nature of the mainstream
reality once – that is, up front in and across the study. Essentially, to most
quantitative researchers, observed measurements of key variables detail
the world as it really exists. In contrast, the qualitative researcher customarily needs to consider these matters of local and situational ontology on a

regular, or repeated basis, perhaps even 'every single time' they enter the given research settings (Holliday 2002: 8, 47–68). The ontological effort to gauge the contours of these various intersecting lived worlds of experience will be very muddy and rather untidy. The identification and profiling of held communal notions of being, becoming and meaning is inevitably a many-sited, slippery business (Fine 1994: 71–75). Ontologically, the emic fit of resultant qualitative interpretations is often richer or more pertinent where the researcher generates 'open-ended' and 'contingent' evocations of being and meaning, rather than yielding totalised, clean and tidy, non-complex classifications of lived reality. And since the border parameters which surround a particular realm of human activity are commonly not known in qualitative research settings, the qualitative researcher is obliged to concern themself with depictions of what is *actual* rather than judging what is *typical*, as under quantitative or probability-based techniques of analysis. Epistemologically, the commonplace messiness of data and settings teaches experienced qualitative researchers not to assimilate their new ontological subjects or their ontological settings too quickly (Marcus 1994: 567). The combined ontological and epistemological craft of 'fifth-moment' qualitative research is thereby generally enriched by a pervasive sense of uncertainty. 'Incompleteness', *ipso facto*, is necessarily a studied and a positive (not positivist!) interpretive attribute. Almost all qualitative analyses can only ever be partial, and therefore open-ended, forms of inquiry; many researchers believe they can only ever yield 'findings' tentatively held, and never 'results' firmly concluded.

Qualitative research in the ontological service of research paradigms

New liberation in the nature of knowledge

In examining ontological questions around the social science of tourism, reference ought to be made to the recent reconceptualisations which have come to the nature of knowledge about human being and societal meaning (Schwartz and Ogilvy 1979; Bernstein 1983). In the past two decades, considerable change has occurred in terms of what is deemed to be appropriate and/or legitimate with regard to sound social science (Slife and Williams 1995: 71–93). Indeed, it is increasingly recognised in the social sciences that considerations of *method* ought to come second after considerations of *methodology*, where questions of a methodological nature are fundamentally 'paradigmatic' issues. Hence, methodological issues are those broader matters which need to be addressed with reference to their epistemological and ontological bearings. In fact, in certain senses after the pioneering thought of Kuhn (1970), paradigms themselves may be seen to be ontological doctrines as to how the world is (i.e. as to *what is important within the imagined world*), or how it ought to be (Slife and

Williams 1995: 163–164). Consonantly, views on the character and role of knowledge have been reconsidered, as new options have emerged among the paradigms. During the past two decades all sorts of new understandings about paradigmatic thought have unfolded to challenge received orthodoxies about *positivism*, the conventional master paradigm of yore, and the philosophical foundations of research methodology in the social sciences have lately been quite well lubricated by them. Again, all sorts of new options within sound social science have been recognised as being valid, justifiable or 'permissible'.

In a chapter of this limited length, space is at a premium for the discussion of the new legitimacies that, during the 1980s and 1990s, have been espoused to help determine what is bona fide within social science. Thus, readers who are keen to appreciate something of the scale and the scope of the magnitude of the changes which have unfolded in knowledge accumulation under the paradigmatic pluralism of this current momentous research liberation era are encouraged to digest Guba's landmark text *The Paradigm Dialog* (1990), which stands as a watershed work in accessible understanding about the twin matters of ontology and epistemology.

The identification of paradigmatic bias under the old and new options for inquiry

In *The Paradigm Dialog*, a paradigm is simply seen to be the basic belief systems which drive disciplined inquiry (Guba 1990: 9, 18). Certainly, particular philosophers and humanists may quibble with Guba over which areas of understanding constitute the major *persuasive* worldviews within the human inquiry of our age (Schwandt 1994: 118, 132). But many social science critics champion the judicious, yet reader-friendly, manner in which Guba and his contributors dissect the philosophical and methodological biases in what are conceivably the four lead 'basic belief systems' which rule contemporary science. In *The Paradigm Dialog* these lead paradigms are:

- *positivism* – with its realist (or 'naive realist'?) ontology (Guba 1990: 19);
- *post-positivism* – with its critical-realist worldviews (Guba 1990: 20–21);
- *critical theory* – with its ideologically oriented standpoints (Guba 1990: 23–24);
- *constructivism* – with its dialectical outlook on the world's multiple realities (Guba 1990: 26).

In clarifying for social scientists the nature of the underlying principles which govern how the world is assumed to be and how the world is assumed to be known (under terms of these four paradigms), Guba (1990) clarifies the premises which tend to regulate contemporary social science. In providing helpful elementary insight on the sorts of methodological bias which govern each of the four organising worldviews, he and his contributors help investigators to recognise the strengths and weaknesses of their own operating presuppositions. He does this by generating a simple critique of the basic paradigmatic belief system as it is discernible in terms of matters of ontological rigour, epistemological rigour and methodological rigour for each of those nominated forms of 'disciplined communal inquiry'. In this regard, Guba supports key philosophical concepts outlined by Phillimore and Goodson in the opening chapter of this book as follows:

- *Matters of ontology* are defined as those concerns and outlooks which help determine or designate the nature of the knowable (or otherwise, the nature of reality in terms of concerns of 'being', 'becoming' and 'meaning', etc.).
- *Matters of epistemology* are defined as those insights and questions which help understand the relationships between knower (the inquirer) and the known (the knowable).
- *Matters of methodology* are defined as those preferred practices and operational partialities (as predicated by the above ontological and epistemological issues) which the inquirer should respect as they go out to find knowledge via the use of particular 'methods' or 'approaches' to inquiry (Guba 1990: 18).

Dispensing with positivism on account of the certitudes and the over-coherencies it is argued to peddle, Table 4.2 now proffers a table which synthesises for the reader the fundamental differences in operating premises between what Guba (1990) deems to be the three chief paradigms of our age. The table delineates what Guba (1990) considers to be the paramount characteristics of the ontology, the epistemology and the methodology of postpositivism, critical theory and constructivism, *ipso facto* at the general level.

Guba's (1990) edited text only reaches into the procedures of social science pertaining to four 'family-type' paradigms, but it is loaded with penetrating insight on the meaning and conduct of human inquiry. Yet it is a comprehensive mapping of the contours of the hidden ideas and the under-recognised conjectures of contemporary human inquiry, and thereby of the deep contestations within social science in our era. Hence, Guba (1990) confirms the judgement of Rorty (1979) that there is no single set of independent criteria available for all social science, which a neutral observer (if such a person existed!) could ever use to determine which was the best or the most appropriate paradigm for each and every social scientist to work within (see Skrtic 1990: 128). Moreover, the pivotal-moment Alternative Paradigms Conference in California in 1989 (upon which *The Paradigm Dialog* is based) did generally support the

Table 4.2 The building blocks of scientific inquiry: a comparison of three lead structural worldviews (paradigms), namely post-positivism, critical theory and constructivism

Truth issues	Post-positivism	Critical theory	Constructivism
Ontology	Realist: reality exists but can never be fully apprehended, only incompletely understood – therefore critical realist	Realist: critical realist (as per post-positivism)	Relativist: realities exist in the form of multiple mental constructions – socially and experientially based, local and specific, dependent for their form and content on the persons who hold them
Epistemology	Dualist/objectivist: objectivity remains a regulatory ideal, but can only be approximated	Interactive/subjectivist: values immediate inquiry which is participative and/or which reflects the values of human players	Interactive/subjectivist: inquirer and inquired are fused into a singular (monistic) entity. Findings are the creation of a process of interaction between the two
Methodology	Interventionist: modified experimental/manipulative methods emphasizing 'critical multiplism'. Redresses imbalances by doing inquiry in more natural settings	Participative: dialogic and transformative – seeking the elimination of false consciousness and the facilitation of a transformed world	Hermeneutic/dialectic: individual constructions are elicited and refined hermeneutically, and are compared and contrasted dialectically – with the aim of generating one (or a few) constructions on which there is general consensus

Source: After Hollinshead (1993: 57); originally adapted from Guba (1990: 20, 23, 25, 27 and 78).

view of Habermas (see Bredo and Feinberg 1982 cited by Firestone 1990: 122) However, in support of Habermas's view, it is argued that paradigms are indeed highly distinguishable by the under-suspected cognitive interests which propel them (Firestone 1990: 122). For instance:

- Under positivism (and postpositivism), ontological thought tends to be constrained by the dictates of technical efficacy and the imperatives of and for instrumental action within organisations (Guba 1990: 20).
- Under critical theory, ontological thought tends to be channelled towards particular emancipatory interests, as are identified through the singular value-window perspectives (or 'standpoints') adopted by action-demanding groups and by ordinarily involved populations (Guba 1990: 25).
- Under constructivism (or constructionism, or social constructivism), the ontological thought is relativist, seeking to understand the identities of, the meanings attributed by and the experiences of different populations, against a background of competing perspectives on life and the world, within the setting being investigated (Guba 1990: 27; see also Skrtic 1990: 135).

Issues concerning the fit between research paradigm and research method today

Whichever paradigmatic views of reality the researcher chooses to address, they are increasingly encouraged these days to make use of quantitative and qualitative methods in tandem (Guba and Lincoln 1994: 105), for the selection of approach(es) is fundamentally *a methods-level consideration* of secondary concern to the initial paradigmatic determinations (i.e. of *methodological-level considerations*) made *vis-à-vis* questions of ontology and epistemology. (See Decrop in Chapter 9 for a further discussion of mixed-method approaches.) In making their choices at the twinned levels of methodology and methods, the researcher must remain alive to the fact that power and politics always play an important (if, often, a below the surface) role in the existence, creation, and/or development of receptive audiences for research projects. While matters of the individual accommodation of the single researcher to particular views of reality are likely to differ idiosyncratically, matters of institutional accommodation are likely to be much more important, always having an underlying internal and external political dynamism to them. Indeed, in many senses, the very choice to use qualitative methods is itself a highly political act. The selection of qualitative research at the methods level can re-establish particular ideas about the nature of the community in question, and thereby whom might be called in to activate the resultant recommendations of the research project. As such, the methods-level choice *to go qualitative* frequently presages ontological judgements about the nature of the actual, or the latent, or the negotiable social order that is being examined (Firestone 1990: 121).

Accordingly, methods-level thinking which is inherently ontological ought to be made within a paradigmatic framework that respects those same ontological impulses and directions at the methodological level. Qualitative research methods are frequently not just about the gain of incremental knowledge. In their routine use to render the world perhaps more humanistic, perhaps more holistic, and perhaps more relevant to the lives of certain disenfranchised populations, qualitative research methods are thereby commonly tied to issues of societal consciousness and emancipation. They tend to have an inherent ontological connection to forms of emancipatory or remedial action empowering 'voices that were previously silenced [to] speak as agents of social change and personal destiny' (Denzin and Lincoln 1994: 207).

Once more, it needs to be stressed that going qualitative is frequently therefore a matter of acting politically. To Denzin, nothing speaks for itself in social science. There is only ever interpretation. The age of a supposedly value-free social science has ended. In his view, what increasingly counts is how the findings of a research study are legitimated, and what epistemological claims as to its own authority over the ontological concerns it addresses. Thus, after the interrogations of postmodernity, the methodological gain of truth is nowadays more commonly taken to be a creative and political act of social discovery, rather than a necessary act of clean and neutral detachment. In Denzin's judgement, such is the loud, or the quiet, ideological truth implicit within the textual production of each and every social science investigation, whatever the driving paradigm may be. To him, a sound, critical and emancipatory qualitative research study is actually 'one that is multivocal, collaborative, naturalistically grounded in the worlds of lived experience, and organised [in terms of demonstrably relevant] critical interpretive theories' (Denzin 1994: 509). To others, however, this very recognition of the so-called ever-present character of the 'political' serviceability of qualitative research only conveniently generates for researchers with emancipative and interpretive interests the new right to over-correctively impose their own voices and values upon the inquiry settings being studied (Quantz 1992: 471). There will always be massed and hidden ranks who will prefer the comfortable sureties of the old and largely empirical conventional scientific objectivities to the emergent and plainly messy irregular social discourses which are involved in the new interpretivist/postmodern ways of knowing.

Summary

In this chapter the following principal insights have been gained:

- That the selection of the most appropriate ontological line of investigation in qualitative research is generally a methodological matter rather than a mere methods-level judgement.

- That efforts to secure insights generally require not only that the researcher is thoroughgoing in their efforts to map and monitor the held 'cultural warrants' of individuals being investigated, but that they are situationally creative in their capacity to deploy salient means of assessment and confirmation to capture (or thickly and appropriately describe) those pertinent 'world-making' realities.
- That within the past couple of decades or so, a new profusion of liberating 'fifth-moment' approaches to inquiry have become available for 'soft science' (i.e. qualitative researchers) to use to pry into ontologies of being – where some of those refreshed and unshackled approaches are 'naturalistic' and others are 'progressivist'.
- That the efforts to work with emergent 'critical theory' family and 'constructivist' family lines of ontological inquiry variously tend towards an emancipatory, action-oriented and admittedly 'engaged' or 'political' outlook on the world.

Questions

1   Recently there has been a considerable liberalisation in the recommended procedures by and through which social science is conducted. For instance, qualitative researchers are not nowadays exclusively encouraged to begin their study with a *formal hypothesis* (which might drive the whole of the study from start to finish); many of them are encouraged to consider the merit of conducting an *emergent study* where a hypothesis might be tenable only at the conclusion of their particular investigation. Under what ontological conditions, therefore, is it still useful or advantageous to commence an investigation with a hypothesis? Under what ontological conditions is it useful or advantageous to conduct a study via an unfolding/emergent design? Compare and contrast the benefits of these two sorts of study design approaches.

2   In his book *Doing and Writing Qualitative Research*, Holliday notes that every qualitative researcher has to work with some notion or other of *culture*. He warns, however, that 'researchers must be wary of reducing reality to the "culture" [which] they themselves construct' (2002: 45). What might Holliday mean by this? How indeed can researchers guard against this ontological (?) problem? Is it just an ontological problem?

3   In his book *The Paradigm Dialog*, Guba doubts whether it is possible for researchers to bring competing paradigms into a uniform joint, complementary or 'ecumenical' position within a singular study. Why might he think that? Is it at all easy for an individual researcher to be effectively trained in the investigative skills which are appropriate to (for instance) 'post-positivist', 'critical-theory', 'constructivist' understanding? Draw up a simple cultural or societal study problem which

needs to be investigated in the tourism development or management of some place or region, and show how *post-positivists*, *critical theorists* and *constructivists* might each respectively tend to want to explore that matter. Likewise, how might *feminists* seek to explore that same problem?

4   Consider one of the following statements about tourism from Selwyn's book *The Tourist Image: Myths and Myth Making in Tourism*:

- 'Tourist sites have always been crucibles for nationalist constructions' (Selwyn 1996: 23).
- 'Tourism is [fundamentally] about the invention and reinvention of tradition. It is about the production and consumption of myths and staged inauthenticities. It has also far-reaching economic, political and social consequences at levels ranging from the household to the nation'. (Selwyn 1996: 28).
- 'Much contemporary tourism consists of] the wholesale Disneyfication of one part of the world built on the wasteland of the other [part]' (Selwyn 1996: 30).

What important ontological issues of SEEING, EXPERIENCING, MEANING, KNOWING or BEING do you think are implied by or raised in the statement you have chosen? Select a tourist destination site/city/region which you are interested in. How might an alert and aware 'fifth-moment' qualitative researcher – that is, a multi-talented *bricoleur* – set up a triangulated (or a crystallised) study to inquire into those nominated issues? What aspects/angles/activities would he or she certainly want to explore or concentrate upon in looking at the role of particular sorts of tourists or the specific function of tourism on or within that matter?

References

Bernstein, R. (1983) *Beyond Objectivism and Relativism: Science, Hermeneutics and Praxis*. Philadelphia, PA: University of Pennsylvania Press.

Bredo, E. and Feinberg, W. (1982) 'The critical approach to social and education research'. In E. Bredo and W. Feinberg (eds) *Knowledge and Values in Social and Educational Research*. Philadelphia, PA: Temple University Press.

Buck, E. (1993) *Paradise Remade: The Politics of Culture and History in Hawai'i*. Philadelphia, PA: Temple University Press.

Chadwick, B. A., Blair, H. M. and Albrecht, S. L. (1984) *Social Science Research Methods*. Englewood Cliffs, NJ: Prentice Hall.

Cohen, E. (1984) 'The sociology of tourism: approaches, issues and findings', *Annual Review of Sociology*, 10: 373–392.

Converse, J. M. (1987) *Survey Research in the United States: Roots and Emergence, 1890–1960*. Berkeley, CA: University of California Press.

Crick, M. (1989) 'Representations of international tourism in the social sciences: sun, sex, sights, savings and servility', *Annual Review of Anthropology*, 18: 307–344.

Denzin, N. K. (1989) *Interpretive Interactionism.* Newbury Park, CA: Sage.

Denzin, N. K. (1994) 'The art and politics of interpretation'. In N. K. Denzin and Y. S. Lincoln (eds) *Handbook of Qualitative Research.* Thousand Oaks, CA: Sage.

Denzin, N. K. and Lincoln, Y. S. (eds) (1994) *Handbook of Qualitative Research.* Thousand Oaks, CA: Sage.

Echtner, C. M. and Jamal, T. B. (1997) 'The disciplinary dilemma of tourism studies', *Annals of Tourism Research,* 24 (4): 868–884.

Featherstone, M. (1995) 'Travel, migration and images of social life'. In M. Featherstone (ed.) *Undoing Culture: Globalization, Postmodernity and Identity.* London: Sage.

Fine, M. (1994) 'Working the hyphens: reinventing self and other in qualitative research'. In N. K. Denzin and Y. S. Lincoln (eds) *Handbook of Qualitative Research.* Thousand Oaks, CA: Sage.

Firestone, W. A. (1990) 'Accommodation: toward a paradigm–praxis dialect'. In E. Guba (ed.) *The Paradigm Dialog.* Newbury Park, CA: Sage.

Fjellman, S. M. (1992) *Vinyl Leaves: Walt Disney World and America.* Boulder, CO: Westview Press.

Guba, E. (ed.) (1990) *The Paradigm Dialog.* Newbury Park, CA: Sage.

Guba, E. and Lincoln, Y. S. (1994) 'Competing paradigm in qualitative research'. In N. K. Denzin and Y. S. Lincoln (eds) *Handbook of Qualitative Research.* Thousand Oaks, CA: Sage.

Gubrium, J. F. (1988) *Analysing Field Reality.* Qualitative Research Methods series 8. Newbury Park, CA: Sage.

Gubrium, J. F. and Holstein, J. A. (1997) *The New Language of Qualitative Research.* New York, NY: Oxford University Press.

Hall, C. M (1994) *Tourism and Politics: Policy, Power, and Place.* Chichester: Wiley.

Harr, R. (1980) *Social Being: A Theory for Social Psychology.* Totowa, NJ: Rowman and Littlefield.

Holliday, A. (2002) *Doing and Writing Qualitative Research.* London: Sage.

Hollinshead, K. (1993) 'The truth about Texas: a naturalistic study of the construction of heritage', doctoral dissertation, Texas A & M University, College Station [Dept of Recreation, Park, and Tourism Sciences].

Hollinshead, K. (1996) 'The tourism researcher as *bricoleur*: the new wealth and diversity in qualitative inquiry', *Tourism Analysis,* 1 (1): 67–74.

Hollinshead, K. (1998) 'Tourism and the restless peoples: a dialectical inspection of Bhabha's halfway populations', *Tourism, Culture and Communication,* 1 (1): 49–77.

Hollinshead, K. (2002) 'Tourism and the making of the world: the dynamics of our contemporary tribal lives'. Honors College Excellence Lecture, The Honors College, Florida International University, Miami, Florida, April 2002. Honors Excellence Occasional Paper Series vol. 1, no. 2.

Horne, D. (1992) *The Intelligent Tourist.* McMahon's Point, NSW: Margaret Gee Holdings.

Howard, D. R. and Crompton, J. L. (1980) *Financing, Managing and Marketing Recreation and Public Resources.* Dubuque, IA: Wm C. Brown.

Jamal, T. B. and Hollinshead, K. (2001) 'Tourism and the forbidden zone: the underserved power of qualitative inquiry', *Tourism Management,* 22 (1): 63–82.

Jennings, G. (2001) *Tourism Research*. Milton, Queensland: Wiley.

Kirshenblatt-Gimblett, B. (1998) *Destination, Culture: Tourism, Museums, and Heritage*. Berkeley, CA: University of California Press.

Kuhn, T. S. (1970) *The Structure of Scientific Revolutions*, 2nd edn. Chicago, IL: University of Chicago Press.

Lanfant, M.-F. (1995) 'Internationalization and the challenge to identity'. In M.-F. Lanfant, J. B. Allcock and E. M. Bruner (eds) *International Tourism: Identity and Change*. London: Sage

Lidchi, H. (1997) 'The poetics and the politics of exhibiting other cultures'. In S. Hall (ed.) *Representation: Cultural Representations and Signifying Practices*. London: Sage.

Marcus, G. (1994) 'What comes (just) after post: the case of ethnography'. In N. K. Denzin and Y. S. Lincoln (eds) *Handbook of Qualitative Research*. Thousand Oaks, CA: Sage.

Meethan, K. (2001) *Tourism in Global Society: Place, Culture, Consumption*. Basingstoke: Palgrave.

Nachmias, D. and Nachmias, C. (1987) *Research Methods in Social Sciences*. New York, NY: St Martin's Press.

Quantz, R. A. (1992) 'On critical ethnography (with some postmodern considerations)'. In M. D. Le-Compte, W. L. Millroy and J. Preissle (eds) *The Handbook of Qualitative Research in Education*. New York, NY: Academic Press.

Rorty, R. (1979) *Philosophy and the Mirror of Nature*. Princeton, NJ: Princeton University Press.

Rothman, H. K. (1998) *Devil's Bargains: Tourism in the Twentieth-Century American West*. Lawrence, KS: University Press of Kansas.

Schwandt, T. A. (1994) 'Constructivist, interpretivist approaches to human inquiry'. In N. K. Denzin and Y. S. Lincoln (eds) *Handbook of Qualitative Research*. Thousand Oaks, CA: Sage.

Schwartz, P. and Ogilvy, J. (1979) *The Emergent Paradigm Changing Patterns of Thought and Belief*. Values and Lifestyle Program, Analytic Report 7. Menlo Park, CA: SRI International.

Selwyn, T. (ed.) (1996) *The Tourist Image: Myths and Myth Making in Tourism*. Chichester: Wiley.

Skrtic, T. M. (1994) 'Social accommodation: towards a dialogical discourse in educational inquiry'. In E. Guba (ed.) *The Paradigm Dialog*. Thousand Oaks, CA: Sage.

Slife, B. D. and Williams, R. N. (1995) *What's Behind the Research? Discovering Hidden Assumptions in the Behavioral Sciences*. Thousand Oaks, CA: Sage.

Vidich, A. J. and Lyman, S. M. (1994) 'Qualitative methods: their history in sociology and anthropology'. In N. K. Denzin and Y. S. Lincoln (eds) *Handbook of Qualitative Research*. Thousand Oaks, CA: Sage.

Weber, M. (1978) *Economy and Society: An Outline of Interpretive Sociology*, vol. 1, G. Roth and C. Wittich (eds). Berkeley, CA: University of California Press.

5 Ontological craft in tourism studies

The productive mapping of identity and image in tourism settings

Keith Hollinshead

Aims of the chapter

- To apply in tourism/travel settings many of the general ontological concerns which were given in Chapter 4.
- To draw attention to a number of particular ontological issues which are frequently encountered in tourism studies – such as occur with difficulties involved in contesting or interpreting 'values', 'realities' and/or 'cultural being'.
- To highlight a number of readings in the recent literature of tourism studies where commentators on tourism have already ventured capably into coverage or critique of ontological matters. These pioneer observers include Golden (1996), Sardar (1998), Wilson (1997) and McKay (1994).

Introduction: methodological thinking in tourism

In Chapter 4 it was emphasised that in qualitative research it is rarely ever sufficient or proper in making selections about the adoption of research approaches to make those choices upon *methods-level* decisions based upon 'technical accuracy' or 'instrumental procedure'. Rather, it was stressed that the choice of approach in most qualitative research settings necessarily involves investigators making their decisions at a foundational or *methodological level* where *ontological* concerns of being, meaning and identity are taken in concert with *epistemological* concerns of knowing. Thus, the preceding chapter underscored the point that ontological craft is fundamentally a situationally creative effort to use forms of reality-aware and context-appropriate 'human instrument' understandings to uncover 'the real cultural world' of the individual realm or the societal spectrum being explored.

This succeeding chapter now explores (at a greater level of scrutiny) the kinds of ontological problems and pressures that are prominent in tourism studies, *per se*. It posits tourism studies as a domain of activity where 'human instrument' forms of assessment are notably in demand because of the broad range of different values which are influential in local and global scenarios. This chapter therefore calls attention to the routine needs for researchers in tourism studies to engage human instrument forms of understanding which are painstakingly 'embedded', 'iterative', 'sustained'. It seeks to do that by, first, delving in some detail into a number of common problematics which characterise the 'locales' in which tourist visitation takes place.

Research insights into such ordinary but deep, messy and ambivalent human and communal importances (and into related investigator reflexivities) are slowly emanating in the field. At present, these sorts of ontological orientations may be unsteady and poorly shared across the domain, but they are steadily surfacing. One hopes that this volume may inspire all sorts of advances and cross-fertilisations in the dedication that inquiry teams in tourism studies show towards these critical human questions of being and becoming.

Commonplace ontological considerations in tourism studies

Now that Chapter 4 has clarified in the kinds of decisions which all social science researchers inevitably have to take regarding the view of the reality they wish to address, and thereby *the methodological-level* paradigmatic framework they want to work within, it is now opportune (for and within tourism studies) to consider the selection of fitting research approach at the *methods level*. To repeat the point, this second chapter on ontological craft is therefore provided on the refined assumption that deliberations over matters of ontology should always precede the choice of particular research method.

In the following pages, ten ontological concerns are listed which commonly occur in studies of international tourism. For comparative clarity, these ten concerns, or problematic issues, are also detailed collectively in Table 5.1.

The first subsection in each case then provides a brief amplification of the ontological concerns which conceivably underpin each of these ten common issues given in the form of a mix of questions routinely implied or generated around the issue.

Thereafter, a second subsection provides a short comment on the sorts of approach or the kinds of methods-level application which can shed relevant light on that sort of ontological poser or problem. These recommended qualitative techniques/recommended qualitative methods-level constructs are provided in terms of the strategic gain they appear to offer the researcher in tackling or handling the important ontological matter

Table 5.1 Common ontological issues in tourism studies

Sample issue in tourism settings/ scenarios	*Sample associated areas of uncertainty in seeing/experiencing/meaning/knowing/ being*
1 The obscure nature of cultural ways in foreign places	Cultural being: What does it entail being an inner member of a culture or society? What important aspects of being and acts of knowing (about a particular place) should be known by overseas businesses operating in the promotion of it? What should intending visitors to that place know? What should landing/arriving visitors there know?
2 The varied and often incoherent meaning of sites and experiences to tourists who visit foreign places	The held identities of places: Why do specific sorts of tourists wish to visit and celebrate particular cultural, particular heritage and/or particular natural sites?
3 The deep and often hidden meaning, in foreign places, of sites and stories to resident host populations there	The intensity of held identities of place: Which particular identities of belonging (out of the found or potential range of contesting associations with place) are currently dominant, and thereby supported ardently/triumphantly in the given locale?
4 The shadowy and indistinct 'unique ways' in which foreign peoples differ from each other	Values: What are the cherished values of the projected culture of a given location? What is ontologically true (and/or 'real') and what is ontologically false (and/or 'faux') for that population? How do those unique ways vitally differ from those of cousin/neighbouring populations who might superficially appear to be 'same'/'similar'?
5 The small and knotty ways in which tourists and/or the tourism industry commits quotidian acts of ethnocentric misinterpretation in foreign/distant/removed locales	Critical reality: Are specific tourism practitioners (or specific tourism studies researchers) conscious of their own possible/actual role in the small but additive everyday objectification of destination societies?
6 The large, long-standing and highly disempowering ways in which the tourist-producing West has continued to Other specific populations	The character and form of different worldviews: How do the nature of doxa (i.e. the held cosmologies or the esteemed societal worldviews) differ between population blocks? How does each individual cultural warrant (i.e. each singular instance of a found doxa/worldview/cosmology) differ between particular national peoples? Historically, what has each population tended to misconstrue about its 'other' populations?

Table 5.1 Continued

Sample issue in tourism settings/ scenarios	Sample associated areas of uncertainty in seeing/experiencing/meaning/knowing/ being
7 The frequently enigmatic and inconstant symbolic significance of signs/markers/objects in different societies, as serviced in international tourism	The symbolic meaning of places (objects/ events): What do objects/places/events 'authentically' or 'precisely' mean for their owning populations?
8 The perplexing manner in which seemingly long-cherished traditions in alien populations are suddenly replaced by new/ transitional practices or consumed within new/transformed pursuits	Contextuality: Just which cultural traits and/or which ontological beliefs have changed (within the given society) and which local circumstances/scenarios has that change accelerated or otherwise retarded?
9 The nebulous and ambiguous ways in which emergent/hybrid populations – particularly in developing nations – freshly seek to reposition themselves through the imagery of tourism, distinguishing themselves markedly from the previously colonised projections of the populations of that locality	Nature of the knowable: Where is the tourism industry indeed engaging in dialogue with new/liberated groups in the post-colonial world to help issue 'new sense' articulations of selfhood … and where is it closed to such emergent populations, thereby perhaps continuing to deal in stale, monologic, colonialist narratives?
10 The complex and amorphic ways in which the tourism industry (and all its public- and private-sector players) generally act internationally/globally to market/ de-market/re-market places	Reality: Whose vision of how the world *is* being (or ought to be?) unashamedly peddled through tourism? Whose 'preferred version' of social reality is being subconsciously or unexpectedly advanced through a particular tourism development? Whose rendition of things is being collectively/collaboratively advanced through the assumed logics and/or the performed narratives of tourism?

which is being critiqued. This last subsection thereby offers up a number of emerging or consolidating advances in interpretive/phenomenological inquiry which have considerable practical worth within tourism studies in helping the researcher gain access into the contexts they must confront. As such, they tend to be context-responsive applications which strategically or tactically help the investigators optimise the deployment of *their 'human instrument' skills*, given the temporal and geographic limits of the study. Once more, readers should recognise that since international tourism can throw up all sorts of different cultural, political, psychic, socio-economic and other ontological problematics, the recommended qualitative actions and constructions offered in the final subsection can only ever be suggestive: they cannot meaningfully be offered in a comprehensively indicative or universalist fashion to immediately suit each and every scenario.

Ten commonplace problems/problematics in international tourism

Sample issue 1: the obscure nature of cultural ways in foreign places

Tourism is a large and complex international activity in which businesses in one tourist-generating country have to deal with a welter of tourist-receiving countries in distant corners of the globe. Those operators have scarce opportunity to know those distant societies well, particularly since consumer interest in visiting those distant and exotic places can alter very suddenly. Tourism practitioners in tourist-generating countries are inevitable outsiders to so many places and cultures they promote.

Area of ontological difficulty

Cultural being: What does it entail being an inner member of a culture or society? What important aspects of being and acts of knowing about a particular place should be known by overseas businesses operating in the promotion of it? What should intending visitors know? What should arriving visitors know?

Suggested qualitative research gain

Many qualitative research approaches are designed purposefully to help the tourism studies researcher become *ontologically embedded* within the removed or distant population they seek to investigate. See Pearce and Chen (1989) for a critique of James Clifford's views on the need for *embedded dialogue* over time with key representatives of such a population, particularly where the researcher seeks to gain credibility for the precision and truthfulness of the emic insight that they claim to have gained.

Sample issue 2: the varied and often incoherent meaning of sites and experiences to tourists who visit foreign places

There are many types of tourists, and there are many differing motivations which pull or push tourists to particular destinations. Little work has currently been done, however, on how individual tourists 'receive'/'interpret' the special places or the sacred sites which they visit in removed/ distant lands.

Area of ontological difficulty

The held identity of places: Why do specific sorts of tourists wish to visit and celebrate particular cultural, heritage and/or natural sites?

Suggested qualitative research gain

Many qualitative research approaches can help the tourism studies researcher towards deeply held ontological identifications which they can take confidence in over time through studied processes of steadily spiralled *confirmability*. The work of Lincoln and Guba (1985) on 'naturalistic inquiry' (which became 'social constructivism') is based on this dovetailed ontological–epistemological approach. In naturalistic inquiry, the call for *confirmability* is parallel to (and replaces) the objectivity of positivism, just as *credibility* replaces internal validity, *transferability* replaces external validity, and *dependability* replaces reliability (Guba 1981; Lincoln and Guba 1985). Taken together, the four *italicised* terms above constitute the *trustworthiness criteria* of naturalistic inquiry, to parallel the four conventional methodological pillars of conventional linear research.

Sample issue 3: the deep and often hidden meaning, in foreign places, of sites and stories to resident host populations there

In distant/removed places around the world, the tourism industry is not particularly 'intelligent' in the manner or the degree to which it seeks to understand significances held by local society concerning the precious places it deals in (Horne 1992). Moreover, the meaning of those culturally esteemed places is often indeterminate or mythically 'scrambled', anyway, notably where different interest groups in that host population may each uphold differently 'interpreted' or differently 'invented' meanings.

Area of ontological difficulty

The intensity held identities of place: Which particular identities of belonging, out of the found or potential range of contesting associations with place, are currently dominant, and thereby supported ardently/triumphally in the given locale?

Suggested qualitative research gain

Many qualitative research approaches can help the tourism studies researcher towards understanding about what local/indigenous/removed populations consider themselves to be currently 'up to'. Many of these approaches are richly demanding of interpretive skill not only to capture those local ontologies, but to be able to broadcast them faithfully (yet with communicative force) to external urban-industrial/metropolitan/ 'Western'/other audiences. For a discussion on the use of 'writing' (*ipso facto*) as a vital qualitative approach to faithful and effective communication, see the Pearce and Chen (1985) critique of *interpretive narrative* in terms of their analysis of Geertzian *'storytelling'*. This chapter examines the interpretive authority of Geertzian styles of narrative (after Geertz) as a plausible research mode of both inquiry and communication. In contrasting the ethnographic authority of Clifford Geertz with the ethnographic authority of a James Clifford, Pearce and Chen (1985) examine the bona fides of Geertzian storytelling as a technique capable of creatively but justly capturing and then richly but justly translating the held ontologies and cosmologies of culturally distant populations.

Sample issue 4: the shadowy and indistinct 'unique ways' in which foreign peoples differ from each other

The distinctions and boundaries which exist between neighbouring nations or between proximal populations may be difficult to fathom because of the differential effects of a multiplicity of factors such as culture, religion, ethnicity, history and contemporary governance, among others. Currently, tourism managers/operators/researchers are rarely trained in cultural studies or in policy studies so as to be discerning on such overlapping matters of identification, matters which, to outsiders, can often appear to be mercurial or perverse.

Area of ontological difficulty

Values: What are the cherished values of the projected culture of a given location? What is ontologically true (and/or 'real') and what is ontologically false (and/or 'faux') for that population? How do those unique ways vitally differ from those of cousin/neighbouring populations who might superficially appear to be 'the same'/'similar'?

Suggested qualitative research gain

Many qualitative research approaches can help the tourism studies researcher towards *multiple or coterminous 'different interpretations'* about the perceived authenticity or the perceived legitimacy of the cultural warrants (i.e. of the held truths) owned or supported by the given population being studied. Quayson (2000) has recently produced a valuable analysis of postcolonial studies as a critical qualitative approach itself, which can be practically applied to examine the dynamisms of truth-making/truth-talking that conceivably exist about a population that is being misrepresented in some way, which has been essentialised over time in some regular fashion, or which has been stereotyped as indistinctly part of a larger 'people'.

Sample issue 5: the small and knotty ways in which tourists and/or the tourism industry commit quotidian acts of ethnocentric misinterpretation in foreign/distinct/removed locales

Tourists who travel on long-haul vacations often visit many countries on their trip and generally receive very little 'grounding' from the travel industry on the new overseas locations they visit. Frequently they are forced to make hurried judgements about the country they are delivered to, and they are inclined to react negatively against cultural practices which at first inspection appear to them to be unacceptable or unsavoury. Indeed, the cross-cultural tourism encounter has a high likelihood of generating ethnocentric understandings and poor presentations of local cultures in this manner (Hollinshead 1993a,b). Many of these ethnocentric judgements on the part of the industry and the tourist tend to be petty and seemingly inconsequential, but can in fact be hugely cumulative through time and unthinking repetition. To a certain extent they are natural sociocentrisms (or the sorts of raw and inexorable misinterpretations) which all in-group populations deal in with regard to outsiders.

Area of ontological difficulty

Critical reality: Are specific tourism practitioners, or researchers, conscious of their own possible/actual role in the small but additive everyday objectification of destination societies?

Suggested qualitative research gain

Many qualitative research approaches can help the tourism studies researcher towards the gain of the kind of *reflexive and transformative self-awareness* that Rojek (1997) demands, and which is called for by Sampson (1993). Fine (1994: 72–75, 79–81) offers an informed critique of how unfolding qualitative approaches can help researchers (and practitioners) comprehend how they themselves so easily and axiomatically participate in day-to-day constructions of the *politics of location* which ontologically imprison both 'the misrepresented outsider' and 'the misrepresenting self'! Where those petty (i.e. ordinary, everyday, *natural* sociocentric, community-focused or ethnocentric, ethnic-focused) representations mount up over time to accrete into large single-rationality-based articulations about an outsider population, that particular population is deemed to have been 'Othered', or at least 'objectified' (see issue 6).

Sample issue 6: the major, long-standing and highly disempowering ways in which the tourist-producing West has continued to 'Other' specific populations

Not only does the tourism industry have a high propensity to yield small-time day-by-day ethnocentric cognitions about the world (see issue 5), but it has a high propensity to serve as an ongoing vehicle for the continued projection/renewed articulation of centuries-old understandings which potently Other the peoples, places and pasts of distant locales.

Area of ontological difficulty

The character and form of different worldviews: How do the nature of doxa (i.e. the held cosmologies or esteemed societal worldviews) differ between population blocks? How does each individual cultural warrant – that is, each singular instance of a found doxa – differ between particular ethnic groups/between particular national peoples? Historically, what has each population tended to misconstrue about its Other populations?

Suggested qualitative research gain

Many qualitative research approaches can help the tourism studies researcher towards more robust understandings of the different 'sustained' or 'continental' pre-appreciated ontologies held about the world. Frequently, this will involve the effort to gain a number of different outlooks on the held doxa or the newly encountered worldviews, secured via a mix of groups/subgroups and/or from a mix of standpoints/locations. In recent years, some progress has been made in the area of *triangulation*, whereby investigators are encouraged to conduct social science inquiry from several interfeeding and compensatory fronts in this fashion, so as to help nullify the intrinsic bias of any singular mode of investigation upon the target population or data source (for further discussion on triangulation, see Chapter 9). Importantly, some authors object to the use of *triangulation* here, however, deeming it to be a fixed-point and 'positivist' notion. Such protestors prefer qualitative researchers to think and operate in terms of '*crystallisation*', where that sought mix of interfeeding and compensatory crystallising approaches is actually selected on account of the found enhanced and aggregate suitability of those methods-in-tandem to differentially, yet collectively, capture a multiple range of interpretations about the issue being explored. Under crystallisation, it is recognised that many of the target areas being explored can only ever be known in a 'fuzzy' rather than a 'fixed' fashion. Hopefully, some of that fuzziness disappears over time as the researcher begins to refine their awarenesses about the locations and the contexts being probed. Thus, crystallised approaches tend to stretch more comfortably across an untidy and unpredictable assemblage of non-isomorphic dimensions (Miller and Crabtree 1994: 345; Richardson 1994: 522), rather than being contained on one single pre-formulated conceptual plane of understanding, as is generally the case with triangulated approaches.

Sample issue 7: the frequently enigmatic and inconstant symbolic significance of signs/markers/objects in distant societies, as serviced in international tourism

As stated under issue 1, on p. 87, those who work continually in inter-national tourism generally have scant opportunity to get to know all of the continents and countries they send tourists to. Accordingly, relatively few travel-trade practitioners have anything more than a superficial knowledge of the cultures/ethnicities/cosmologies they 'deal in' within their tourism programming and packaging activities. Many of the interpretations of distant societies they take up may be idiosyncratically 'absorbed', or are otherwise subject to elastic or variable forms of contextualisation. Quite frequently, under such loose circumstances of interpretative decoding, the processes of people/place projection in tourism become cryptic and mercu-rial (Horne 1992).

Area of ontological difficulty

The symbolic meanings of places (objects/events): What do objects/places/events 'authentically' or 'precisely' mean to owning populations?

Suggested qualitative research gain

Many qualitative research approaches can help the tourism studies researcher towards in-depth analyses of ontologies of meaning, such as the nature of national identity or the character of gendered understandings of place. Qualitative research techniques such as narrative analysis/semiotic analysis are often selected because they potentially afford the investigator rich opportunity to capture *thick descriptions* (Holliday 2002: 77–78) of the settings being examined. Recent advancement in qualitative research revolves around the emergent role of the researcher as *bricoleur* (Denzin and Lincoln 1994: 2–3, *et passim*; Hollinshead 1996), where a *bricoleur* is a multi-skilled craftsperson who is gradually able to use a variety of increas-ingly fine-tuned approaches as they immerse themselves in the research. Through such painstaking and deepening *bricolage*, the researcher seeks to win appropriate thick descriptions. Hence, the thick description of high-in-context interpretive/constructivist work is the antithesis of the sorts of orthodox *thin description* which is customarily favoured by low-in-context conventional linear styles of research, and which are prone to focusing exclusively upon the cause-and-effect power of single variables.

Sample issue 8: the perplexing manner in which seemingly long-cherished traditions in alien populations are suddenly replaced by new/transitional practices or consumed within new transformed pursuits

All cultures are dynamic, and change is the constant character of all societies. All traditions evolve or alter over time, for it is difficult to keep stable and constant all the factors that have produced a given 'tradition'. Tourism can itself play a large role in changing the levels of cultural appreciation which exist both within or about a society. But the storylines in vogue in tourism can also suspectingly/unsuspectingly ring-fence or contain societies within fixed images about unchanging traditions, denying the chance for that depicted population to be seen in modern-day light, or the opportunity to become 'contemporary'.

Area of ontological difficulty

Contextuality: Just which cultural traits and/or which ontological beliefs have changed (within the given society) and which local circumstances/scenarios has that change accelerated or otherwise retarded?

Suggested qualitative research gain

Many qualitative research approaches can help the tourism studies researcher towards deep-seated inspections of the ontological considerations involved in settings where evolutionary change is suspected. In recent decades there have been important procedural advances in both observational techniques and ethnomethodology, notably with regard to rendering their conduct less intrusive on local populations under investigation (Alder and Alder 1994). Both techniques are demanding in terms of the timescale and the iterative processes they rely on, in order that theoretical saturation is obtained (Glaser and Strauss 1967). Observation techniques often yield rich rigour when used in triangulated (or rather, in crystallised) combination with other techniques. This typically might occur where the 'soft evidence' of researcher observation is assessed in tandem with the 'hard evidence' of interviews with figures of authority from the study scenario (Denzin 1989; Alder and Alder 1994: 382).

Few tourism studies researchers around the world have yet been accorded the right to engage in multi-sited/longitudinal work of this kind where they have been able to deploy triangulated or crystallised inspections on a multiple operational inquiry basis. Such plural but closely inter-worked approaches might, for instance, involve the programming of hard and soft forms of investigation (in harness strategically) at those multiple sites over sustained time. Brewer and Hunter (1989: 13) call for the use of multi-method approaches in the kinds of complex cultural scenarios commonly examined by tourism studies researchers. They make the case for the dovetailed use of complementary sorts of multiple methods where qualitative (ql) approaches are used mutually to offset weaknesses in quantitative (qn) approaches. Brewer and Hunter (1989) especially recommend the reciprocal use of *fieldwork* (qn or ql), *survey research* (qn or ql), *experimentation* (routinely qn) and *non-reactive research* (routinely ql) in these sorts of singular, but multi-pronged, investigations. Such a co-ordinated panoply or interfusive diversity of methods can, in their view, ontologically 'cross-fertilise' and epistemologically 'cross-validate' the understandings gathered. While some readers might find that the thinking of Brewer and Hunter (1989) is littered with small-'p' positivisms/small-'p' postpositivisms, their text on the necessity for and the potential of carefully triangulated investigative approaches yields a healthy ontological scepticism about the strengths of singular method, *ipso facto*. Their work is a well-argued call for methodological diversity, and also for interactivity between theory and method.

Sample issue 9: the nebulous and ambiguous ways in which emergent/hybrid populations freshly seek to reposition themselves through the imagery of tourism, markedly vis-à-vis the previously colonialised projections of the population of that locality

As cultural, societal, economic and political factors change, subcultures and subpopulations rise and fall in their interest in relating to the larger state, and in their capacity to 'be free' from received or externally enforced representations of their supposed selfhood (Bhabha 1994). Under the post-colonial moment, a large number of new or revived such populations have emerged or are emerging around the world who are keen to redefine themselves *vis-à-vis* their in-state overlords, or their so-called colonial 'masters' (Thomas 1994). Many of them are turning to the new articulative opportunities of tourism to make those corrective or emanative declarations of selfhood/statehood on an international or 'big league' stage (Hollinshead 1998a,b).

Area of ontological difficulty

Nature of the knowable: Where is the tourism industry indeed engaging in dialogue with new/liberated groups in the post-colonial world to help issue 'new sense' articulations of selfhood? Where is it closed to such emergent populations, thereby perhaps continuing to deal in stale, monologic, colonialist narratives?

Suggested qualitative research gain

Many qualitative research approaches can help the tourism studies researcher towards much more carefully tracked and deciphered *dialogic understandings* about the difficult psychic and ontological circumstances in which the world's emergent, halfway or institutional populations ambivalently find themselves to be caught (Hollinshead 1996). Indeed, it is not just some small restricted set of freedom-fighters that are contained in ambivalent and difficult psychic positions, for in many senses most of the world's subpopulations and subcultures (and indeed, most of the world's cultures) are nowadays located within difficult psychic locations and uncertain restless spaces regarding the ways in which their assumed and established mainline cultural inheritances are or are not in balance with their new intercultural coalescences and their often admixed and venturesome new worldviews (Bhabha 1994). It is expected that the twenty-first century will demand high numbers of interpreters and textualists who have been sensitively trained in the iterative decoding of immanent and contemporary rhetorical statements and in the local-political translation of fresh and fashionable nationalist expressions of being (Deleuze and Guattari 1987). Already at the start of the twenty-first century, it is hard to tell where the inventive and imaginal representational repertoire of the tourism industry differs from inventive and imaginal significatory systems of the 'public culture' industry (Horne 1992; Kirshenblatt-Gimblett 1998; Rothman 1998): both are becoming heavily propagandist. Ontologically speaking, it appears that the realm of tourism is inherently merging seamlessly into the declarative realm of public culture placemaking.

Sample issue 10: the complex and amorphous way in which the tourism industry and all its public- and private-sector players generally act globally to market, de-market and re-market places

The tourism industry is a large and an ultra-complex one consisting of many subsectors such as 'transportation', 'attractions management', 'entertainment', 'education/interpretation', 'accommodation' and 'travel marketing'. It is difficult for any one management body to have effective and sustained control over all these sectors. Management bodies have to know and relate to the intentionalities of a complex web of other players and partners, along the long, long chain of distribution in tourism. Thus, the stuff of travel management is needfully the stuff of vertical integration, horizontal integration and diagonal integration (Poon 1989), as these players in the complex international architecture of the industry constantly seek to set up improved relationships with these other players and these potential partners. Each operator seeks to maximise its degree of control over its volatile (fast-changing) operational milieu, its dynamic collaborative and connective plane of action.

Area of ontological difficulty

Reality: Whose version of how the world is, or ought to be, peddled through tourism? Whose 'preferred version' of social reality is being subconsciously or unsuspectingly advanced through a particular tourism development? Whose rendition of things is being collectively/ collaboratively advanced through the assumed logics and/or the performed narratives of tourism?

Suggested qualitative research gain

Many qualitative research approaches can help the tourism studies researcher towards the kinds of descriptive analyses which Hall (1994) calls for to complement the field's apparent current reliance on quantifi- able and prescriptive understanding. In this respect, the need for sustained *qualitative critique case studies* to supplement or compensate for the domain's strong existing predilection for 'quantological' case studies is imperative (Stake 1994) for a clarification of the differences between qualitative case studies compared to quantitative case studies). Tourism studies researchers who might need a sound exemplar of a qualitative case study to follow are encouraged to inspect Rodriguez's (1990) protracted study of the politics of power and place in Taos (New Mexico), or the elongated Cranz (1982) inquiry into the development of parks, as mechan- isms of social control, throughout the United States. Moreover, since so many different players are involved in the business of tourism, future tourism researchers must learn how to differentially audience their find- ings (see Fiske 1994 on the qualitative technique of *audiencing*) – where audiencing constitutes an action-orientated 'technique' (Hollinshead and Jamal forthcoming).

The above ten exemplars have been drawn up to illustrate customary or prevailing difficulties over matters of uncertain being, contested meaning, unclear identity, etc. which ubiquitously crop up in international tourism development and/or tourism management scenarios. Each of these prob- lematic areas will have a welter of its own local institutional place- situational, time-contextual and intersubjective factors. No tourism studies researchers should expect there to be a prescriptive list available to deter- mine which single qualitative method or which sort of soft-science tech- nique has the 'best fit' for their own investigative setting. Each ontological problem area will be idiosyncratic to the singular dynamics of its time, its place, its history, and to the standpoint perspectives on that setting which the given researchers (or better still, the transdisciplinary research team) recognise to be salient. Hence, the researchers must be painstakingly *reflective* about all the above local or situational factors which constitute the tourism development or the travel-trade operational setting in question. And to repeat that important point from Chapter 4, they must also be pointedly *reflexive*. An investigator should only ever issue strong findings from qualitative research methods or constructs with which they are comfortably familiar, and from value standpoints with which they are particularly at ease. That is the reflective and the reflexive craft of the conduct of qualitative research, where emergent ethnographies of local truth-finding necessarily recognise and position the researcher as the critical human and social interpreter of all sorts of aspects of aspiration, voice, signature and audience (Patel 1997). Crafted considerations of ontological fit must thereby be taken in full and acknowledged cahoots with crafted considerations of epistemological suitability.

It inevitably takes a long, long time for qualitative researchers to become familiar with the large cosmologies of being, meaning and becoming, and the small and accented variations of lived experience they have to work under, ontologically, in found settings (Whatmore 2002). And it then takes a long, long time to become familiar with the epistemological techniques and methodological tools appropriate for that found scenario (Haraway 1991). There are nowadays so many emergent techniques and new tools to learn about. But if, in tourism studies, we are to cover the lived world of travel experience, and we are regularly to inspect who is making/re-making/de-making the imagined world of social/mythic experience (through the selection of esteemed 'natural' sites and the celebration of favoured 'tribal' storylines), we do need that large lexicon of versatile and vernacular soft-science choices.

Summary

In this chapter, the following principal insights have been gained:

- That ontological issues stand common and tall in tourism studies, particularly where the internationalities of tourism/travel render it difficult for those who work in the industry (and for those who travel!) quickly to come to terms with the local/emic ways of different/distant populations (in general), and with the specific meaning or importance of revered places to those different/distant communities.

Questions

1   Take one of the ten issues in tourism studies outlined in this chapter. Design a one-year-long study in a resort, city or island state of your choice to show how you would endeavour to deal satisfactorily with the issue selected, or how otherwise would endeavour to gain sound and useful insight on that particular area of ontological difficulty. Ensure that you explain how would you seek to build *ontological rigour* into your proposed study.

2   The French sociologist Lanfant maintains that 'tourism indirectly causes the different national societies to become gradually interlinked in economic, social, and cultural networks that are organised internationally on the basis of a central decision-making body and at the same time cut across national reference systems' (Lanfant 1980: 22). What ontological issues, questions or uncertainties do you think are implied within Lanfant's statement? Explain why you think that those who work in tourism in 'management', and/or 'planning' and/or 'research' positions are or are not well trained to understand these implied or inherent ontological matters.

3   Horne (1992) considers that tourism is the greatest continuing mass movement of peoples in human history. What does all that mobility of people do to the meanings and the values that particular populations carry/cherish? Can you find any substantive journal articles on the subject? In particular, can you locate – in any recent issues of the six journals below – at least three articles which directly or indirectly cover ontological issues concerning conveyance or change in 'meaning' or 'value' in tourism? The default list of journals is:

Annals of Tourism Research   *Current Issues in Tourism*
Tourism Analysis   *Tourism, Culture and Communication*
Tourism Geography   *Tourism Management*

Compare and contrast these three articles. What do they appear to agree and to differ on in terms of the ontology of seeing/experiencing/meaning/knowing/being/becoming? What do the articles say (if anything) about the role, the function or the power of tourism today?

4   Hyperreality is an important feature of many contemporary commercial forms of tourism (Rojek 2002) – an important aspect of the commodified ontology of sites and attractions in tourism. To cultural studies thinkers, hyperreality is the deliberate construction of new leisure forms of experience based upon large doses of simulation, spectacle and simulation, where 'absolute unreality is offered as a very real presence' (Eco 1986: 7). For instance, Hollinshead defined hyperreality as follows:

> Hyperreality: during recent decades, distinctions between 'the real' and 'the represented' have become problematic as one travels and experiences places. To Baudrillard, one increasingly consumes signs or copies rather than real entities themselves. When the simulation of things is particularly ubiquitous or spectacular, hyperreality results, constituting a state of manipulated discourse where referential reason disappears – a realm where illusion is not achievable because the real itself is no longer knowable or attainable. Images projected in tourism frequently habituate people to hyperreal *replica*

worlds which are surreal and exhilarative replica worlds to some, but depthless, inauthentic, and commoditised to others.
(Hollinshead's definition, commissioned in Jafari 2000: 292)

What do hyperreal conditions and/or hyperreal environments and/or hyperreal performances do to the ways in which 'tourists' and 'local/resident/host populations' understand the world? Relate the subject of hyperreality to the ontological questions and issues which arise in tourism, as detailed in the current chapter.

References

Alder, P. A. and Alder, P. (1994) 'Observation techniques'. In N. K. Denzin and Y. S. Lincoln (eds) *Handbook of Qualitative Research*. Thousand Oaks, CA: Sage.

Bhabha, H. (1994) *The Location of Culture*. London: Routledge.

Brewer, J. and Hunter, A. (1989) *Multimethod Research: A Synthesis of Styles*. Newbury Park, CA: Sage.

Cranz, G. (1982) *The Politics of Park Design: A History of Urban Parks in America*. Cambridge, MA: MIT Press.

Deleuze, G. and Guattari, F. (1987) *A Thousand Plateaus: Capitalism and Schizophrenia*. Minneapolis, MN: Minnesota University Press.

Denzin, N. K. (1989) *Interpretive Interactionism*. Newbury Park, CA: Sage.

Denzin, N. K. and Lincoln, Y. S. (eds) (1994) *Handbook of Qualitative Research*. Thousand Oaks, CA: Sage.

Eco, U. (1986) *Faith in Fakes*. London: Secker and Warburg.

Fine, M. (1994) 'Working the hyphens: reinventing self and other in qualitative research'. In N. K. Denzin and Y. S. Lincoln (eds) *Handbook of Qualitative Research*. Thousand Oaks, CA: Sage.

Fiske, J. (1994) 'Audiencing: cultural practice and cultural studies'. In N. K. Denzin and Y. S. Lincoln (eds) *Handbook of Qualitative Research*. Thousand Oaks, CA: Sage.

Glaser, B. and Strauss, A. (1967) *The Discovery of Grounded Theory*. Chicago, IL: Aldine.

Golden, D. (1996) 'The Museum of the Jewish Diaspora tells a story'. In T. Selwyn (ed.) *The Tourist Image: Myths and Myth Making in Tourism*. Chichester: Wiley.

Guba, E. (1981) 'Criteria for assessing the trustworthiness of naturalistic inquiry', *Educational Communication and Technology Journal*, 29: 75–92.

Hall, C. M. (1994) *Tourism and Politics: Policy, Power, and Place*. Chichester: Wiley.

Haraway, D. (1991) 'Situated knowledges: the science question in feminism and the privilege of partial perspective'. In *Simians, Cyborgs and Women: The Reinvention of Nature*. San Francisco, CA: Free Association Books.

Holliday, A. (2002) *Doing and Writing Qualitative Research*. London: Sage.

Hollinshead, K. (1993a) 'Encounters in tourism'. In M. A. Khan, M. D. Olsen and T. Var (eds) *VNR Encyclopaedia of Hospitality and Tourism*. New York, NY: Van Nostrand Reinhold.

Hollinshead, K. (1993b) 'Ethnocentrism in tourism'. In M. A. Khan, M. D. Olsen and T. Var (eds) *VNR Encyclopaedia of Hospitality and Tourism*. New York, NY: Van Nostrand Reinhold.

Hollinshead, K. (1996) 'The tourism researcher as *bricoleur*: the new wealth and diversity in qualitative inquiry', *Tourism Analysis*, 1 (1): 67–74.

Hollinshead, K. (1998a) 'Tourism, hybridity and ambiguity: the relevance of Bhabha's third space cultures', *Journal of Leisure Research*, 30 (1): 121–156.

Hollinshead, K. (1998b) 'Tourism and the restless peoples: a dialectical inspection of Bhabha's halfway populations', *Tourism, Culture and Communication*, 1 (1): 49–77.

Hollinshead, K. and Jamal, T. B. (forthcoming) 'Tourism and "The Third Ear": further prospects for qualitative research'. Manuscript submitted to *Tourism Management*.

Horne, D. (1992) *The Intelligent Tourist*. McMahon's Point, NSW: Margaret Gee Holdings.

Jafari, J. (2000) *Encyclopedia of Tourism*. London: Routledge.

Kirshenblatt-Gimblett, B. (1998) *Destination, Culture: Tourism, Museums, and Heritage*. Berkeley, CA: University of California Press.

Lanfant, M.-F. (1980) 'Introduction: tourism in the process of internationalization', *International Social Science Journal*, 32: 7–43.

Lincoln, Y. S. and Guba, E. (1985) *Naturalistic Inquiry*. Beverly Hills, CA: Sage.

McKay, I. (1994) *Quest for the Folk*. Montreal: McGill and Queens University Press.

Miller, W. L. and Crabtree, B. F. (1994) 'Clinical research'. In N. K. Denzin and Y. S. Lincoln (eds) *Handbook of Qualitative Research*. Thousand Oaks, CA: Sage.

Patel, G. (1997) 'Gender and voice, signature and audience in north Indian lyric traditions'. In R. Hertz (ed.) *Reflexivity and Voice*. Thousand Oaks, CA: Sage.

Pearce, W. B. and Chen, V. (1989) 'Ethnography as Germanic: The rhetorics of Clifford Geertz and James Clifford'. In H. W. Simons (ed.) *Rhetoric in the Human Sciences*. London: Sage.

Poon, A. (1989) 'Competitive strategies for a "new tourism"'. In C. P. Cooper (ed.) *Progress in Tourism, Recreation and Hospitality Management*. London: Belhaven Press.

Quayson, A. (2000) *Postcolonialism: Theory, Practice or Process?* Cambridge: Polity Press.

Richardson, L. (1994) 'Writing: a method of inquiry'. In N. K. Denzin and Y. S. Lincoln (eds) *Handbook of Qualitative Research*. Thousand Oaks, CA: Sage.

Rodriguez, S. (1990) 'Ethnic reconstruction in Taos', *Journal of the Southwest*, 3 (2/4): 541–555.

Rojek, C. (1997) 'Indexing, dragging and the social construction of tourist sights'. In C. Rojek and J. Urry (eds) *Touring Cultures: Transformations of Travel and Theory*. London: Routledge.

Rojek, C. (2002) 'Fatal attractions'. In D. Boswell and J. Evans (eds) *Representing the Nation: A Reader – Histories, Heritage, and Museums*. London: Routledge.

Rothman, H. K. (1998) *Devil's Bargains: Tourism in the Twentieth-Century American West*. Lawrence, KS: University Press of Kansas.

Sampson, E. E. (1993) *Celebration of the Other: A Dialogic Account of Human Nature*. New York, NY: Harvester Wheatsheaf.

Sardar, Z. (1998) *Postmodernism and the Other: The New Imperialism of Western Culture*. London: Pluto Press.

Stake, R. E. (1994) 'Case studies'. In N. K. Denzin and Y. S. Lincoln (eds) *Handbook of Qualitative Research.* Thousand Oaks, CA: Sage.

Thomas, N. (1994) *Colonialism's Culture: Anthropology, Travel and Government.* Princeton, NJ: Princeton University Press.

Whatmore, S. (2002) *Hybrid Geographies: Natures, Cultures, Spaces.* London. Sage.

Wilson, C. (1997) *The Myth of Santa Fe: Creating a Modern Regional Tradition.* Albuquerque, NM: University of New Mexico Press.

6 (Dis)embodied experience and power dynamics in tourism research

Margaret Byrne Swain

Aims of the chapter

- To provide an overview of embodiment issues in qualitative tourism research methods.
- To give examples of good practice drawing from feminist analysis of embodiment in tourism studies.
- To raise questions about power dynamics in the intersections of researcher and the researched in terms of embodiment and reflexivity.

Introduction

This chapter addresses questions of embodied experience and power in qualitative tourism research, drawing on a litany of embodied power relations often recited in literature on the body: gender, sexuality, ethnicity, race, nationality, class, age and (dis)able-bodiness. Naturally I see these questions from my own positions as a woman, straight, Anglo, white redhead, American middle-class and middle-aged, semi-dyslexic, left-handed researcher. I rarely discuss all these aspects of my persona at once, and have written about my position in terms only of gender in past work. However, all of what I am affects the problems I see and the power dynamics I experience as a researcher. Don't worry, this chapter is not all about me. Qualitative (and I would argue quantitative as well) research must acknowledge the researcher's positionality intersecting with that of the researched in order to understand how our understandings are constructed. This is not a unique observation, as we all know that knowledge is not created in a vacuum. My question here is what might we learn from such observations to promote more equitable human conditions in the course of our research about tourism. Power relations in research, in terms of who controls what, how hierarchies are built, maintained and changed, and how equality occurs, all emanate out of these intersections within multiple contexts.

A focus on embodiment in qualitative research acknowledges the corporal selves of the researchers' as well as the researched 'subjects' as primary factors in the research process. How a researcher incorporates embodiment into their work is more than a personal choice of episteme. Research contexts of local landscapes and global systems must also be taken into consideration. Intersecting power dynamics emanating from ethnic/racial hierarchies, patriarchal systems including academia, and disability stereotypes, for example, as well as ideologies of equity and inclusiveness, shape qualitative research agendas and results.

In this chapter I will briefly discuss qualitative research using embodiment theory in general before turning to the field of tourism studies. It is quite possible to research embodied topics without being informed by embodiment theory, as we shall see in work presented in several international tourism conferences. Within tourism studies, one of the richest areas of work on embodiment is found in articulations of feminist theory and tourism studies. This feminist perspective on embodiment in tourism is expanded on in this chapter as an example of best practices. I draw from my experience in editing several collections representing a number of scholars writing on gender in tourism (Swain 1995, 2002), and my own work on gender and ethnicity issues for women producers of ethnic arts (1993, 2001). I conclude with references to work by university students, heavily influenced by ideas of embodiment, which provides a look into the next wave of tourism research. As I comment on other researchers' work, as well as reflexively on my own, I want to be aware that conditions embodied in myself such as my gender, class, age and ethnicity help structure what I understand. It is a far cry from the disembodied authoritative perspective of objective science. Not wilfully unscientific or subjective, this approach embraces embodiment theory as a way of knowing with research principles applicable to any issue.

Ethical questions about speaking for others and the interpretation of research results also need to be addressed in terms of the researcher's objectives. Within tourism studies, we have ample opportunities to act from an ethical position, to engage the oppressed, identify possibilities for agency and resistance. A basic ethical question we confront in qualitative tourism research is, why does difference often mean inequality and domination rather than co-operation and survival? Difference based in political and social power must be transformed into the valuing of diversity as resources (Bunch 1992). Tourism studies is an ideal vehicle for promoting the richness of diversity, in cultural tourism for example, and the justness of equality in resource access and opportunities for tourism providers and tourist consumers. If the researcher is writing from a feminist perspective, basic beliefs in feminist ethics start with the conditions of women's lives. Women are oppressed and mistreated, and this is wrong. Analysis and transformation of this or any form of unjustified domination is necessary for the creation and promotion of good societies (Como 1998: 55).

Terms of embodiment

'Embodiment' in the dictionary sense means to give something a concrete or discernible form, a tangible body. The term's use in contemporary scholarship on the body has come to mean 'individuals' interactions with their bodies and through their bodies with the world around them' (Davis 1997: 9). Davis (1997: 1), in a comprehensive essay on embodiment theory, notes an explosion of academic and popular culture interest in 'the body' during the late twentieth century. She discerns three interconnected explanations for why this occurred:

- a reflection of culture at large;
- theoretical development;
- the impact of feminism as political action.

A culturally based analysis of the body, for example the work of Featherstone (1983), reflects changes in cultural landscape of late modernity from industrial capitalism to a consumer culture that encourages leisure and, we should note, tourism. Bodies then became understood as a means for self-expression, an aspect of individual identity, and/or cyborg intersections with machines, rather than as representations of the social order. This new theorising, however, is 'masculinist and disembodied as it ever was' (Davis 1997: 14), reflecting back to Cartesian dualities and ideas about the separation of mind and body that have dominated Western thought.

At the same time, there has been a paradigmatic shift across disciplines to a distinct focus on the body, interrogating philosophical preference of mind over body, theological preference for the soul, and social science relegation of the body to biology, desire or 'nature'. This embodiment theory has grown out of philosophy (phenomenology), linguistics and cognitive science, and ecology, influencing all the humanities and social sciences. Embodiment is understood as the existential conditions of human life (Csordas 1994), an integrative sensate and cognitive reality that never privileges mind over body or body over mind. It provides ways of thinking about humanity that makes it impossible to commoditise bodies as objects to be simply bought and sold. Fluidity and temporality are other aspects of embodiment. As Weiss (1999: 43) writes, 'the very notion of embodiment suggests an experience that is constantly in the making ... being constituted and reconstituted from one moment to the next ... changing in significance and form'.

Embodiment theory thus challenges deeply rooted Western constructs about the duality of mind and body, and virtually all life experience as opposite pairs rather than integrative wholes. Cognitive linguists Sinha and Jensen de Lopez (2000: 19) warn that embodiment theory may break with only half of the dualistic framework of the Cartesian paradigm. It may successfully challenge the mind–body divide, but 'it leaves intact the dualism or opposition between the individual and society.... [It] tends to see cognitive mapping in terms of a one-way street from individual (embodied domains) to society (abstract and social domains).' An integrative approach is an ongoing challenge for embodiment researchers. Disabilities studies provides another challenging area of embodiment theory, raising questions on established ideas about having a body. It destabilises concepts of 'the normal' and frames embodiment as theories about the subjectivity of the able-bodied, not engaged with the corporeality of the disabled body (Breckenridge and Vogler 2001).

Foucault (1978, 1988) is often lauded as having greatly influenced social theory with his understanding of the body as constructed through processes of discipline and normalisation by various regimes of domination. Researchers taking this perspective have sought to understand how people's bodies are subject to power. Ideas about construction of the body fuel some theorists, while others point to the body as fact, 'the only constant in a rapidly changing world ... the final arbiter in what is just and unjust, human and humane, progressive and retrogressive' (Frank 1990: 133). Such contradictory understandings of the body, as material fact and always in ideological flux, fuel research about it.

In tourism studies, as we shall see, it is quite possible to research embodied topics without being informed by embodiment theory. If we look at tourism research that is not concerned with embodiment, we can find that an embodiment perspective exists but is under-theorised and often ignored. Why this is so raises interesting questions about the heritage of tourism studies that are taken up by Veijola and Jokinen (1994) in work on feminist research on tourism, as discussed in the next section of this chapter. This 'stealth-embodiment' was the case with some papers offered at two international conferences in 2001–2002 that I attended. Here we can find a bit of work on embodied topics without reference to embodiment theory, but offering something other than the usual disembodied 'scientific' analyses. My comments come from my own position as a fellow presenter at these meetings, with reference to published conference reports by Aramberri (2002a,b). Our first stop is the Macao 2001 meeting of the International Academy for the Study of Tourism (IAST). Aramberri (2002a: 871) comments on the 'clubby' atmosphere of this group with members from 23 countries, but he does not note that these participants/bodies are almost all male, which makes for a very masculine research community quite reminiscent of the one constructed by Veijola and Jokinen (1994). He focuses on the conference theme of globalisation, and comments how uncertainty about the future 'causes a funky feeling of transciency that also affects the discipline [tourism studies]'. Empirical papers dominate the proceedings, and Aramberri (2002a: 873) raises the question of why there seemed to be so little theory. His explanation has to do with how an implosion of tourism studies over the past few years reflects conditions in academe, where conflicting discourses of theoretical frameworks talk past each other with no lingua franca. While Aramberri bemoans this, he also draws attention to several researchers writing about the Balkans, China and outer space (the next tourism frontier) in what could be seen as a common currency of researcher embodiment in distinct positions that shape their questions. Certainly in many of the conference papers, researchers spoke from personal experience or, in the case of outer space, desire. These researchers presented work cross-cutting disciplinary boundaries, engaging embodied differences in tourism volunteerism, ecotourism, authenticity practices and nationalism. My point is that research called empirical is often written from an unacknowledged perspective of embodiment and positionality that is rarely if ever theorised.

The second conference visited was the international gathering 'Tourism Development, Community and Conservation: Shaping Eco-Tourism for the Third Millennium', held in 2002 at Jhansi, India. This was an open event, rather than by IAST membership invitation only as in Macao, so that the participation was self-selected and relatively diverse in terms of nationality, academic status and gender. Aramberri (2002b) makes the point that most papers did not pay attention to the theme of ecotourism in the third millennium, but rather retreated to case studies describing specific events or issues. However, there were indeed theoretical perspectives running through the papers, including the silent question of embodiment. Some papers were offered from the disembodied position of scientific investigation, but others were written from the perspectives of international development workers, national tourism analysts, and – yes – a stray feminist or two. Non-government organisations were analysed from engaged, embodied positions in presentations on pro-poor sustainable tourism in Nepal, and other tourism development projects. Tourism health issues, a clear case of the body in tourism, were explored, and the perspective of ecofeminism, inserting gender into the study of ecotourism, was raised (for a further discussion on feminist standpoint research in ecotourism, see Humberstone's chapter, Chapter 7). Papers from these conferences show that embodiment theory is often in the closet, but present and useful in tourism studies.

There has of course been overt writing on embodiment in tourism, growing especially in response to the seminal work of Urry (1990), which proposed an ocular approach focused on the tourists' 'gaze'. Franklin and Crang (2001) have noted a growing trend to a fuller embodied analysis beyond the gaze, and Urry himself (2002) has now reflected on the relationship between tourism and embodiment. The work of Edensor (2001) raises some very interesting questions about two aspects (dare we say a duality) of embodiment that he has discerned in his research on ways in which tourism is staged and performed. He notes (2001: 60) culturally coded patterns of tourist behaviour evolving around class, gender, sexuality, ethnicity, and other 'dispositions' that he argues are grounded in 'the specific habitus of tourists, distinct "common sense" unreflexive ways of being (Bourdieu 1994). But besides these dispositions, there are also expectations.' Thus, forms of tourist habitus are also determined by unreflexive, embodied shared assumptions about appropriate behaviour in particular contexts. Edensor (2001: 63) develops his argument that 'the everyday can partly be captured by unreflexive habit, inscribed on the body, a normative unquestioned way of being on the world'. He stresses that 'tourist practices abound with their own habitual enactments, and tourism is never entirely separate from the habits of everyday life, since they are unreflexively embodied in the tourist' (Edensor 2001: 63). However, as we have already seen in previous discussion of embodiment, everyday life is full of both routine and change, fluid and the unpredictable. This other side of tourism embodiment Edensor calls 'reflexive' and comments that (2001: 62) 'where reflexive improvisation and a critical disposition are mobilised, the resulting ambiguity can threaten the sense of well-being that is one of the main aims of tourism'. He (2001: 78) thus highlights the 'ambivalent nature of tourism, as the desire to escape and the pleasures of conformity clash'. This tourist performance may be understood 'as intentional and unintentional, concerned with both being and becoming, strategically and unreflexively embodied'. Naming two aspects of embodiment as reflexive and unreflexive is a bold move, but it elides the original coda of 'dispositions' which are also set and continually contested as both life conditions and the body itself change. Can embodiment ever be fully unreflexive, or would it be more profitable to see reflexivity as a continuum, sometimes less felt and cognisant than others? It would also be interesting to apply Edensor's ideas to the full spectrum of tourism practice, to tourism providers as well as tourists. In the following section I discuss how feminist researchers in tourism studies have integrated human 'dispositions' of race, class, gender, sexuality, age, ethnicity into their work on embodiment.

Feminist questions of embodiment in tourism studies

Feminism, simply defined as 'a movement to end sexism, sexist exploitation and oppression' (hooks 2000: 1), has politicised women's bodies and led to analysis of all forms of difference that feed social inequalities. Feminist theory on the body focuses on differences, domination and subversion as necessary ingredients for understanding the terms of embodiment in contemporary culture. Individual differences are seen to intersect, giving meaning to embodied conditions and experiences that are systematically organised by patterns of control. Human differences can be explained in terms of biological bodies and/or culturally constructed meanings. Wary of privileging biology, which could legitimise inequalities as normal or immutable, feminist scholars asking about women's experience separated sex and gender as analytical constructs, with some scholars viewing the body as a 'coat rack' on which traits are hung. Butler (1990) made a major theoretical move by insisting that distinctions made between female and male bodies are entirely arbitrary. Such a radical stance had engaged a good deal of critique about the physical realities of bodily functions that shape human experience, for example the unavoidable fact so far that female bodies bear children (Nussbaum 1999).

In another approach, difference is seen as an essential concept for understanding embodiment, culturally formed and physically marked conditions such as gender, race and sexuality (Davis 1997: 7). Dominant powerful groups name differences embodied in each individual, privileging bodies that are most like those in control (Moore 1994). Difference, however, may also be seen as the basis of individual and group agency, as people negotiate the limitations of their embodied experience (Davis 1997: 12) through compliance, resistance and transformation. Experimentation with alternative meanings and identities is a basis, for example, of queer theory that questions all forms of binary thinking including dualistic concepts of sex and gender. Whether a researcher's focus is on body practices, regimes or discourses about the body, power is a constant factor that needs to be taken into account. One dimension of power imperative to be acknowledged in qualitative research is the embodiment of the researcher. Davis (1997: 14) notes that 'by assuming that the theorist is also embodied, feminist theory opens up possibilities for exploring new ways of doing theory ... which uses embodiment as a theoretical resource for an explicitly corporal epistemology or ethics'.

The term 'embodiment' gained use in feminist writing on 'the body' in the 1990s to indicate the physical reality of intersecting power relations such as gender, sexuality, ethnicity, race, nationality, class, age and (dis)able-bodiness within individuals and among people. Qualitative research in geography (Rose 1993) and anthropology (Moore 1994) has begun to acknowledge that the researcher's embodied positionality intersecting with that of the researched affects how our understandings are constructed. This direction has been embarked on in tourism studies, raising questions about how embodiment theory may be useful for analysing differences and diversity in tourism practice, regimes and discourses. Tourism is an industry built on differences (Swain 2002) in landscapes, peoples and experiences. In tourism research we study the processes and results of 'Othering', including stereotypes, prejudice, self-promotion, entrepreneurship, and forced exploitation of bodies. By combining embodiment theory with feminist-type ethics, we can strive to understand differences and promote more equitable human conditions in the course of our research about tourism.

Veijola and Jokinen's seminal article 'The body in tourism' (1994) opened up this line of inquiry in tourism studies. They burlesque the all-male star scholars of tourism studies, inserting their feminine selves into an imagined vacation (probably an international conference) where they interact with these shapers of tourism theory. Veijola and Jokinen (1994: 129) playfully ask 'would it not be possible to thematise the embodiment, radical Otherness, multiplicity of differences, sex, and sexuality in tourism?' They draw from feminist theory to process their imagined conversations with some of the heavy hitters of tourism studies, including MacCannell (1989), Urry (1990) and Rojek (1993) (see Chapter 8 for a further discussion on the influence of tourism gatekeepers on research). Citing Butler (1990) and Braidotti (1991), Veijola and Jokinen reinforce the point that gender and sexuality are both socially constructed and culturally encoded into diverse bodies. Butler found that 'gender trouble' could be transformative. As Veijola and Jokinen emphasise (1994: 142) 'Butler advises us to resignify ... instead of just asserting the alternatives given by culture we can create new ones for gender/sex by using the prevalent rules governing signification. The rules need not be only restrictive.' Braidotti in Veijola and Jokinen (1994: 141) asks 'how to rethink the body in terms that are neither biological nor sociological?'. Pettman (1997: 99), in her subsequent work on body politics in international sex tourism, illustrates how this perspective from Braidotti (1991) and Veijola and Jokinen (1994) in thinking about embodiment is a critical analytical step. Grosz (1992), a theorist not included by Veijola and Jokinen, has also had significant influence on embodiment studies through her arguments against Cartesian dualism, or the separation of mind and body, and advocating for the integrated analysis of gender and sexuality located in corporeal entities. An excellent example of this separation has been the focus in tourism studies on 'the gaze' (Urry 1990), an ocular perspective that tells us a great deal about the mind but very little about what the tourist feels or experiences.

Veijola and Jokinen (1994: 149) conclude their contribution with a powerful summary, still current, of the status (corpus) quo:

> So far the tourist has lacked a body because the analyses have tended to concentrate on the *gaze* and/or structures and dynamics of *waged labour societies*. Furthermore, judged by the *discursive postures* given to *the writing subject* of most of the analyses, the analyst himself has, likewise, lacked a body. Only the pure mind, free from bodily and social subjectivity, is presented as having been at work when analysing field experiences, which has taken place from the distance required by the so-called scientific objectivity, from the position-in-general.

I would like to trace how Veijola's and Jokinen's (1994) challenge has and has not been taken up in the ensuing years by drawing from my experience in editing several collections representing a number of scholars writing on gender in tourism (Swain 1995, 2002), and my own work on gender and ethnicity issues for women producers of ethnic arts (Swain 1993, 2001), as well as some recent work by my students. In the special issue of *Annals of*

Tourism Research on gender that I edited in 1995, I had bodies, unconsciously, on my mind when I argued for 'thinking about tourism issues as gendered relationships between individuals, groups, social categories, types of tourism, and nations in First/Third Worlds' (Swain 1995: 247). The collection's articles explored gender in terms of consumption practices, perceptions of tourism development, identities, sexuality and nationalism, in the political economy of tourism, in relations between tourism providers and consumers, and in the reframing of ideologies through tourism leisure practice. In these articles, sexed bodies, social sexuality and assorted gender hierarchies are clearly significant dimensions of tourism and its study. This is to be expected in a collection on gender in tourism, but the larger point is that any qualitative research on tourism deals with people, and thus has a gender dimension whether it is analysed or not.

While I drew great inspiration from Veijola and Jokinen (1994) in framing this collection, I did not follow their lead to interrogate why individual tourism researchers choose to study what they do in terms of personal body politics, and how that affects their interactions with their subjects and research results. What I wrote about Veijola and Jokinen (1994) was on their humorous insertion of 'the body in tourism', taking up their project in terms of what we study. At the time, I thought it was rather obvious that what we are influences what we study, but I missed an opportunity there to explore the meaning of who was writing what, of embodiment, in this edited collection. Researchers positioned in varied degrees from their subjects wrote the articles in the Special Issue. There was no 'native' ethnography, but there were embodied connections. Virtually all the contributors engaged in some form of participant observation, with the exception of those who collected quantitative data based on surveys and other instruments. Long-term relationships of researcher and subjects (Cone 1995; Meisch 1995) and co-nationalist identities (Garcia-Ramon *et al.* 1995; Anderson and Littrell 1995; Harvey *et al.* 1995) shaped the authors' embodied perspectives. They wrote from their situatedness (Haraway 1988), collections of relationships and histories played out in their own bodies as gendered, raced, sexed, classed, aged, etc. beings.

A number of articles focused specifically on issues of sexuality, but there was no mention of sexualities or sexual difference in tourism. As Johnston has noted (2001), this was a significant gap, which would have been even more obvious if I had used a more developed understanding of embodiment in the Introduction (Swain 1995). Other significant gaps were the lack of any focus on disabilities or age differences and gender. The fact that I did not think to write about sexualities, age or other aspects of the body was reinforced at the time by the lack of submissions about them for the Special Issue. Times have changed, and we are beginning to see more inclusive scholarship in tourism studies, as evidenced by some of the following examples in this book.

Gibson (2002) provides a nuanced review of work to date on gender in tourism, drawing on a wide range of materials, including the Special Issue and important work by Kinnaird and Hall (1994, 1996). Using varied theoretical perspectives, Gibson (2002) analyses this body of work but does not invoke theory on embodiment. Rather, she cites work in leisure studies by Wearing and Wearing (1996), who:

> critique the mainstream, male-oriented tourism theories that focus on a subject–object relationship between host and guest.... [Rather, they] adopt an interactionist perspective to proffer the idea that tourism should be conceptualised as a dynamic set of social relationships and meanings for the people who interact in the tourism space.
>
> (Gibson 2002: 25)

A focus on both intersectionality and embodiment would help us understand the dynamics of tourism, its problems and potentials for human interaction. It would engage researchers in thinking beyond an empirical description by embodying their selves within the conditions they are studying. Gibson embraces the Platonic ideas of 'chora' and 'choraster' as developed by Wearing and Wearing (1996), who in turn were influenced by feminist philosopher Grosz. Applying Plato's 'chora', a space between being and coming, to tourist destinations, the tourist then can be understood as a 'choraster', the one who interacts with the chora, including host peoples. The idea is that the people who make use of space give it meaning. While Gibson sees this space as inhabited by host people who are then intersected by the choraster tourists, I would push this a bit further from an embodiment perspective. Host bodies physically take up space and interact with visiting bodies. This chora destination is given meaning in tourism by the intersection of many choraster bodies, primarily of the tourist, the toured (workers and resident hosts) and the tourism researcher who thinks about such things. Analysis of gender in tourism provided a focus on what could be called embodiment in tourism studies, accompanied by rethinking of ethnicity and class, and new work on sexualities, age, nationality and corporal experience in tourism encounters (Swain 2002). While in past work, bodies had often been conceptualised as blank slates on which socially constructed identities were written, an embodied approach asked how these identities are personally experienced as complex, interacting whole beings. Veijola and Jokinen (1994: 148) ask, 'how are we going to change our research practices and tourist practices in a way that prevents us from constituting the Other *outside* of ourselves?' In the intervening years, other researchers have picked up Veijola and Jokinen's challenge, and have asked about sex bodies, sexualities and gender politics in the academy (Pettman 1997; Johnston 2001).

Writing about her own conversion to this newer perspective, Pettman (1997: 94–95) notes that bodies erupted into her text, inscribed with differences that mattered.... Sex – as desire, danger, eroticized bodies, transgressions, violations – came through my writing too, including women's experiences in identity conflicts, as boundary markers or community possessions, as women warriors, as commodified cheap labour on the global assembly line, as labour migrants, 'foreign' domestic workers, and international sex workers.

Franklin and Crang (2001: 13–14) note Johnston's (2001) work on Gay Pride parades as erotic spectacles for straight audiences as an example of the move in tourism studies away from the 'gaze' only to a more fully embodied analysis. In sum, it seems that tourism studies must investigate the sensual, embodied and performative dimensions of change in tourism cultures. Johnston (2001: 196) concludes in her article that for researchers, 'writing from one's located and embodied position is one way that may unsettle positivist, rational, and masculinist constructions of tourism researchers'. In terms of subject matter, 'bringing the gendered/sexed body into tourism epistemologically challenges the distinctions between mind and body, Self and Other, tourist and host'. While she sets these parameters well in her study of queer event tourism, I missed the articulation of researcher embodiment with that of her subjects: tourists, performers and hosts. What did Johnston feel, literally, about her research, and how did that shape her results? If Johnston had written herself into the text, we would have a better understanding of her own embodied position.

The intersections of researched and researcher bodies are approached, but rarely looked at as sites of knowledge, in current tourism research. Pettman (1997: 95) comments that most work on tourism remains disembodied, with neither the writers nor their subjects in visible bodies. However, she argues, 'the body you are/are in clearly makes an enormous difference – it places you, or me, on one side or the other of boundaries that mark both power relations and entitlements'. This perspective explicitly appears in several chapters in the *Gender/Tourism/Fun(?)* collection from revised conference papers originally given in late 1997 that I edited with Momsen (Swain and Momsen 2002). These authors foster an intimacy of embodiment, and chose to speak in their own voice (rather than the usual disembodied type of analysis), to acknowledge some somatic empathy as well as an outsider perspective. Selänniemi (2002) takes us into the sunlust liminoid zone of Finnish tourists at Playa del Anywhere, while Small (2002) engages the age-graded good and bad memories of Australian women and girls on holiday. Meisch (2002: 172) explores her 'gender radical' self as a woman trek guide in the highly masculinised Andean trade, and evokes the perspective of 'halfie', someone who is both native to the culture being studied, and an outside analyst. Hottola (2002) has expanded on questions of the researcher's embodiment in his study of sexual harassment, stereotypes and misunderstandings between Indian male locals and Western women tourists. This perspective of the researcher being mindful of one's own embodiment enriches any qualitative research project, in terms of results and applications of our work.

My own work on gender and ethnicity issues for women producers of ethnic arts (Swain 1993, 2001a) has been devoid of overt body analysis while taking a firm perspective on gender issues as culturally and socially constructed. My training as an anthropologist in the United States first led me to study Native peoples in the Americas, but I simply could not work in North America, and lasted only a few years in Panama owing to what might be called 'group colonialization guilt'. This had nothing to do with my specific ancestors or actions and was all about histories of colonisation and what I looked, sounded and acted like. I embody many signs of imperialism I could not conceal: I'm white for a start. I have been infinitely more comfortable in south-west China, where the race politics are distinct and my white foreignness feels more like an advantage than a badge of historically tainted power. Granted, I would not even have been allowed to work in China 25 years ago, but once the doors opened, I embodied many desirable traits as a researcher. I was drawn to the people in the place that I now work, the Sani Yi in the Stone Forest of Yunnan, China, in part because of the 'Martian effect'. To the Sani, I appeared in their life a bit like a Martian; I was as different from them as possible in many ways. All those personal characteristics that I listed earlier set me apart from my 'subjects', and what did not – often gender, sexuality and age, rarely class or left-handedness – was heavily coded in distinct norms. What I focused most on, with due respect to the 'white western feminist misappropriates third world women of colour' perspective (Mohanty 1987) was what I found most familiar, the lives of women. My Otherness was an entrée into the field. My unexpected arrival, full of foreign entertaining sights, sounds and actions, made me quite attractive for a while. Once the mutual charm wore off, I was still an anomaly, not expected to measure up to local standards or manipulate power in particular ways, which was a great advantage. This, I hasten to add, was how I felt, that what I embodied both limited my research possibilities and made others very possible. I am not saying of course that I think that men should not study women, or brown should not study white. Rather, the point is that what we embody must affect how we feel, see, interpret and understand. This resonates with an observation by Grosz (1992: 21) about how the corporal and the psychic combine to shape the body, which 'thereby defines the limits of experience and subjectivity'. Reality is in my body, including my head. The interactions I have had with numerous Sani and Han people in Stone Forest over the years were predicated on how we perceived each other from bodily clues, actions and circumstances, as well as any relationships we developed.

The Sani Yi are an ethnic minority in their nation-state, living in a region highly exploited for tourism based on natural scenic attractions and, more recently, Sani culture. The Han majority population of China colonised their ancestors many generations ago. In Sani eyes, the Han are

Other, the ones to be feared, distrusted and stereotyped. I did not feel hostility for what I embody in terms of race, ethnicity and nationality. The gender dimension, however, was curiously maddening and inciting. Yes, I was restricted in my living arrangements and time in the field by state officials – because I am a woman, they said. Rather perversely, this gave me something to complain about with my Sani friends, who readily saw some fellow victimhood. While I did not use the rhetoric of 'embodiment' at the time, I was acutely aware that my body and theirs made an enormous difference in how I conducted my research, what I was able to learn about Sani women tourism entrepreneurs, and how I might affect their lives. The ethics of inserting myself/my body into their daily routines to extract knowledge are something I tried to deal with in the field and have written about (Swain 2001a). For me, the classic qualitative research method of participant observation led me to collaborate in the creation of new tourism souvenirs for my Sani companions to sell. Appropriately enough, these were little representations of Sani bodies: cloth dolls. As we have seen in my work and a very limited review of some tourism researchers writing from varied disciplinary and theoretical perspectives, an embodied approach has some currency. The future of this approach in qualitative research may be indicated by pre-publication sites in the classroom, which we will briefly visit next.

Embodiment in the classroom

In a feminist studies course on tourism, one would expect to find overt attention to bodies and embodiment analysis, and we do in the following excerpts from students' research projects. The students' own agendas and interest in their subjects' embodied experiences led them to investigate issues of subjugation and empowerment in tourism ranging from the sex trade to specific sites of tourist encounters and experiences. They were primarily influenced by third-wave feminism that explicitly asks about women in relationship to men, and they found intriguing recent tourism research such as Ryan and Martin's (2000) study of stripper tourism, clearly written from an embodiment perspective.

A poignant paper written from a position of family relations described sex tourism in Nigeria, asking about the web of economic forces and sexist cultural practices that virtually enslave women as sex workers – women like the student's grandmother, who died in the trade while still supporting a transnational network of children in diaspora. Another student enrolled in a sex tourism Web site, http://WorldSexArchives.com, to obtain his data about Western masculine motives in Asian romance and sex tourism. His insider/outsider perspective as a Western straight male of Asian descent led him to conclude that the masculine culture of the Web site encourages displays of physical over emotional desire, reinforced by a 'gender double standard' in both Asia and the West. Western sex tourists themselves perceive sex tourism as a male prerogative, and romance tourism as a female pursuit. A woman student from Japan focused on a developing form of domestic tourism of the body called in Anglo-Japanese 'New Half Show Pubs' – tours to view male-to-female transgender drag performances. The student began her paper presentation with the question, 'What do you think I am, a man or a woman?' She grounded her research in website advertisements for bus tours to 'new-half' performances, and an email survey of tour participants with 106 respondents. Her gender analysis of the tourist gaze and somatic consumption showed that the female majority of responding tourists enjoyed seeing the show, while feeling sympathy for the performers who had to show their bodies as commodities. Male respondents (40 per cent of her sample) consumed the 'new-half' bodies without critically questioning their sexuality, but rather looked for authenticity and experiencing titillation in the 'Other'. She also interviewed on email a 'new half' performer about 'hir' life as an entertainer and as a member of an increasingly visible minority in Japanese culture and society. 'New-half' tourism performers are constructed by both their own and their consumers' desires. Worldwide, from sex tourism to street performances, 'in the informal sector that feeds off the tourism industry, how the body is marked is integral to the commodity exchanged' (Sanchez Taylor 2000: 50).

Summary

Disembodied disciplines, wrapped in the mantle of objectivity, have denied the existence of bodies for researchers and their subjects, taking the dominant male position to be the norm, as 'the tourist', for example, noted by MacCannell himself (1989: xi). From a perspective of hegemonic masculinity, all the messy bits, in 'sites of desire and economies of pleasure' (Jolly and Manderson 1997), become coded as women's or disempowered (minority, 'less developed country') men's issues. An understanding of power dynamics – how bodies interact and influence each other in multiple relationships, creating these unequal relations – is needed to move us forward into new solutions. I have framed this discussion in terms of feminist ethics (Bunch 1992), questioning the researchers' respect and understanding, while speaking for others and interpreting research results. Discourse about the body may inform all qualitative tourism research in useful ways, not just feminist research. As we have seen so far in tourism studies, most embodiment research has been from a feminist perspective, but there is much potential for any qualitative research, especially research engaged in praxis. An embodiment approach offers tourism studies opportunities to change and adapt to more engaged analysis in terms of our subjects, and our qualitative research methods. The approach has grown out of feminist theory, expanding an emphasis beyond women, gender and sexuality analysis to embrace all embodied differences that physically mark and channel experience. In terms of methodologies, an embodied approach acknowledges that the researcher's own embodiment affects research questions and results. If bodies are invisible or silenced in qualitative research, we are missing a very rich source of data and denying a method of investigation that acknowledges the researcher's complicity in knowledge-building.

Questions

1   How might an embodied approach in qualitative tourism research result in more equitable conditions in tourism power dynamics?
2   What has feminist analysis added to the discussion of embodiment in tourism studies?
3   What are some examples of how the embodiment of tourism researchers and tourism studies' subjects may be articulated in research work?
4   How might embodiment be seen as reflexive or non-reflexive in tourism research?

References

Anderson, L. and Littrell, M. (1995) 'Souvenir-purchase behaviour of women tourists', *Annals of Tourism Research*, 22: 328–348.

Aramberri, J. (2002a) 'International academy for the study of tourism', *Annals of Tourism Research*, 29: 871–873.

Aramberri, J. (2002b) Conference report, *Newsletter of the International Academy for the Study of Tourism*, 12 (3): 6. Online. Available www.tourismscholars.org (accessed 1 October 2002).

Bourdieu, P. (1994) *Distinction.* London: Routledge.

Braidotti, R. (1991) *The Patterns of Dissonance.* London: Polity Press/Basil Blackwell.

Breckenridge, C. and Vogler, C. (2001) 'The critical limits of embodiment: disability's criticism', *Public Culture*, 13 (3): 349–357.

Bunch, C. (1992) 'A global perspective on feminist ethics and diversity'. In E. Cole and S. Coultrap-McQuin (eds) *Explorations in Feminist Ethics: Theory and Practice.* Bloomington, IN: Indiana University Press.

Butler, J. (1990) *Gender Trouble: Feminism and Subversion of Identity*, London: Routledge.

Como, C. (1998) *Feminism and Ecological Communities: An Ethic of Flourishing.* London: Routledge.

Cone, C. (1995) 'Crafting selves: the lives of two Mayan women', *Annals of Tourism Research*, 22: 314–327.

Csordas, T. (ed.) (1994) *Embodiment and Experience: The Existential Ground of Culture and Self.* Cambridge: Cambridge University Press.

Davis, K. (1997) 'Embody-ing theory: beyond modernist and postmodernist readings of the body'. In K. Davis (ed.) *Embodied Practices: Feminist Perspectives on the Body*, London: Sage.

Edensor, T. (2001) 'Performing tourism, staging tourism: (re)producing tourist space and practice', *Tourist Studies*, 1: 59–81.

Featherstone, M. (1983) 'The body in consumer culture', *Theory, Culture and Society*, 1 (2): 18–33.

Foucault, M. (1978) *The History of Sexuality*, vol. 1. New York, NY: Pantheon.

Foucault, M. (1988) *Politics, Philosophy, Culture: Interviews and Other Writings 1977–1984*. New York, NY: Routledge and Chapman and Hall.

Frank, A. (1990) 'Bringing bodies back in: a decade review', *Theory, Culture and Society*, 7: 131–162.

Franklin, A. and Crang, M. (2001) 'The trouble with tourism and travel theory?', *Tourist Studies*, 1: 5–22.

Garcia-Ramon, M. D., Canoves, G. and Valdovinos, N. (1995) 'Farm tourism, gender and the environment in Spain', *Annals of Tourism Research*, 22 (2): 267–282.

Gibson, H. (2002) 'Gender in tourism: theoretical perspectives'. In Y. Apostolopoulos, S. Sönmez and D. Timothy (eds) *Women Producers and Consumers of Tourism in Developing Regions*. Westport, CT: Praeger.

Grosz, E. (1992) 'Bodies–cities'. In B. Colomina (ed.) *Sexuality and Space*. New York, NY: Princeton Architectural Press.

Haraway, D. (1988) 'Situated knowledges: the science question in feminism and the privilege of partial perspective', *Feminist Studies*, 14 (3): 575–599.

Harvey, M., Hunt, J. and Harris C. (1995) 'Gender and community tourism dependence level', *Annals of Tourism Research*, 22 (2): 349–366.

hooks, b. (2000) *Feminism Is for Everybody: Passionate Politics*. Cambridge, MA: South End Press.

Hottola, P. (2002) 'Amoral and available? Western women travellers in South Asia'. In M. Swain and J. Momsen (eds) *Gender/Tourism/Fun(?)*. Elmsford, NY: Cognizant.

Johnston, L. (2001) '(Other) bodies in tourism studies', *Annals of Tourism Research*, 28 (1): 180–201.

Jolly, M. and Manderson, L. (1997) 'Introduction'. In L. Manderson and M. Jolly (eds) *Sites of Desire, Economies of Pleasure: Sexualities in Asia and the Pacific*. Chicago, IL: University of Chicago Press.

Kinnaird, V. and Hall, D. (eds) (1994) *Tourism: A gender analysis*. New York, NY: Wiley.

Kinnaird, V. and Hall, D. (1996) 'Understanding tourism processes: a gender-aware framework', *Tourism Management*, 17 (2): 95–102.

MacCannell, D. (1989) *The Tourist: A New Theory of the Leisure Class*, 2nd edn. New York, NY: Schocken Books.

Meisch, L. (1995) 'Gringas and Otavalenos: changing tourist relations', *Annals of Tourism Research*, 22 (2): 441–462.

Meisch, L. (2002) 'Sex and romance on the trail in the Andes: guides, gender and authority'. In M. Swain and J. Momsen (eds) *Gender/Tourism/Fun(?)*. Elmsford, NY: Cognizant.

Mohanty, C. (1987) 'Under Western eyes: feminist scholarship and colonial discourses', *Feminist Review*, 30: 65–88.

Moore, H. (1994) *A Passion for Difference: Essays in Anthropology and Gender*. Bloomington, IN: Indiana University Press.

Nussbaum, M. (1999) 'The professor of parody', *The New Republic*, issue 02-22-99.

Online.   Available   www.tnr.com/archive/0299/022299/nussbaum022299.html (accessed 18 October 2002).

Pettman, J. (1997) 'Body politics: international sex tourism', *Third World Quarterly*, 18 (1): 93–108.

Rojek, C. (1993) *Ways of Escape: Modern Transformations in Leisure and Travel*. London: Macmillan.

Rose, G. (1993) *Feminism and Geography: The Limits of Geographical Knowledge*. Cambridge: Polity Press.

Ryan, C. and Martin, A. (2001) 'Tourists and strippers: liminal theater', *Annals of Tourism Research*, 28 (1): 140–163.

Sanchez Taylor, J. (2000) 'Tourism and "embodied" commodities: sex tourism in the Caribbean'. In S. Clift and S. Carter (eds) *Tourism and Sex: Culture, Commerce, and Coercion*. London: Pinter.

Selänniemi, T. (2002) 'Couples on holiday: (en)gendered or endangered experiences?'. In M. Swain and J. Momsen (eds) *Gender/Tourism/Fun(?)*. Elmsford, NY: Cognizant.

Sinha, C. and Jensen de Lopez, K. (2000) 'Language, culture and the embodiment of spatial cognition', *Cognitive Linguistics*, 11: 17–41.

Small, J. (2002) 'Good and bad holiday experiences: women's and girls' experiences'. In M. Swain and J. Momsen (eds) *Gender/Tourism/Fun(?)*. Elmsford, NY: Cognizant.

Swain, M. (1993) 'Women producers of ethnic arts', *Annals of Tourism Research*, 20 (1): 32–51.

Swain, M. (1995) 'Gender in tourism', *Annals of Tourism Research*, 22 (2): 247–267.

Swain, M. (2001) 'Ethnic doll ethics: tourism research in southwest China', in V. Smith and M. Brent (eds) *Hosts and Guests Revisited: Tourism Issues of the 21st Century*. Elmsford, NY: Cognizant.

Swain, M. (2002) 'Gender/tourism/fun(?): an introduction'. In M. Swain and J. Momsen (eds) *Gender/Tourism/Fun(?)*. Elmsford, NY: Cognizant.

Swain, M. and Momsen, J. (eds) (2002) *Gender/Tourism/Fun(?)*. Elmsford, NY: Cognizant.

Urry, J. (1990) *The Tourist Gaze*, London: Sage.

Urry, J. (2002) *The Tourist Gaze*, 2nd edn. London: Sage.

Veijola, S. and Jokinen, E. (1994) 'The body in tourism', *Theory, Culture and Society*, 11: 125–151.

Wearing, B. and Wearing, S. (1996) 'Refocusing the tourist experience: the flâneur and the choraster', *Leisure Studies*, 15 (4): 229–243.

Weiss, G. (1999) *Body Images: Embodiment as Intercorporeality*, London: Routledge.

7 Standpoint research

Multiple versions of reality in tourism theorising and research

Barbara Humberstone

Aims of the chapter

- To identify, critique and develop thinking that recognises the Other, insider perspectives and the partial nature of knowledge.
- To examine the constructs of standpoint and 'post'-standpoint research.
- To examine the connections between standpoint perspectives and ecological perspectives.
- To point out the need for and usefulness of standpoint research for tourism research.

It has been argued for some time by critical commentators (see, for example, Foucault 1980; Women and Geography Study Group 1997) that dominant ideologies[1] in the construction of knowledge have tended to convey a single truth. This production of knowledge and development of theory was in the past largely based upon Eurocentric research and the ideas of mainly white middle-class men (Haraway 1991). Even today, in the field of tourism there are charges of lack of attention to post-modern thinking (Aitchison 2002). There appears to be reluctance in the field to engage with current debates ongoing in emancipatory and cultural theories (Deem 1999). As Aitchison (2002: 21) argues, 'postmodernism and feminism present a challenge to the codification of power and knowledge, the power of knowledge and the challenge of power within the subject field'. Tourism research, then, has largely failed to acknowledge issues raised through standpoint epistemology, which dialectically, through the synthesis of often opposing ideas, informed the thinking of developing postmodern thinking and emancipatory frameworks. Consequently, it has tended to neglect the perspectives and experiences of marginalised groups. These groups include women, black people, people with disability, and so forth.

This chapter examines the development of standpoint epistemologies to the position of 'post'-standpoint research. Tourism studies are nothing if they are not about relations between the visitor, the Other (the host) and the locale – cultures and contexts intermingling. The environmental context as non-human is part of the equation of oppression of marginalised people. 'Post'-standpoint research is important for tourism research and theory since it explicates the importance of deconstruction (questioning) from the inside of a school of thought and the interrogation of assumptions that there can be one Truth regime, be it from dominant discourses or even those in their ascendancy. 'Post'-standpoint epistemologies question the authority not only of the researcher but also that of the 'text'. The challenge for tourism research is aptly posed by Soper (2003): 'Can we find ways of living rich, complex, non-repetitive lives without social injustice and without environmental damage?' This chapter argues that tourism researchers need to ask critical questions about the effects of tourism both upon those whose environment is a tourism destination and upon the environment itself. Standpoint epistemology provides ways in which tourism researchers can think about such questions, raise awareness and perhaps bring about change.

Standpoint approaches to research

Standpoint research was first 'named' by North American critical feminists to identify feminist research approaches, then emerging in the 1980s, that challenged traditional social science that sought to replicate positivistic scientific method associated with the traditional period (Lincoln and Denzin 1998).[2] The writings of Collins, Haraway, Smith and Harding have helped to legitimate and give name to the voices/work of researchers who took the starting point of research to be the lived experience of women. This 'naming' of a 'feminist standpoint' epistemology by feminist philosophers was significant in gaining credibility for women researchers and their work (Stanley and Wise 1990). It reflected and acknowledged longstanding feminist criticisms of the absence of women or marginalised reports of women in social research accounts. Research undertaken from the perspective of standpoint feminist theories stressed a particular view that builds on and from women's experiences.

Standpoint epistemology may be understood in simple terms as a move towards local, contextualised, situated knowledge as represented by fifth-moment research (see Chapter 1 for a discussion of the five 'moments' of qualitative research), which draws upon the experience of subordinate groups away from a universalised, value-neutral knowledge. Ward (1997: 774), a critic of standpoint epistemology, sees 'standpoint epistemologies' as varying 'considerably in scope' … but, 'they generally contend that gender, ethnicity, culture, sexuality or some other group- or site-specific element of fact-production are ultimately responsible for both the form and content of knowledge' (see also Hartsock 1983; Collins 1985; Harding 1993b)'. According to Harding (1993a), standpoint theories start from the lives of marginalised peoples and are historically and contextually situated (see Haraway 1991).

Consequently, early research from standpoint perspectives raised fundamental questions around epistemological, ontological and methodological elements of research. Taken-for-granted epistemological assumptions around the nature of knowledge and the relation of the researcher to knowledge in conventional scientific discourse were discarded. For Smith (1992: 91), the rejection of 'standard good social scientific methodologies' was imperative, because these methodologies 'produce people as objects', for if sociologists 'work with standard methods of thinking and inquiry, they import the relations of ruling into the texts they produce'. Questions regarding the ideal of value-neutral social science research emerged. That is, social scientists who were adopting critical and feminist perspectives began to recognise that researchers brought their own values, presumptions and histories to their research which could not be erased from the research process (see Hammersley and Atkinson 1983; Richardson 1991). They acknowledged and expounded that knowledge and its production were both complex and contestable.

> The production of 'valid' knowledge from research takes place through its legitimisation by various dominant ideologies. Knowledge constituted by research becomes acceptable/unacceptable, valid/ invalid depending upon whether it 'fits' with the values, assumptions and ideologies of those in a position to legitimate its credibility. Many critical sociologists and feminist researchers are concerned with the question, 'what constitutes valid knowledge and in whose interests does it operate?'.
>
> (Humberstone 1997: 201)

Thus, from such perspectives ontological, epistemological and methodological dimensions have implications for the choice of methods or techniques and the ways in which the research is assessed as legitimate and the findings taken to be acceptable knowledge. Those implications are examined in more detail in the next section, on the characteristics of standpoint research.

Tourism as a field of study was slow in recognising and responding to these paradigmatic shifts in social scientific thought. Phillimore and Goodson argue in Chapter 1 that much tourism research remains influenced primarily by the traditional period. Recent tourism studies, though concerned to better understand the lived experience of host communities or the tourist him or herself, have often focused on the interpretation by 'experts' of the community view.[3] In doing this, tourism studies has more often sought to look for or apply universalising or totalising explanation for social phenomena. This was exemplified in studies that developed typologies of tourists and so forth (e.g. Cohen and Taylor 1976; Smith

1994). Nevertheless, it should be acknowledged that some tourism research inspired by neo-anthropological perspectives adopted some of the beliefs of interpretive or critical theoretical perspectives, sharing similar epistemological positions and a common set of methodological commitments as 'post'-standpoint research (Strathern 1987). Exemplary recent tourism research evidenced in Duffy (2002) and Hall and Ryan (2001), which is concerned with the dark side of tourism (sex tourism) and the ecological implications of ecotourism respectively, represents the type of research that is underpinned by these forms of standpoint epistemologies. This sort of research questions assumptions that tourism is necessarily a wholly good or wholly bad thing, and highlights the damage that it can do to marginalised people, cultures and to the local and global environment even when the objective of tourism is to provide ecotourism experiences with some positive spin-offs for the community. For tourism research to gain further insight into the latter, it needs to engage with issues around the nature of knowledge and its production, which has been central to much social science and philosophical discussion. Standpoint research is central to these debates. Ecotourism takes a standpoint approach in that it focuses on the experiences and perspectives of disadvantaged/marginalised groups or, in tourism terms, the host perspective, but in addition emphasises the importance of the environment within which these groups live.

The 'standpoint' debate is significant, as it systematically sets out characteristics of feminist and critical epistemology and highlights the ways in which these are linked with issues such as 'objectivity' that are continually interrogated within social science research, in particular critical research and, more widely, in sociology. There continue to be areas of contention as 'post'-standpoint positions emerge. These characteristics and the development of standpoint epistemologies, together with the contentions around standpoint research, are considered in the next section.

Characteristic elements of standpoint research

The main characteristics of research can be identified in terms of ontology, epistemology and methodology, and it is around these characteristics or sets of beliefs that questions emerged through and for standpoint research.

Ontology and epistemology

Ontology has been described as 'a theory which claims to describe what the world is like – in a fundamental, foundational sense – for authentic knowledge of it to be possible' (Barnes and Gregory 1997: 511; see also Hollinshead's discussion on ontological issues in Chapters 4 and 5 of this book). It can be understood as assumptions about the nature of reality. At a taken-for-granted, lived-experience level of individual authenticity, it is a state of being. Standpoint perspectives hold that reality lies in the lived experience of people within their situations and contexts. For tourism research, this means exploring the lived experiences of the host community, its environment and the tourist. For Stanley (1990a: 21) and other critical feminists, 'woman is a socially and politically constructed category, the ontological basis of which lies in a set of experiences rooted in the material world'. Ecological perspectives would add 'in the natural world', and this is clearly significant in understanding the effects tourism may have not only on local communities but also on local and global environments. Since men's and women's ways of knowing, particularly in developing countries, where people may have close contact with the land to sustain their lives, may be constituted by different experiences and practices, such lived experiences constitute forms of knowledge which are constructed by gendered categories (Warren 1997: 88).

Epistemology in its general form is a study of knowledge through which 'rules' can be established to identify what is to count as 'true' (Barnes and Gregory 1997: 505). It answers questions such as Who can be the knower? What 'truth' test must beliefs pass to be legitimated as knowledge? What kind of things can be known and what counts as valid knowledge? (Harding 1987: 3). Most important to standpoint research is the epistemological question: what is the relationship of the knower to the known? (Skeggs 1994: 77). We can therefore ask from whose perspectives –those of the local community, the tourist or the tourist business – is 'reality' legitimated? That is, for whom are particular ways of seeing and doing privileged?

Therefore, the epistemological stand taken depends on how reality is understood by the researched and by the researchers. This compounds the complexity of research. Local, situated reality, ways of doing and being may be incomprehensible and unintelligible to the distant researcher who understands knowledge to be 'objectifiable' and 'out there'. Arguably, such positivistic approaches disadvantage marginalised groups and local communities, as they do not fully recognise the lived experiences and subjective interpretations of local communities, or the 'baggage' that the Western tourism researcher may carry with them. Proponents of standpoint epistemology drawing upon Harding's early work placed women's experience at the heart of the research project. Women's and men's experiences are constituted by particular actions and practices. These are fundamental ways in which we live in the world and thus create ways of categorising, experiencing and valuing the world. In turn, the requirements of practice generate distinctive forms of knowledge. Epistemic attitudes thus derive from the 'doing'. Hartsock (1983) argued that there are typically women's ways of categorising, experiencing and valuing the world. Furthermore, Curtin (1997: 95) draws attention to the interconnection and relations of this locatedness in relation to the natural world and the movements that value and promote the sustainability of the natural world:

[F]eminist epistemology derives from women's lived experience, centred on the domains of inter-connectedness and affectual rationality. It emphasised holism and harmonious relationships with nature, which is why feminism has links with that other major social movement of our time, ecology. Eco-feminist standpoint values and organises the world through a revaluation of those practices that dominant discourses marginalise, practices that are pushed to the periphery of what has counted as important.

Epistemologies of ecological feminisms are fundamental to arguments which lead to the questioning of the notion of one-standpoint epistemology (a single Truth) and to an understanding of the complex nature of the connections between the environment (nature) and human nature. This human–nature connection was recognised and legitimised through events such as the Brundtland Report (World Commission on Environment and Development 1987) and the 1992 Earth Summit, which highlight a holistic approach to sustainable development (see Leslie and Muir 1997). Standpoint research and its various underlying epistemologies help us to understand and recognise the needs of marginal groups and their environmental contexts, and in so doing find new ways of encouraging sustainability in the interest of local groups. This is clearly of interest to the tourism researcher.

Ecofeminism, epistemology, standpoint research and tourism

Ecofeminist standpoint perspectives have considerable potential and application for tourism research. Since tourism is fundamentally about going to some other place, and frequently somewhere 'exotic', it is about place as well as people. For the tourist and their hosts, the natural environment is as significant as the cultural context. Eco-feminist epistemology is related to standpoint research in that not only does it make central the lived experience of marginalised people, but also it centralises the needs of the environment. These are dialectically interrelated and come together in praxis through ideas around 'sustainable' development in tourism (see Scheyvens 2000, and also Belsky's chapter on ecotourism in this book (Chapter 16)). Middleton and Hawkins (1998: 247) attempt to define the concept of sustainable tourism, highlighting the difficulties in measuring 'sustainability' since they suggest that the process of evolution means that the environment is constantly in a state of change. However, significantly they note that sustainability implies 'a state of equilibrium in which the activities of the human population co-exist in harmony with their natural, social and cultural environment'. Tourism researchers examining sustainable tourism might benefit from exploration of ecofeminist epistemologies.

Antecedents of ecofeminism

Sustainability and tourism research can be understood if we examine the ways in which thinking began to question taken-for-granted Western ideas about human beings' place in the world. Since the 1980s there has been a call by many environmentalists, feminists and proto-poststructuralists for a foundational re-conceptualisation of the woman/nature, man/culture binaries and non-human exploitation so prevalent in early Western thought. Current environmental feminism argues that this reconceptualisation is premised upon the acknowledgement and understanding of the significance of the connections between the oppression of women and other subalterns (oppressions as a consequence of race, class, sexuality, and so forth) and the exploitation of 'nature'. Consequently, given the link between oppressed peoples and nature, in order to challenge and change such webs of unequal power relations, a synthesis of feminist and ecological approaches is required (Nesmith and Radcliffe 1997; Warren 1990).

Environmental feminism(s) seek to draw attention not only to the material reality of nature as substance which is subjected to exploitation, but also to the symbolisation of 'nature' as a legitimator of various forms of oppression (Diamond and Orenstein 1990; Soper 1995). Drawing upon Ortner's (1974) work, which suggests that the predominant belief in all societies is that women, as a consequence of their biology, are closer to nature than men, environmental feminists highlighted the puissant links between 'nature' and women's and men's practices and ways of being, and the common oppression of both 'nature' and women. Nature (aligned with woman) is devalued compared with Culture (associated with man). Underpinning the ecofeminist standpoint is a recognition of this common oppression of both woman and nature and the concern to transform this double subjugation (Merchant 1990). Soper (2002) draws attention to the ways through which women have found themselves more closely identified with nature, and nature has itself been feminised by the sexual polarisation of the nature–culture divide.

This sexual polarisation took hold in Western societies during the Enlightenment period of the eighteenth century. The Enlightenment tradition emerged as a 'rational' counter to faith and superstition through the exercise of reason. The Enlightenment created an idea of humanity whose purpose was to uncover universal moral codes and realisation of self. The scientific paradigm emerged as a consequence of rational philosophy, which was to provide the foundation of objective and reliable universal knowledge. It spawned in the United Kingdom the humanistic project that promoted the legal entitlement of the 'Rights of Man'. Women were, however, excluded. 'Woman' as a category became defined as Other, and opposite to 'Man'. As men became more involved in public life, women were assigned to the home, to the private sphere, to birth and symbolically aligned with nature. Through the construction of sets of binaries, masculinity was equated with science, rationality, objectivity and culture, while femininity was equated with emotionality, subjectivity irrationality and nature. This dualism constituted opposites, and those most closely associated with the 'feminine' became devalued.[4] Thus was this double bind, and its role in reinforcing the sexual division of labour and its domestication of women, constituted. We see this from research into the division of labour within the tourism industry (Adkins 1992, 1995; Jordan 1997) and the considerable profits made from the sex industry in developing countries.

Standpoint epistemologies incorporated into tourism research and thinking could provide the philosophy and underlying research processes by which lived experiences, within local environments, can be sensitised to destructive factors, providing local, contextualised evidence and support for locally driven sustainable tourism development. Such a move would represent a shift from tourism research based in the traditional period to more of a fifth-moment approach.

Ecofeminism, difference and 'post'-standpoint research

Important for tourism research is an understanding and awareness of 'difference', since it is frequently the white, middle-class, able-bodied who are the researchers, and the black and/or indigenous people who are those 'being' researched. Ecological and post-colonial perspectives raised issues around difference and 'otherness' that began to inform standpoint epistemology (see Stanley and Wise 1990). Such imperatives as 'race' and diversity became central to poststructuralist standpoints (see Donald and Rattansi 1995; Kobayashi and Peake 1997).

Collins (1986) drew attention to and interrogates the dominance of whiteness in positioning Afro-American women as 'outsider' to society. Their life experiences are different from those who are not black and female. Thus, it is important for tourism researchers to acknowledge and so become aware of their particular identities in relation to those whom they research. As a white male researcher, you might begin to reflect upon your own whiteness and male embodiment and how this affects your understanding of your data. Similarly, white female researchers need to question their whiteness and female embodiment, and the unconscious influence these have on their actions and interpretations. (In Chapter 6, Swain too highlights the importance of acknowledging embodied characteristics and their impact on the research process.) Further, even when such reflexivity is embraced, the reflective tourism researcher should be aware that much anthropology that has informed some schools of tourism research has been charged with silencing these marginalised peoples (subalterns) in its pursuance of the unified Other (Spivak 1988). That is to say, difference has been seen not only as exotic but also simplistically, as a monoculture. Work of women of difference demonstrates a diversity of standpoint(s) (see Anzaldua 1987; Chow 1993; Smith 1997).

The work of ecofeminists such as Warren (1994, 1997) and Mies and Shiva (1993) in bringing together the voices of women of difference is highly relevant to the development of standpoint epistemology and to its potential relationship with tourism research.

For early standpoint theories/research, therefore, the focus on one standpoint became problematic and contentious. Attention was drawn to the importance of the fact that an emphasis upon lived experience might mean not only the lived experience of white middle-class heterosexual women, but more particularly the diversity of lived experience, based on social positionings and attributes such as class, 'race', sexuality, ethnicity, disability, and so forth. Arguably, tourism research tends only in a few exemplar cases (Kinnaird and Hall 1996; Aitchison 2001) to enter into critical interrogation highlighted through standpoint epistemological dilemmas. Pertinently, prevailing discourses in tourism show evidence of 'seeing' only through the 'spectacles' shaped by the characteristics of a non-disabled academic community. This point is made by Aitchison (2002: 27) in her analysis of leisure and tourism studies when she refers to tourism discourses:

> Indeed there is a sense in which both mobility and sight are deemed to be prerequisites for engaging with tourism as conventionally constructed by the leisure and tourism industry and leisure and tourism studies. This emphasis on mobility to sites and ability to see sights is evident in the research of many leisure and tourism scholars. Urry, for example, has focused on 'touring cultures' and the 'tourist gaze' signifying the importance of sight and mobility within tourism.

It is clear that tourism studies and tourism research can benefit from engaging with the sorts of debates initiated through and by standpoint epistemologies. To do so would require tourism studies and research to look through varied different standpoint 'spectacles' (that is, take 'post'-standpoint perspectives) at phenomena under examination. The tourism researcher examining the impact of tourism on particular destinations would benefit considerably from asking pertinent questions about the lived experiences of local marginalised or minority groups, their environment and how they interrelate.[5] This suggests that tourism researchers need to engage minority groups collaboratively in their research and to take a reflexive approach to their research, critically reflecting upon their own gender identities, ethnicity, nationality, and so forth.

Ecotourism and 'post'-standpoint research

The ecotourism industry provides a pertinent arena in which to demonstrate the appropriateness of standpoint research and the importance of centralising gender in such approaches. It is an industry that is projected to expand by some 25 per cent from 1996 to 2005 (Herliczek 1996). It is also Third World destinations to which tourists are largely attracted to enjoy nature and, on occasions, the local cultures. As Scheyvens (2000) points out, the label of 'ecotourism' has frequently been used as a marketing tool with little consideration for indigenous community involvement or the effects of tourist impact upon the environment or local culture. Even where these are taken into account, gender perspectives have frequently been ignored. Kinnaird and Hall (1996: 97) cited in Scheyvens (2000: 234) point out that:

> ... while some critics of mass, large-scale tourism development have advocated the pursuit of small-scale, 'sustainable', 'alternative', 'responsible' or 'appropriate' tourism which is locally controlled, sensitive to indigenous cultural and environmental characteristics and directly involves and benefits local population, gender considerations have yet to be placed centrally within such a debate.

Clearly, then, the tourism researcher informed by 'post'-standpoint epistemologies would be better placed to uncover the experiences of the marginalised, as they would be attuned to the need to recognise gender and other 'differences' in the collection and analysis of their data and in assessing how their work may impact upon the local population. As suggested earlier, indigenous women may have greater connections with nature and so concern for sustainable use of natural resources. Thus, uncovering their experiences and coming to understand local traditions could not only benefit the local community but also increase the sustainability of ecotourism.

Standpoint epistemology, ethnography and tourism research

Standpoint research requires the researcher to relinquish positivistic notions of 'objectivity' in the research process through adopting an interpretive stance. The researcher no longer becomes the adjudicator for competing worldviews but the interpreter speaking for and with the host community and its environment (Bauman 1987). Examples of 'good practice' in which the voices of the indigenous women have been heard in ecotourism initiatives are provided by Lama (1998), who describes mountain-based tourism, and Hemmati (1999), who discusses equal participation in sustainable tourism initiatives.

Questions remain for tourism research. For example, if objectivity is rejected, how then can research findings be considered 'valid', acquire legitimacy and be drawn upon to develop theory? Clearly, the notion of removing researcher 'bias', so well beloved of positivistic methodology, becomes unintelligible; rather, it becomes imperative to move towards a fifth-moment approach and make visible the researcher's values, ideological and philosophical underpinnings and their interpretations. There may also be a role for Decrop's measures of trustworthiness. How then does standpoint research manage these developments and respond to such questions as what 'ways of finding things out' can address such issues as inter-subjectivity, the invisibility of women's and Others' diversity of lived experiences and the unequal power relations in research. How can we utilise these developments in tourism research? One important methodology which purports to uncover the lived experience is that of ethnography (see Denzin 1997).

Around the late 1980s, feminists undertaking ethnographic research were in something of a dilemma. Through 'doing' research, these ethnographers were living the contradictions between interpretive ethnography and feminist standpoint theory. The notion of one feminist standpoint epistemology – that is, one Truth claim to represent all women – contradicted the developing notions of the partial nature of knowledge explored earlier and the experiences of ethnographers informed by feminism and critical theory in the field (see Humberstone 1997: 202–205). A number of texts were published by women engaged in ethnographic and anthropological research exploring such questions as 'Can there be a feminist ethnography?' (see Abu-Lughod 1990).

Strathern (1987), looking at this 'awkward relationship between anthropology and feminism', argued that neither can challenge the other but, rather, their perspectives are central to their practices so they can only 'mock' each other. Stacey (1988: 22) concluded that ethnography is compatible with the feminisms that had begun to emerge from the critiques of the traditional feminist standpoint. She argued that both ethnography and feminism focus on lived experiences, the experiential. Ethnography was concerned with the day-to-day nature of human practice and agency. And important for both is the centrality of the researcher as the research 'instrument' who must draw upon their empathy and concerns in the research process. Both methodological approaches, she argued, provide for respect for and power to the participants of the research. This is imperative for the tourism researcher adopting post-standpoint approaches, where the perspectives, experiences and values of local culture are recognised, and Otherness is questioned (see Bell *et al.* 1993).

Intellectual biography/reflexivity

Stanley (1990b: 209), interested in developing feminist theory, critically engaged with standpoint theorists and identified the importance of the researcher in creating/recreating knowledge, suggesting that this inter-subjectivity and integration could be addressed through detailed researcher accounting during the research process. Like Strathern and the ethnographers, she centres experience, and argued that '[f]eminist theory would be directly derived from "experience" whether this is experience of survey or interview or an ethnographic research project, or whether it is experience of reading or analysing historical or contemporary documents'. She continues that through an 'intellectual autobiography', the credibility of the research is made visible. 'Intellectual autobiography' may provide for detailed accounting of the research process through which the reader can judge the credibility and authenticity of the research. This has commonality with the notion of reflexivity, much considered in the field(s) of ethnography and discussed in some depth in Hall's chapter. Marcus (1998) points to the equivalence of autobiographic accounting with reflexivity in critical research, highlighting the deconstruction from within, the principle of contextuality and partiality of truth claims.

> The feminist version of the highly valued, powerfully evoked base-line form of subjectivity, experiential reflexivity has more recently been discussed and theorised as the practice of positioning, which is not that different from the politics of location that gives shape to reflexivity in critical ethnography.... Positioning (of standpoint epistemologies) as a practice in feminism is most committed to the situatedness and partiality of all claims to knowledge, and hence contests the sort of essentialist rhetoric and binarism (male/female, culture/nature) as a cognitive mode that has so biased towards rigidity and inflexibility questions of gender or 'otherness' in language use.
>
> (Marcus 1998: 401)

Feminist standpoint epistemologies have much in common with recent ethnographic work which encourages careful and public interrogation of the researcher's background, assumptions and values. It is through this and the personal account of the research experience that the social construction of the research process can be explored. Through this process, researcher–research participant relations can be highlighted and issues around power in the research may be uncovered. The cultural practices that create the situations of the research are made visible. Researcher–participant relations are recorded in much literature on ethnographic research in the 'field', where feminist standpoint accounts emphasise the social construction of the interactions and events. There is little evidence to suggest that tourism has addressed, let alone embraced, these ideas, uncertainty and practices. However, Scheyvens (2000) points to the possibilities of such approaches in her analysis of ecotourism and women's experiences in Third World communities.

'Post'-standpoint research, credibility and tourism research

Denzin (1997: 55), following Smith (1987), draws attention to what may be conceived of as 'post'-standpoint approaches which are organised around 'the experience of women, of persons of colour, postcolonial writers, gays and lesbians, and persons who have been excluded [from] the dominant discourses in the human disciplines' (Smith 1993) and which challenge the very 'notion of a single standpoint from which a final overriding version of the world can be written' (Smith 1993: 184). 'Post'-standpoint epistemologies, importantly, mix together various theoretical and methodological perspectives. In doing this, Smith moves beyond widely accepted understandings of the importance of inter-subjectivity in qualitative work such as ethnography, participant observation and interview towards the fifth moment of research.

This rejection of objectivity and emphasis upon inter-subjectivity in 'post'-standpoint epistemologies make them vulnerable to positivists' criticisms around 'validity' and credibility. Further, as Denzin (1997: 63) points out, standpoint and critical research, with their specific interests which foreground praxis (theory into empowering practice), frequently 'left unclear' the methodological aspect of the research process. The notion of validity in legitimating research speaks to assumptions that have largely been rejected by 'post'-standpoint epistemologies. 'Post'-standpoint researchers' concern is for making aware the reflexive awareness or 'practical reasoning' of participants rather than any attempt at verification of 'facts' (see Humberstone 1997: 207–208) – the questions raised around positivistic criteria such as What is validity? and Valid in whose terms? For the tourism researcher this means Whose experiences are legitimate? Those of the host community? Those of the tourist? Those of the researcher? Further, as Lather (1993) points out, a call to validity in research is partial, and is 'the researcher's mask of authority, which allows a particular regime of truth within a particular text (and community of scholars) to work its way on the world and the reader' (Lincoln and Denzin 1998: 415). The set of assumptions about the research process that are signalled by the notions of trustworthiness, credibility and authenticity are different from those signalled by the traditional criteria inherent in constructs such as validity or reliability. These issues are discussed in further detail by Decrop in his chapter on trustworthiness (Chapter 9). Critical and 'post'-standpoint researchers recognise credibility in texts that are plausible to those who constructed them, the participants and the researcher(s); the host community and the tourism researcher. If tourism research is to address the challenges posed by Soper (2003) ('Can we find ways of living rich, complex, non-repetitive lives without social injustice and without environmental damage?'), then tourism researchers need to engage with the dilemmas both in practice and theory in construction, as identified in this chapter.

Summary

This chapter has looked at the development of 'post'-standpoint research and the importance of the contradictions and dilemmas of its associated epistemologies and methodologies to research in tourism. Its key points are as follows:

- Standpoint research developed through criticisms of taken-for-granted assumptions about the construction of knowledge and ways of knowing.
- Standpoint research centres the lived experience of people within their situations and contexts.
- Gender is central to standpoint research.
- Tourism research, through taking standpoint perspectives, can come to better understand and recognise the needs of host communities and their environmental contexts.
- Ecofeminist standpoints developed as a criticism of Western binary thinking and the exploitation of human and non-human nature.
- Sensitivity to difference such as class, race, disability, and so forth is central to 'post'-standpoint research.
- Ecotourism highlights the significance of applying 'post'-standpoint perspectives to tourism research to support sustainability and local communities.
- Ethnography is a relevant methodological approach for tourism research informed by 'post'-standpoint perspectives.
- Reflexivity is necessary in tourism research informed by 'post'-standpoint research.
- 'Post'-standpoint tourism research replaces positivistic research criteria such as 'validity' with a different set of assumptions: trustworthiness, credibility and authenticity.

Questions

1 Discuss your understanding of the concepts ideology and dualistic thinking. How can dualistic thinking contribute to the oppression of disadvantaged groups and the environment?
2 Examine literature on tourism destinations in three 'lifestyle' magazines.

   - How do they represent the host communities?
   - How is the environment represented?
   - How is diversity represented?

3 Identify an 'exotic' destination which you have decided to research. What theoretical and practical consideration do you need to take into account if you adopt (a) a standpoint approach; (b) a post-standpoint approach?

Notes

1 'Ideologies are sets of ideas, assumptions and images by which people make sense of society, which give a clear sense of social identity, and which serve in

some way to legitimise relations of power in society' (McLennan 1995: 126). See also Humberstone (2003), who discusses ideologies associated with masculinities and femininities.

2 It is worth noting Harding's (1993a: 53–54) comments regarding the intellectual preconditions of feminist standpoint theory, which, she asserts, 'is conventionally traced to Hegel's reflections on what can be known about the master/slave relationship from the standpoint of the slave's life versus that of the master's life and to the way Marx, Engels, and Lukacs subsequently developed this insight into the "standpoint of the proletariat" from which have been produced Marxist theories of how class society operates. In the 1970s, several feminist thinkers began to reflect upon how the structural relationship between women and men had consequences for the production of knowledge.'

3 But see Goodson (2003), who does address some of these issues.

4 The cultural ecofeminist standpoint is represented by work published by Mies and Shiva (1993). See also Diamond and Orensten (1990). Social ecofeminists' work includes that of Merchant (1990), King (1990), Stanley (1990a) and Warren (1994). Spirituality in relation to an ecofeminist standpoint is discussed by Daly (1973). See also Humberstone (2003).

5 This type of approach has highlighted the effects of globalised tourism upon the First Nation peoples of North America, the Aboriginals of Australia and the Sami peoples of Lapland. See Pedersen (2003), whose research highlights the effects of globalisation on young Sami Norwegians.

References

Abu-luighod, L. (1990) 'Can there be a feminist ethnography?', *Women and Performance: Journal of Feminist Theory*, 5 (1): 7–27.

Adkins, L. (1992) 'Sexual work and employment of women in the service industries'. In M. Savage and A. Witz (eds) *Gender and Bureaucracy*. Oxford: Blackwell/Sociological Review.

Adkins, L. (1995) *Gendered Work: Sexuality, Family and the Labour Market*. Bristol: Open University Press.

Aitchison, C. (2001) 'Theorising other discourses of tourism, gender and culture', *Tourist Studies*, 1 (2): 133–147.

Aitchison, C. (2002) 'Leisure studies: discourses of knowledge, power and theoretical "crisis"'. In L. Lawrence and S. Parker (eds) *Leisure Studies: Trends in Theory and Research*. Leisure Studies Publication 77. Eastbourne: Leisure Studies Association.

Anzaldua, G. (1987) *Borderlands/La Frontera*. San Francisco, CA: Aunt Lute.

Barnes, T. and Gregory, D. (eds) (1997) *Reading Human Geography: The Poetics and Politics of Inquiry*. London: Arnold.

Bauman, Z. (1987) *Legislators and Interpreters: On Modernity, Post-modernity and Intellectuals*. Cambridge: Polity Press.

Bell, D., Caplan, P. and Karin, W. J. (1993) *Gendered Fields: Women, Men and Ethnography*. London: Routledge.

Chow, R. (1993) *Writing Diaspora: Tactics of Intervention in Contemporary Cultural Studies*. Bloomington, IN: Indiana University Press.

Cohen, S. and Taylor, L. (1976) *Escape Attempts*. Harmondsworth: Penguin.

Collins, P. (1985) *Changing Order: Replication and Induction in Scientific Practice*. London: Sage.

Collins, P. H. (1986) 'Learning from the outsider within: the sociological significance of black feminist thought', *Social Problems*, 33: 4–32.

Curtin, D. (1997) 'Women's knowledge as expert knowledge: Indian women and ecodevelopment'. In K. J. Warren (ed.) *Ecofeminism: Women, Culture, Nature.* Bloomington, IN: Indiana University Press.

Daly, M. (1973) *Beyond God the Father: Towards a Philosophy of Women's Liberation.* Boston, MA: Beacon Press.

Deem, R. (1999) 'How do we get out of the ghetto? Strategies for research on gender and leisure for the twenty-first century', *Leisure Studies*, 18: 161–177.

Denzin, N. K. (1997) *Interpretive Ethnography.* London: Sage.

Diamond, I. and Orenstein, G. (1990) *Reweaving the World: The Emergence of Ecofeminism.* San Francisco, CA: Sierra Club Books.

Donald, J. and Rattansi, A. (eds) (1995) *'Race', Culture and Difference.* London: Sage.

Duffy, R. (2002) *A Trip Too Far: Ecotourism, Politics and Exploitation.* London: Earthscan.

Foucault, M. (1980) 'Truth and power'. In *Power? Knowledge: Selected Interviews and Other Writings, 1972–1977.* Brighton: Harvester.

Goodson, L. (2003) 'Tourism impacts, community participation and gender: an exploratory study of residents' perceptions in the city of Bath'. Unpublished PhD thesis, Buckinghamshire Chilterns University College, Brunel University.

Hall, M. and Ryan, C. (2001) *Sex Tourism: Marginal Peoples and Liminalities.* London: Routledge.

Hammersley, M. and Atkinson, P. (1983) *Ethnography: Principles in Practice.* London: Tavistock.

Haraway, D. (1991) 'Situated knowledges: the science question in feminism and the privilege of partial perspective'. In *Simians, Cyborgs and Women: The Reinvention of Nature.* London: Free Association Books.

Harding S. (1987) *Feminism and Methodology.* Milton Keynes: Open University Press.

Harding, S. (1993a) 'Rethinking standpoint epistemology: what is "strong objectivity"?'. In L. Alcoff and E. Potter (eds) *Feminist Epistemologies.* London: Routledge.

Harding, S. (ed.) (1993b) *The 'Racial' Economy of Science: Towards a Democratic Future.* Bloomington, IN: Indiana University Press.

Hartsock, N. (1983) 'The feminist standpoint: developing the ground for a specifically feminist historical realism'. In S. Harding and M. Hintikki (eds) *Discovering Reality.* London: D. Reidel.

Hemmati, M. (ed.) (1999) *Gender and Tourism: Women's Employment and Participation in Tourism.* Report for the United Nations Commission on Sustainable Development, 7th Session, April 1999, New York. London: United Nations Environment and Development Committee of the United Kingdom.

Humberstone, B. (1997) 'Challenging dominant ideologies in the research process'. In G. Clarke and B. Humberstone (eds) *Researching Women and Sport.* London: Macmillan.

Humberstone, B. (2002) 'Femininity, masculinity and difference, what's wrong with a sarong?'. In A. Laker (ed.) *The Sociology of Sport and PE: An Introductory Reader.* London: Falmer Press.

Humberstone, B. (2003) 'Gender transgressions and contested natures'. In K. Pedersen, S. Mathisen and A. Viken (eds) *Nature and Identity.* Kristiansand: Høyskoleforlaget/Norwegian Academic Press.

Jordan, F. (1997) 'An occupational hazard? Sex segregation in tourism employment', *Tourism Management*, 18 (8): 525–534.

King, Y. (1990) 'Healing the wounds'. In I. Diamond and G. Orenstein (eds) *Reweaving the World: The Emergence of Ecofeminism.* San Francisco, CA: Sierra Club Books.

Kinnaird, V. and Hall, D. (1996) 'Understanding tourism processes: a gender-aware framework', *Tourism Management*, 17 (2): 95–102.

Kobayashi, A. and Peake, L. (1997) 'Unnatural discourses: "race" and gender in geography'. In T. Barnes and D. Gregory (eds) *Reading Human Geography: The Poetics and Politics of Inquiry.* London: Arnold.

Lama, W. B. (1998) 'CBMT: Women and CBMT in the Himalaya'. Mountain-Forum Discussion Archive article from 'Community-Based Mountain Tourism' conference 5 August 1998. http.//mtnforum.org/emaildiscuss/cbmt/cbmt4/050898d.htm.

Lather, P. (1993) 'Fertile obsession: validity after poststructuralism', *Sociological Quarterly*, 34: 673–694.

Leslie, D. and Muir, F. (1997) *Local Agenda 21, Local Authorities and Tourism: A United Kingdom Perspective.* Glasgow: Glasgow Caledonian University.

Lincoln, Y. S. and Denzin, N. K. (1998) 'The fifth moment'. In N. K. Denzin and Y. S. Lincoln (eds) *The Landscape of Qualitative Research: Theories and Issues.* London: Sage.

McLennan, G. (1995) 'The power of ideology in Block 4'. *Power and Politics.* Society and Social Science: A Foundation Course. Milton Keynes: The Open University.

Marcus, G. E. (1998) 'What comes (just) after "post"? the case of ethnography'. In N. K. Denzin and Y. S. Lincoln (eds) *The Landscape of Qualitative Research: Theories and Issues.* London: Sage.

Merchant, C. (1990) 'Ecofeminism and feminist theory'. In I. Diamond and G. Orenstein (eds) *Reweaving the World: The Emergence of Ecofeminism.* San Francisco, CA: Sierra Club Books.

Middleton, V. and Hawkins, R. (1998) *Sustainable Tourism: A Marketing Perspective.* Oxford: Butterworth-Heinemann.

Mies, M. and Shiva, V. (1993) *Ecofeminism.* London: Zed Books. (Halifax, Nova Scotia: Fernwood Publications.)

Nesmith, C. and Radcliffe, S. (1997) '(Re)mapping Mother Earth: a geographical perspective on environmental feminisms'. In T. Barnes and D. Gregory (eds) *Reading Human Geography: The Poetics and Politics of Inquiry.* London: Arnold.

Ortner, S. (1974) 'Is female to male as nature is to culture?'. In M. Rosaldo and L. Lamphere (eds) *Women, Culture and Society.* Stanford, CA: Stanford University Press.

Pedersen, K. (2003) 'Discourses on nature and ethnicity among youth in northern Norway'. In B. Humberstone, H. Brown and K. Richards (eds) *Whose Journeys? The Outdoors and Adventure as Social and Cultural Phenomena,* Penrith: Institute for Outdoor Learning.

Richardson, L. L. (1991) 'Value constituting practices, rhetoric, and metaphor in sociology', *Current Perspectives in Social Theory*, 11: 1–15.

Scheyvens, R. (2000) 'Promoting women's empowerment through involvement in ecotourism: experiences from the Third World', *Journal of Sustainable Tourism*, 8 (3): 232–249.

Skeggs, B. (1994) 'Situating the production of feminist ethnography'. In M. Maynard and J. Purvis (eds) *Researching Women's Lives from a Feminist Perspective*. London: Taylor & Francis.

Smith, A. (1997) 'Ecofeminism through an anticolonial framework'. In K. J. Warren (ed.) *Ecofeminism: Women, Culture, Nature*. Bloomington, IN: Indiana University Press.

Smith, D. E. (1987) *The Everyday World as Problematic: A Feminist Sociology*. Evanston, IL: Northwestern University Press.

Smith, D. E. (1992) 'Sociology from women's perspective: a reaffirmation', *Sociological Theory*, 10: 88–97.

Smith, D. E. (1993) 'High noon in textland: a critique of Clough', *Sociological Quarterly*, 34: 183–192.

Smith, V. (1994) *Hosts and Guests: The Anthropology of Tourism*, 2nd edn. Oxford: Blackwell.

Soper, K. (1995) *What Is Nature? Culture, Politics and the Non-human*. Oxford: Blackwell.

Soper, K. (2003) 'Naturalised identities and the identification of nature: a realist and humanist perspective'. In K. Pedersen, S. Mathison and A. Viken (eds) *Nature and Identity*. Kristiansand: Høyskoleforlagat/Norwegian Academic Press.

Spivak, G. C. (1988) 'Can the subaltern speak?' In C. Nelson and L. Grossberg (eds) *Marxism and the Interpretation of Cultures*. Urbana, IL: University of Illinois Press.

Stacey, J. (1988) 'Can there be a feminist ethnography?', *Women's Studies International Forum*, 11: 21–27.

Stanley, L. (ed.) (1990a) *Feminist Praxis: Research, Theory and Epistemology in Feminist Sociology*. London: Routledge.

Stanley, L. (1990b) 'Feminist autobiography and feminist epistemology'. In J. Aaron and S. Walby (eds) *Out of the Margins: Women's Studies in the Nineties*. Lewes: Falmer Press.

Stanley, L. and Wise, S. (1990) 'Method, methodology and epistemology in feminist research process'. In L. Stanley (ed.) *Feminist Praxis: Research, Theory and Epistemology in Feminist Sociology*. London: Routledge.

Strathern, M. (1987) 'An awkward relationship: the case of feminism and anthropology', *Signs*, 12: 276–292.

Ward, S. (1997) 'Being objective about objectivity: the ironies of standpoint epistemological critiques of science', *Sociology*, 31 (4): 773–800.

Warren, K. J. (1990) 'The power and promise of ecological feminism', *Environmental Ethics*, 12: 125–146.

Warren, K. J. (ed.) (1994) *Ecological Feminism*. London: Routledge.

Warren, K. J. (ed.) (1997) *Ecofeminism: Women, Culture, Nature*. Bloomington, IN: Indiana University Press.

Women and Geography Study Group (1997) *Feminist Geographies: Explorations in Diversity and Difference*. Harlow: Addison Wesley Longman.

World Commission on Environment and Development (1987) *Our Common Future* (the Brundtland Report). Oxford: Oxford University Press.

8 Reflexivity and tourism research
Situating myself and/with others

Michael Hall

Aims of the chapter

- To identify issues relating to reflexivity and undertaking tourism research.
- To identify how 'the rules of the game' may affect tourism research.
- To discuss how so-called private life may influence the public conduct of tourism research.

Introduction

The growth of a high degree of 'reflexivity' or self-consciousness among the populations of contemporary industrial societies tends to be regarded as one of the hallmarks of postmodernity (Gergen 1991; Lash and Urry 1994). By this is meant that modern societies have reached a position where not only are they forced to reflect on themselves but also they have the capability of reflecting *back* on themselves. For Giddens (1990, 1991), this has meant the capacity for greater personal, individual self-reflexivity, while for Beck (1992) it is societal self-reflexivity, through social monitoring and social movements (Beck *et al.* 1994). For researchers, this means that via the principle of reflexive explanation, 'each of us as members of society are able to participate via certain roles and come to reflect on the products of that participation' (Evans 1988: 2000). However, whether the condition of modern societies is branded as reflexive modernity or postmodernity, the vagaries of the postmodern condition are virtually unavoidable in contemporary examinations of social science and the worlds from which social research are formed, including our own.

Ironically, this is itself a product of the nature of postmodernity, which 'does not offer itself as a theory to be tested and assessed in the usual fashion. In a peculiar way, post-modernity has to be assessed not from the detached viewpoint of the external observer but from within, from inside its own discourse' (Kumar 1995: 184). Arguably, the critical culture of postmodernity has established new spaces, opportunities and languages of debate, with such debates becoming the proof of postmodernity's own existence. 'The battle of the books is also an ontic battle against death' (Hassan 1985: 120).

Nevertheless, the contemporary economic, cultural and political sphere of industrialised economies has undoubtedly undergone change in recent years as old 'certainties' have seemingly given way to uncertainty, fluidity and flux. Yet despite an increase in reflexivity in an era of 'disorganised capitalism' (Lash and Urry 1987), what the present mainly displays is the constant flux of capitalism, and its tendency to transform itself over space and time – particularly through the growth of individualisation and privatisation, which are redrawing the boundaries between state and society, the public and the private spheres, society and the individual (Kumar 1995), and, given the contemporary debate over migration and identity in many industrialised societies, the accepted and the unaccepted in a society. Indeed, the increase in reflexivity, or self-consciousness and awareness of social processes, that many have seen as characterising contemporary societies and subcultures, such as those of the academy, might be regarded as an expression of this heightened individualism (Kumar 1995). Nevertheless, while there does appear to be greater individualism, the growth of reflexivity also creates new possibilities for social relations in a wide variety of spheres, 'for intimate relations, for friendship, for work relations, for leisure and for consumption' (Lash and Urry 1994: 31). Significantly, recognition of increased reflexivity is also related to increased academic interest in the development of social capital, networks and collaboration in wider forms of planning (Healey 1997) as well as in tourism planning in particular (Hall 2000), because of its emphasis on participation, observation and communication (Evans 1988).

All this is not to say that I am an unabashed postmodernist – far from it. Instead, I would agree with Nederveen Pieterse (1992: 26) that postmodernity means that modernity can now be seen 'in the rear-view mirror'. Or, as Calinescu (1987: 278) put it, postmodernity is not 'a new name for a new "reality", or "mental structure", or "world view", but a perspective from which one can ask certain questions about modernity in its general incarnations'. Postmodernism therefore expresses the crisis of modernism – a crisis of our truths, our values, and our beliefs. It is a crisis that owes to reflexivity its origin, its necessity and its force (Lawson 1985), but it does not mean the end of modernity, or even of modernism. Postmodernity is modernity recollected, 'if not in tranquillity at least at the end of the working day' (Kumar 1995: 140).

> [Postmodernity] supplies a new and external vantage point, from which some aspects of the world came into being in the aftermath of the Enlightenment and the capitalist revolution (aspects not visible, or allotted secondary importance, when observed from inside the unfinished process), acquire saliency and can be turned into pivotal issues of the discourse . . .
>
> Postmodernity may be interpreted as fully developed modernity

taking a full measure of the anticipated consequences of its historical work . . . modernity conscious of its true nature – modernity for itself.

(Bauman 1992: 102–103, 187)

For Bauman, as for myself, this means that we are now more aware of the limits of modernity as well as the limits of ourselves. For intellectuals this may mean the recognition that the old authorities of the past have now passed. There are no absolute rules and standards for society. No such universal principles exist even though reference may still be made to them (e.g. the Universal Declaration of Human Rights). How can one hold to a belief in a single, objective account in the face of a multiplicity of viewpoints? Claims to truth and light are more reasonably viewed as the constructions of communities with particular interests, values and ways of life (Gergen 1991; Hall and Jenkins 1995). To this end, participation is a refinement of the methods used to reflexively understand and interpret everyday life. As Evans (1988: 201) noted, 'To this end, the measure of adequacy of the articulation of the social phenomena researched is the success of the participation by the researcher in the "collective contract" . . . of the everyday life being studied.' The core of the collective contract is participation and communicability (Evans 1988), an observation parallel to that articulated within the notion of tourism policy as theory, and tourism policy as argument, by Hall and Jenkins (1995) and Hall (2000). This therefore means that intellectuals should accept a more modest role, that of interpreters and brokers of civil societies and cultures. Idealistically, this should provide individuals with the ability and responsibility to shape their own futures. 'The postmodern state of mind is the radical . . . victory of modern (that is, inherently critical, restless, unsatisfied, insatiable) culture over the modern society it aimed to improve through throwing it wide open to see its own potential' (Bauman 1992: viii, 188). In reality it has not, though I subscribe to David Harvey's (2000) observation that there are still 'spaces of hope' (see Hall 1988).

As we face up to a world of uncertainty and risk, the possibility of being quite undone by the consequences of our own actions weighs heavily upon us, often making us prefer 'those ills we have than flying to others that we know not of'. But Hamlet, beset by angst and doubt and unable to act, brought disaster upon himself and upon his land by the mere fact of his inaction. It is on this point that we need to mark well the lessons of capitalist historical geography. For that historical geography was created through innumerable forms of speculative action, by a preparedness to take risks and be undone by them. While we labourers (and philosophical under-labourers) for good reasons 'lack the courage of our minds', the capitalists have rarely lacked the courage of theirs (Harvey 2000: 254).

The above discussion, I would argue, is critical for examining reflexivity as it positions the subject not only in the wider context of social science, and hence tourism research, but also in any identification of self, which is clearly integral to reflexive approaches to research. Accounts of any discipline and of research within that field of study are *situated*. That is, 'they depend on the point of view of the author, which in turn reflects how he/she is positioned intellectually, politically and socially' (Barnes and Sheppard 2000: 6). However, how often does one read research which explicitly recognises its situatedness in tourism? Despite the postmodern recognition of the absence of absolutes, this does not seem to have been widely translated into the representation and reading of tourism research and scholarship. Why?

To understand tourism research and the significance of reflexivity, we need to understand the structures that surround tourism as an area of study and also ourselves (see also Tribe's chapter on the epistemology of tourism (Chapter 3) for a discussion of the nature of tourism studies). This is in itself highly reflexive. In order to do this, the remainder of the chapter will first present a model of the structural framework within which tourism studies operates. The key elements of the model will then be discussed in the light of the author's own experiences and observations. Following this, the chapter will then go on to discuss the challenges of reflexivity in the wider tourism research context.

Looking in a rear-view mirror

The study of tourism and recreation does not occur in isolation from wider trends in the social sciences and academic discourse, or of the society of which we are a part. Tourism academics are 'a society within a society'; academic life 'is not a closed system but rather is open to the influences and commands of the wider society which encompasses it' (Johnston 1991: 1). The study of the development and history of a discipline, and the how, where and why of what we study, 'is not simply a chronology of its successes. It is an investigation of the sociology of a community, of its debates, deliberations and decisions as well as its findings' (Johnston 1991: 11).

'The contents of a discipline at any one time and place reflects the response of the individuals involved to external circumstances and influences, within the context of their intellectual socialization' (Johnston 1983: 4). Grano (1981) developed a model of external influences and internal change in geography that provides a valuable framework within which to examine the field of tourism studies (Figure 8.1). The figure is divided into three interrelated areas:

- knowledge – the content of tourism studies;
- action – tourism research within the context of research praxis; and
- culture – academics and students within the context of the research community and the wider society.

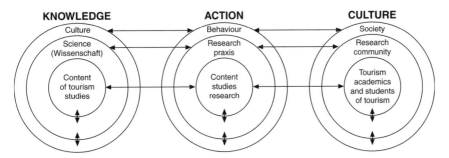

Figure 8.1 The context of academic tourism knowledge, action and culture.

Source: Hall and Ryan 1999: 9. After Grano 1981.

Tourist academics are a subcommunity of the social science community within the wider community of academics, scientists and intellectuals, which is itself a subset of wider society. That society has a culture, including a scientific/academic subculture within which the content of tourism is defined. Action is predicated on the structure of society and its knowledge base: research praxis is part of that programme of action, and includes tourism research. The community of tourism academics is therefore an 'institutionalising social group' (Grano 1981: 26), a context within which individual tourism academics are socialised and which defines the internal goals of their subject area in the context of the external structures within which they operate (after Johnston 1991). The content of a subject area or discipline must, in turn, be linked to its milieu, 'so that disciplinary changes (revolutionary or not) should be associated with significant events in the milieu' (Johnston 1991: 277) such as changes in how universities and research are funded. Similarly, Stoddart (1981: 1), in his review of the history of geography, stated, 'both the ideas and the structure of the subject have developed in response to complex social, economic, ideological and intellectual stimuli'. I would argue that Johnston's and Stoddart's observations on geography also apply to the field of tourism studies. However, unlike in the case of geography and geographers, an intellectual history of tourism studies or sociology of tourism knowledge is still to be written. Yet Grano's model provides a useful first step in understanding some of the elements of what it would look like.

Knowledge, action and culture

Why is the content of tourism studies as it is? In one sense, the disciplinary battleground is often mapped out at the beginning of textbooks, where key definitions, including that of tourism, are identified and laid out. Tourism is full of such boundary-marking. The very nature of the subject matter precludes itself from definitional contestation. Yet such issues are not only matters of social scientific debate, but also ways of mapping out intellectual terrain and resources. For example, by defining a subject matter you also identify what is inside and what remains outside. This becomes important for deciding what is the proper domain of a textbook or journal as well as of academic departments, and those who are not defined in it may not be able to publish, gain resources such as research grants or gain appointments.

Furthermore, the domain of tourism knowledge also includes how such knowledge was itself gained in terms of the acceptability of research methods and, hence, research results and their writing up. Think of a number of tourism journals. Reflect on their style. To what extent does the delineation of journal style affect the ability of manuscripts to be accepted? In this chapter I am using 'I' a number of times, as I believe that it better fits a discussion on reflexivity in tourism. Yet think of how few times such an expression of the first person is included in academic writing as expressed in the majority of tourism journals and texts. Indeed, academic writing in the third person conveys an impression of objectivity and scientific rationality which is almost the antithesis of the realisations of reflexive modernity. If one were to submit an article written solely in the first person to most academic tourism journals, the likelihood is that it would not be accepted or that major modifications would have to occur before acceptance. Under the rubric of convention and style, academic institutions and the culture of academia have therefore greatly influenced what is acceptable or unacceptable in being represented as tourism knowledge.

But we can go further. Arguably, some tourism journals also find certain research methods or approaches more acceptable than others. This may be because of conscious editorial policy or it may have developed over the years as a certain journal style developed – for example, one that favours more quantitative approaches as evidence providing for knowledge assertions – or it may be because of how editorial boards are selected and the particular breadth of views that may be provided for. Nevertheless, such considerations are important, as academics work in an environment in which the reward system is determined in great part by the publications one produces, which may therefore influence not only the selection of journal to which research articles are sent but even the choice of research method. For example, what I would consider one of the most important articles on tourism, and tourism geography in particular, a paper by Stephen Britton (1991) ('Tourism, capital and place: towards a critical geography of tourism'), was posthumously published in *Environment and Planning D: Society and Space*. However, before he died, I had the chance to ask him why he had selected that journal and not *Annals of Tourism Research*, in which he had previously published. He responded that he 'didn't believe' that it would be accepted there. Now in some ways it does not make a difference whether he was right or wrong in this assertion; the important thing was that it reflected a perception regarding the matching of what was being produced and where it would be accepted, and this influenced both the writing up of the ideas behind the article and where it was published.

Yet perhaps we could question further still the role of the gatekeepers of a field of study: the publishers, editors, editorial boards and reviewers as well as those who rank research or who sit in judgement on whether to fund research. These people have enormous effect in determining what is acceptable or not in the publication of research results and in research overall. And in the case of some of the Research Assessment Exercises (RAE) or Performance-Based Research Funding (PBRF) exercises which go on around the world, many of these panels will have little representation from the tourism research field, with the vast majority of panel members having no idea whatsoever about tourism as an area of academic study.

In tourism there is very little discussion of the role that such gate-keepers (of which I would be counted as one) have in determining the scope and direction of tourism knowledge. For example, several years ago I was a member of a panel of reviewers for an article that was submitted to one of the better-known tourism journals. The three reviewers all recommended that the article not be published (I was perhaps the most positive and recommended that it be published as a research note). However, a little over a year later the article was published in full. Now I recognise the rights of an editor to execute their own judgement as to the merits of a manuscript, yet the situation still struck me as somewhat odd. Indeed, I was later told by the author of the article, who had not realised that I had participated in the double-blind review, that upon receiving the referees' reports he had contacted the editor of the journal, of which he was a member of the editorial board, to request that the manuscript be published because he and others, who were not part of the review process but who were on the editorial board, believed it had merit. Again, to use this example is not to say that the article was unworthy of being published. Instead, it is to suggest that the field of tourism studies is partly influenced by the relationships that exist within the research community, rather than depending solely on so-called objective academic merit. Yet the rules of the academic game, including application of a RAE, are increasingly based upon the notion that it is a rational merit-based system in which journal publications are produced totally upon merit!

Similar issues arise with respect to the review of research grants. Historically, tourism has been seen by many social scientists as not a serious area of research. To what extent might this have influenced perceptions of applications for research grant funding and the methods which have been adopted? In several situations, including when I was sitting on the research committee of the University of Canberra, it was often noticeable how difficult it was for colleagues to get funding to travel to study tourists or to undertake comparative research at an international level – with one possible interpretation being that applicants were thought to be using their subject matter as an excuse to be a tourist. At a personal level I would note that, given that I study tourism and tourists, if I want to maximise my leisure space I would normally spend it at home in the garden or in the company of friends rather than travelling, so that I do not suddenly find myself reflexively analysing what's around me! Furthermore, one of the other problems that we tourism researchers often have is that because most of our colleagues are privileged enough to travel on holidays, they often believe that they understand tourism.

Yet even if research funding bodies are made up of tourism researchers, it does not necessarily mean that proposals may be accepted. Again, issues arise of who is on the committee, what their values and interests are, and to what extent applications meet the expectations of the panel. Indeed, in many industrialised countries, such as Australia and New Zealand, what is funded may be substantially dependent on government and industry objectives. In that kind of situation, to what extent might funding be provided for research projects that are likely to produce results that may be critical of government or business policy? Instead, research issues will often be kept within relatively safe boundaries. Here, researchers may face, if they were to think about it, substantial ethical dilemmas about what to study and comment on. Yet given that position confirmation, continued tenure and promotion are often dependent on the attraction of research funding and publication in the 'right journals', little public criticism is usually found of directed research – otherwise a researcher might be biting the hand that feeds. Indeed, increasingly the criteria for gaining professorships seem more geared for success in attracting external monies (and administration) than being a site of trusted academic authority. As Robertson (1999) stated:

> Above all, professors must be steadfast in their defence of academic freedom. This is not optional. The professoriate acts as the guardian of the ethical core of the university. An ethically united professoriate protects society's capacity to call down reliable knowledge for the common good.

Moreover, comments might affect not only the researcher but also those associated with him or her, whether as colleagues or as students. In such circumstances, what would be the most appropriate path to take, and to what effect might these decisions influence not only the immediate decisions over research and publication but also future directions of the discourse of tourism studies?

It is perhaps ironic that in some academic circles, particularly with respect to place marketing and promotion, there has been substantial criticism of public–private partnerships. In reflecting on the economics and politics of postmodernity, Harvey recognised that 'the new entrepreneurialism of the smaller state has, as its centrepiece, the notion of a "public–private partnership" in which a traditional local boosterism is integrated with the use of local government powers to try [to] attract external sources of funding, new direct investments, or new employment sources' (1989: 7). However, in most cases partnership does not include all members of a community: those who do not have enough money, are not of the right lifestyle, or simply do not have sufficient power are ignored. For example, in referring to Derwentside in the United Kingdom, Sadler (1993: 190) argued:

> The kind of policy which had been adopted – and which was proving increasingly ineffective even in terms of its own stated objectives ... rested not so much on a basis of rational choice, but rather, was a simple reflection of the narrow political and intellectual scope for alternatives. This restricted area did not come about purely or simply by chance, but had been deliberately encouraged and fostered.

Yet despite such warnings from the experience of contemporary governance in the industrial world, models of collaboration are increasingly advocated for research endeavours, often as a means to encourage more 'relevant' (begging of course the question of relevant to what and to whom?) and effective research (read targeted and cost-effective). As Clegg and Hardy (1996: 678) remind us, 'We cannot ignore that power can be hidden behind the facade of "trust" and the rhetoric of "collaboration", and used to promote vested interest through the manipulation of and capitulation by weaker partners.'

This is not to say that collaboration either between researchers or between researchers and government and/or industry is a bad thing. Far from it. Instead, I am posing questions as to the transparency of such processes and the potential limiting effect collaboration might have in some circumstances as to the research questions being asked or funded. For example, in the case of New Zealand, applications to one of the main sources of tourism research funding in 2002, the ironically titled Public Good Science Fund, were no longer sent out for external academic review. Instead, research proposals were sent to a reference group primarily from government and industry with one academic member (who was also an applicant for funding), which made recommendations to the Foundation Board with respect to funding. Applicants who did not receive funding received three- to four-line explanations concerning their lack of success, while those who were successful had to negotiate with respect to the research outputs that would be purchased by government. In such an opaque process it is not difficult to feel, cynically, that in such a small country as New Zealand, personal relationships, including relationships with former students, industry and government contacts, and playing the rules of the game may have more to do with research funding success than the inherent qualities of grant applications. Indeed, in such a situation, networks of social relationships such as who one has been in bed with, as much metaphorically as actually, become critical to influencing who does and does not gain research funding through influencing how a research proposal should be both written and assessed. Of course, such a situation is not unique to New Zealand. Again the point is to identify some of the structural and institutional processes which surround tourism research, particularly given the differing research agendas of academics, industry and government (Ryan and Simmons 1999).

As Hall and Jenkins (1995) observed, one of the great problems in examining the role of interest groups, such as business, in the tourism policy and research process is deciding what the appropriate relationship between an interest group and government should be. At what point does tourism industry membership, or academic membership for that matter, of government research advisory committees or of a national, regional or local tourism agency represent a 'closing up' of the research process to others rather than an exercise in consultation, co-ordination, partnership or collaboration? As Deutsch (1970: 56) recognised:

> this co-operation between groups and bureaucrats can sometimes be a good thing. But it may sometimes be a very bad thing. These groups, used to each other's needs, may become increasingly preoccupied with each other, insensitive to the needs of outsiders, and impervious to new recruitment and to new ideas. Or the members of the various interest group elites may identify more and more with each other and less and less with the interests of the groups they represent.

In cases where university places are government funded, the growth in university-level courses in tourism is itself related to broader government and industry perceptions of where jobs may be created or where human capital needs to be focused. The growth in international tourism to Australia and New Zealand in the late 1980s, for example, led to the rapid development of tourism degrees by universities, particularly new universities, because of the availability of government funding and a rapid rise of student interest in these programmes. Arguably, similar processes of economic change and restructuring and government response to the development of a service economy have been behind the growth of university and trade-orientated tourism programmes in the United Kingdom and elsewhere in Europe.

Government policies affect not only the provision of tourism education but also tourism research. Different funding schemes for publication and research lead to different behaviours by institutions and individuals. In recent years a number of countries, such as the United Kingdom, Australia, New Zealand and South Africa, have developed funding schemes to reward research performance. Where this has occurred, universities and individuals have altered their behaviours in order to attempt to maximise returns from such programmes. Under the UK Research Assessment Exercises (RAE), which have produced 'league tables' of research performance that not only lead to funding but may also influence student decisions with respect to where to study, some institutions have sought to attract individuals from other institutions in order to improve their RAE performance. The development of institutional incentives may also lead to other forms of behaviour. For example, in Australia the responsible national government body allocates differential funding in terms of the number of various research publications which are produced. Such may therefore lead to a situation in which the focus is on the quantity of publications produced rather than the quality (however assessed), as well as on producing publications in certain situations and not others. As a conference convenor one of the first questions I am asked by Australians and, increasingly, New Zealanders is, 'Is it a double blind refereed conference?', not, 'Is it going to be a good conference at which to discuss my work?' This is because the potential financial value of a paper presented at a double blind refereed conference means that universities will often only be willing to fund people to attend such conferences. Yet given the rules of the game, this is rational behaviour in economic terms even if it does not meet our ideal of a community of scholars. Given that our performance indicators, and hence our ability to earn a salary and pay our mortgages or rents, are dependent on certain research and publishing behaviour, it is no surprise that our research decisions can be affected by the structures which surround us.

Academic freedom is a mirage. If I do not meet the research (and teaching and service) performance criteria which are set, and which are increasingly set in terms of attracting income and getting a high rating in terms of the submission of four high-quality research outputs, I may not get a pay rise, or perhaps in the longer term I may even lose my job. Furthermore, as a head of department at the time of writing, I have to worry about not only my own research performance but also that of my staff and graduate students, because in the longer term the departmental budget, and therefore their salaries, departmental research support, conference and travel funding, and teaching assistantships, depends on the direction I set and for which I am (and feel) responsible. If the parameters are thus set, then surely playing the rules of the game and directing research in certain directions so as to achieve certain outcomes with respect to research outputs (for that is how the winners of the game are currently decided) becomes a logical response even if I do not actually support or believe in the rules of the game. However, all this makes for high stress, a bad back and too much comfort food in the evening. In such a situation, what is one to do if one cannot in good conscience work within those parameters, especially when so many other universities also now work within those rules? Whether you are a professor or a graduate student, how we play the rules of the academic game is clearly going to influence our research in terms of what, how and where we research and how we produce the outputs. (Is it to be internationally refereed journals only, and forget about books?).

In my own career I have made publishing decisions based on the rewards that are available. If one is looking at the time to get research published and the return on such publications, then it will often influence the type of research one undertakes and how it is written up. Indeed, a major issue for my own work is whether to write or edit a book or several articles from a piece of research. I actually think that most tourism books get read more than the majority of journal articles, therefore I would prefer to write more books so as to convey my research and scholarship, which is supposedly one of the main tenets of academic life. Yet the reality is that if research funding or reward schemes favour articles (in the right journals, of course) over books, to what extent will I change my behaviour so as to adapt to those schemes? Indeed, even writing textbooks becomes an interesting case in point in terms of adaptation and change, whereby at times I may underplay certain opinions or perspectives I hold so as to make the book more readable to specific target audiences and therefore more saleable, while also continuing to follow certain criteria of academically acceptable writing. Moreover, I have recently been told that the rules of the game in New Zealand mean that I cannot count texts as research publications. OK, so entire cohorts of students would be pleased, but frankly it's easier to write a research article than it is to undertake the scholarship to write a university-level text. In addition, as much as I would love to be more 'playful' and reflexive in much of my academic writing, I also know that there is a danger that such work might not be accepted. For example, while my book *Tourism Planning* (Hall 2000) (I would note that the first draft of this chapter read 'text'; however, I am now not wanting it to be found out that I have claimed it as a research output for the past six years) emphasises reflexivity and situatedness within collaboration and networks, much of the personal reflexivity was taken out at proof stage because of concerns over the appropriateness of the material and how it would be received! Such a concern is not only for myself. I also wish the graduate students I supervise and advise to be more reflexive in their own work, as I think it is a very important and valuable part of the research process. Yet taking such positions or making personal value statements in their dissertations may also upset some examiners if they do not support the inclusions of reflexive statements. This therefore affects not only the choice of examiners but also the final composition of the student work.

In terms of why we research what we do, one also cannot ignore the personal. Yet this is almost completely ignored in discussions of tourism research (also see Crouch 1999). One possible exception of which I am very proud is the book by Chris Ryan and myself on sex tourism, where we note the impact of such research on our own notions of identity and sexuality (Ryan and Hall 2001). Yet when I think about it, even here most reflection was confined to the foreword and afterword rather than being in the main body of the book.

The personal subjectivities of our experiences are vital to our choice of research paths, yet typically go unacknowledged. In thinking of my own experiences and why I follow certain lines of research, reasons apart from sheer intellectual curiosity do appear. For example, I ended up doing my doctorate in Perth, Western Australia, primarily because of a woman I lusted after who was there, not because of any rational decision over research supervision (although I was not really prepared to admit that at the time). If I had gone elsewhere, different topics and research directions would have undoubtedly followed. Indeed, partners or potential partners (as well as friends and relatives) may exert an influence not only on where one lives and works but on the type of research one engages in.

There are also certain colleagues one does not trust and whom I would never want to work with again if at all possible. Similarly, there are also some universities and publishers I regard with enormous disdain because of the manner in which they have treated colleagues or myself. Of course, thankfully, there are also colleagues and publishers I work and write with that I find extremely stimulating and which provide a real sense of fun and enjoyment in scholarship. Similarly, there are some places in the world that I would rather go to and conduct work in than others, and this also influences the kind of work that one does. The personal is therefore critical in determining the kind of research lines we follow, though in ways we barely acknowledge apart from the odd foreword or acknowledgement. For example, geography, one of my core discipline areas, was appealing not only intellectually (it was and is) but also because it gave me an excuse to do the surfing and outdoor activities which I so enjoyed (these opportunities are now greatly restricted by administrative and teaching responsibilities!). Tourism was an accident, a sideline that grew from work on national parks, wilderness and environmental history but which provided good career prospects. Moreover, even some of the things I research follow from the personal. For instance, I am passionate about local foods and can do work on wine and food tourism.

Perhaps more deeply, many of my intellectual passions and therefore publications are concerned with issues of mobility, place (and placelessness), peripherality and identity (e.g. Hall and Williams 2002; Hall and Müller 2003). Many of them find their grounding in tourism and geography. In fact, if I could, I would rather primarily place myself as a student not of tourism but of temporary mobility, but then, while I find great meaning in the very inbetweeness of the term, I doubt whether it would be a wise career move, given current disciplinary structures (although they too are highly mobile, of course!). Where does all this come from? I think there is a connection between some of these interests and my life course. I am a migrant who left Margate, England, at the age of 11 to move to Western Australia. I did my master's in Canada, and have moved backwards and forwards for work between the eastern states of Australia and New Zealand since 1988 upon finishing my doctorate. I travel for work reasons to the extent that I have been away from home for over two months a year for most of the past few years. I do currently live on the periphery, after all. When posing the question of where is home, I feel I have multiple places which are familiar and to which I have some sense of belonging. However, interest in place and identity perhaps goes wider than that in terms of family relationships – for example, my mother remarrying, or an older brother who committed suicide. The question of who I am in a time of globalisation and fluid identity is therefore as much a personal interest as it is a research response to external factors. Such autobiography and reflection is not just a desire to have some kind of cathartic experience by writing this for all to see, so as to prove I'm not as fucked up as the next person, or more so! Instead, it is part of an admittedly imperfect desire to locate some of those links between self and action that reflexivity brings and which as we sometimes peel off the onion-like layers of our understanding concerning the nature of our research may sometimes make us cry. To do so is to make oneself vulnerable. Still, we have to start somewhere, and it is easier if I start with me; after all, it is the subject I know best! Yet I would argue that if we are to be serious about utilising reflexivity, this is where we should start, with ourselves, and ask some of those difficult questions as to how we situate our research with ourselves and others before we go and observe and involve ourselves with those others.

Looking forward in a rear-view mirror

From the above, it is perhaps pretty clear that I regard reflexivity as critical to all tourism research practice, even if it is not as well acknowledged as it should be. For some forms of social research, particularly participant observation and other observational and ethnographic techniques, reflexivity is vital. The use of participant observation, for example, leads to the very questioning of the objectivity of the researcher, the status of the observation of social phenomena and the scientific and ontological standing of social research (Evans 1988). Participant observation requires a profound level of introspection on the part of the researcher with respect to his or her relationship with what is being researched and how the process of 'Othering' takes place. The process of giving voice to others is never neutral and works itself through power structures. Participant observation assumes a subjectivity in which the voices are heard from somewhere in particular rather than from some Archimedean point of reference (Doorne 1998). As Said reminds us, the 'scrubbed, disinfected interlocutor is a laboratory creation' (1989: 210).

So where does this lead us? As Gregson (2000: 322) comments:

Conclusions are part of the performance of academic writing, suggesting it is possible to summarize in a comprehensive, synthetic way, which looks forward and stakes out the terrain of 'progress'. This is part of the expectation here, I am sure. Student texts, after all, carry such presumptions, both from their audiences and producers/publishers. But ... ending in this way is problematic.... This is because conclusions of this nature revert to privileging one narrative, one voice; a writing tactic that therefore erases as it summarizes; and a style that is inappropriate for expressing geographies of tension, contradiction, and polyvocality.

Gregson nicely presents the reflexive dilemma: we are aware of the other voices which we can hear, yet how can we represent them? For me the answer lies in considering the nature of what we write as an argument; it is the craft of creating something the acceptability of which is determined by the different audiences which we seek to engage. The completeness, shape, structure and beauty of an argument is determined not just by ourselves but by those who receive it (Majone 1989). For some arguments, these audiences are very small indeed. Indeed, I often criticise much of the cultural turn in the social sciences not because of the spirit behind it or on grounds of the methods employed, but because of the dense, exclusive language which is used. For me, language should seek to be inclusive so that it can be understood by as many people as possible. It is the communicative aspect of reflexivity. And as Evans has observed, 'Validation is ... by the very act of participation, internal to the research in that whilst interpretations must be justifiable in terms of the cited evidence, they are still the product of the ability of the observer to participate meaningfully' (1988: 201). As noted earlier, I have written elsewhere of the notion of tourism planning and policy as argument and persuasion (Hall and Jenkins 1995; Hall 2000). Therefore, this text should be read as an argument.

> We may watch a play, but we now recognise ourselves as actors in another broader drama. That all the world is a stage and no person an island may be trite but goes to the heart of the interpretative approach, for we are part of and contributors to that which we seek to understand.
>
> (Smith 1988: 258)

However, I am not a valueless interpreter. I have great frustration with much of the research and scholarship undertaken in tourism. Often it is competently done, meeting the RAE and PBRF criteria of 'quality', but without reflexion and thought as to whose interests are being served – which is normally those from business and government with access to power. For all the talk of sustainable and alternative tourism, few alternatives have really shown up which explore the potential for other spaces and places which reflexivity may provide. In my more sanguine moments I believe that this is because researchers often take the easier path in tourism research because within current academic structures that is what provides the rewards. And who am I to talk? As much as my present position stresses and frustrates me, I still sit in a highly privileged position. I have a good salary by New Zealand standards, I have reasonable security of tenure, I can pay off the mortgage and I can travel. Many cannot. So perhaps it is easier to take some sort of high ground in this position. However, I also realise that for me it is time for a change. I feel that growing sense of disenchantment and unease in my research and in the structures I am embedded in that, if it cannot be given an outlet, it will lead to further disenchantment. Perhaps others feel this as well in terms of their own situation. Perhaps others, like me, read Sartre at 17 and never came back. I then return to Harvey's (2000: 255) 'spaces of hope':

> The lesson is clear: until we insurgent architects know the courage of our minds and are prepared to take an equally speculative plunge into some unknown, we too will continue to be the objects of historical geography (like worker bees) rather than active subjects, consciously pushing human possibilities to the limits. What Marx called 'the real movement' that will abolish 'the existing state of things' is always there for the making and for the taking. This is what gaining the courage of our minds is all about.

We make our own futures, we conduct our research, we live with the consequences. At the conclusion of *Tourism Planning* (Hall 2000), I refer to my master's thesis (from 1984) in which I quoted a geographer by the name of Gilbert White as a kind of research credo:

> Speaking only as one individual, I feel strongly that I should not go into research unless it promises results that would advance the aims of the people affected and unless I am prepared to take all practicable steps to help translate the results into action.
>
> (White 1972: 102)

For me, that philosophy still resonates, stronger than ever. Indeed, writing this piece on reflexivity has allowed me to reflect deeply on my research, and on the research cultures, structures and institutions which surround tourism and in which I am situated. Will you do the same with your research?

Summary

- The chapter initially places the notion of reflexivity within the context of modernity and postmodernity.
- The chapter presents the means by which the academic reward system,

government and society influence the nature of tourism research. The rules of the academic game are noted, and how gatekeeping may influence the conduct and publishing of research as well as how RAE/PBRF exercises may also affect research.

- The extent to which the researcher's personal life affects research is also noted. In addition, the extent to which the personal and private is not discussed in research is also regarded as important.
- Reflexivity is regarded as an essential ingredient in qualitative tourism research, particularly with respect to participant observation, and readers are challenged to reflect on their own research undertakings and the manner in which they engage with research subjects as well as the production of academic knowledge.

Questions

1 To what extent do you regard the writing of academic papers in the third person as reflecting a supposed objectivity in tourism research?
2 Would you undertake tourism research because it was intrinsically interesting or because it might reward your career or provide financial rewards?
3 Has your personal life ever influenced academic or research decisions? Should every tourism thesis or dissertation include a 'value statement' by the author which indicates how their personal values may have influenced the choice of topic, the research method utilised and the conclusions reached?
4 What are the rules of the academic game?

References

Barnes, T. J. and Sheppard, E. (2000) 'Introduction: the art of economic geography'. In E. Sheppard and T. J. Barnes (eds) *A Companion to Economic Geography*. Oxford: Blackwell.

Bauman, Z. (1992) *Intimations of Postmodernity*. London: Routledge.

Beck, U. (1992) *Risk Society: Towards a New Modernity*, trans. M. Ritter. London: Sage.

Beck, U., Giddens, A. and Lash, S. (1994) *Reflexive Modernization: Politics, Tradition and Aesthetics in the Modern Social Order*. Cambridge: Polity Press.

Britton, S. G. (1991) 'Tourism, capital and place: towards a critical geography of tourism', *Environment and Planning D: Society and Space*, 9: 451–478.

Calinescu, M. (1987) *Five Faces of Modernity*, Durham, NC: Duke University Press.

Clegg, S. R. and Hardy, C. (1996) 'Conclusion: representations'. In S. Clegg, C. Hardy and E. Nord (eds) *Handbook of Organization Studies*. London: Sage.

Crouch, D. (ed.) (1999) *Leisure/Tourism Geographies: Practices and Geographical Knowledge*. London: Routledge.

Deutsch, K. (1970) *Politics and Government: How People Decide Their Fate*. Boston, MA: Houghton Mifflin.

Doorne, S. M. (1998) 'The last resort: a study of tourism policy, power and participation on the Wellington waterfront', unpublished doctoral thesis, Victoria University of Wellington, Wellington.

Evans, M. (1988) 'The researcher as research tool'. In J. Eyles and D. Smith (eds)
Qualitative Methods in Human Geography. Cambridge: Polity Press.

Gergen, K. J. (1991) *The Saturated Self: Dilemmas of Identity in Contemporary
Life*. New York, NY: Basic Books.

Giddens, A. (1990) *The Consequences of Modernity*. Cambridge: Polity Press.

Giddens, A. (1991) *Modernity and Self-Identity*. Cambridge: Polity Press.

Grano, O. (1981) 'External influence and internal change in the development of
geography'. In D. Stoddart (ed.) *Geography, Ideology and Social Concern*.
Oxford: Blackwell.

Gregson, N. (2000) 'Family, work and consumption: mapping the borderlands of
economic geography'. In E. Sheppard and T. Barnes (eds) *A Companion to Economic Geography*. Oxford: Blackwell.

Hall, C. M. (1988) 'The geography of hope: the history, identification and preserva-
tion of wilderness in Australia'. Unpublished Ph.D. thesis, Department of Geography, University of Western Australia, Nedlands.

Hall. C. M. (2000) *Tourism Planning*. Harlow: Prentice-Hall.

Hall, C. M. and Jenkins, J. M. (1995) *Tourism and Public Policy*. London: Routledge.

Hall, C. M. and Müller, D. (eds) (2004) *Second Homes, Tourism and Mobility*.
Clevedon: Channelview Publications.

Hall, C. M. and Page, S. (1999) *The Geography of Tourism and Recreation*.
London: Routledge.

Hall, C. M. and Williams, A. M. (eds) (2002) *Tourism and Migration: New
Relationships between Production and Consumption*. Dordrecht: Kluwer
Academic.

Harvey, D. (1989) 'From managerialism to entrepreneurialism: the transformation
in urban governance in late capitalism', *Geografiska Annaler*, 71B: 3–17.

Harvey, D. (2000) *Spaces of Hope*. Berkeley, CA: University of California Press.

Hassan, I. (1985) 'The culture of postmodernism', *Theory, Culture and Society*, 2
(3): 119–131.

Healey, P. (1997) *Collaborative Planning: Shaping Places in Fragmented Societies*.
Basingstoke: Macmillan.

Johnston, R. J. (1983) 'On geography and the history of geography', *History of
Geography Newsletter*, 3: 1–7.

Johnston, R. J. (1991) *Geography and Geographers: Anglo-American Human Geography since 1945*, 4th edn. London: Edward Arnold.

Kumar, K. (1995) *From Post-industrial to Post-modern Society: New Theories of
the Contemporary World*. Oxford: Blackwell.

Lash, S. and Urry, J. (1987) *The End of Organized Capitalism*. Cambridge: Polity
Press.

Lash, S. and Urry, J. (1994) *Economies of Signs and Space*. London: Sage.

Lawson, H. (1985) *Reflexivity: Problems of Modern European Thought*. London:
Anchor.

Majone, G. (1989) *Evidence, Argument and Persuasion in the Policy Process*. New
Haven, CT: Yale University Press.

Nederveen Pieterse, J. (ed.) (1992) *Emancipations, Modern and Post-modern*.
London: Sage.

Robertson, D. (1999) 'Professing integrity', *Times Higher Education Supplement*, 19 November.

Ryan, C. and Hall, C. M. (2001) *Sex Tourism: Marginal People and Liminalities*. London: Routledge.

Ryan, C. and Simmons, D. (1999) 'Towards a tourism research strategy for New Zealand', *Tourism Management*, 20: 305–312.

Sadler, D. (1993) 'Place-marketing, competitive places and the construction of hegemony in Britain in the 1980s'. In G. Kearns and C. Philo (eds) *Selling Places: The City as Cultural Capital, Past and Present*. Oxford: Pergamon Press.

Said, E. (1989) Representing the colonised: anthropology's interlocutors', *Critical Inquiry*, 15 (2): 202–225.

Smith, S. (1988) 'Constructing local knowledge: the analysis of self in everyday life'. In J. Eyles and D. Smith (eds) *Qualitative Methods in Human Geography*. Oxford: Polity Press.

Stoddart, D. (1981) *Geography, Ideology and Social Concern*. Oxford: Blackwell.

White, G. (1972) 'Geography and public policy', *Professional Geographer*, 24: 101–104.

9   Trustworthiness in qualitative tourism research

Alain Decrop

Aims of the chapter

- To show the relevance of trustworthiness issues in qualitative tourism research.
- To position trustworthiness in the framework of paradigmatic stances in qualitative research.
- To describe the basic criteria of trustworthiness.
- To present practical techniques to achieve trustworthiness, with a few examples.
- To discuss triangulation as one of the most comprehensive techniques with which to achieve trustworthiness.

Introduction

In tourism research, sociologists and anthropologists have been turning to qualitative approaches for a long time. This is not the case for researchers working in other disciplines such as psychology, economics and marketing. Riley notes that 'the majority of tourism marketing research has relied on structured surveys and quantification' (1996: 22). Walle further regrets that 'techniques which bear the imprints of logical positivism, statistical investigation, and the scientific method continue to dominate' (1997: 525). The focus is on that which is general, average and representative in order to allow for statistical generalisation and prediction. Because this is not possible in qualitative research, the approach has been regarded as exploratory and largely unscientific. For many tourism researchers, qualitative research exists only to provide information for developing further quantitative research. The aim of this chapter is not to defend the prevalence of one approach over the other. Indeed, methodological eclecticism is desirable: research questions must direct the choice of appropriate methods and research designs rather than the other way round. However, there is a need to consider the issue of trustworthiness in qualitative research. Trustworthiness refers to scientific inquiry that is able to 'demonstrate truth value, provide the basis for applying it, and allow for external judgements to be made about the consistency of its procedures and the neutrality of its findings or decisions' (Erlandson *et al.* 1993: 29). Addressing the trustworthiness issue is important in helping to make qualitative and interpretive tourism studies more rigorous and more acceptable to quantitative and positivist researchers.

This chapter starts with a discussion on the issue of trustworthiness, which will be treated within the framework of paradigmatic debates in qualitative research because trustworthiness is not relevant to all qualitative tourism researchers. Following this discussion, Lincoln and Guba's (1985) basic criteria of trustworthiness in social research, which parallel positivists' reliability and validity constructs, are outlined. Subsequently, some practical techniques to enhance trustworthiness are presented. Triangulation is one of the techniques discussed in more depth since it offers the most comprehensive means of applying the criteria put forward by Lincoln and Guba. Finally, these techniques are illustrated with the support of a PhD research project on vacation decision-making (Decrop 1999b).

Paradigmatic stances in qualitative research

Frequently, qualitative research is connected with interpretivism as a general philosophy. In contrast to positivism, which strives to explain phenomena in order to predict and control them, interpretivism focuses on understanding and interpretation. Interpretivism emphasises relativism: reality is not objective, single and divisible but socially constructed, multiple, holistic and contextual (Ozanne and Hudson 1989). This ontological premise implies the use of different methodologies. Interpretivism does not suggest a separation but rather an interactive and co-operative relationship between the investigator and the object of investigation. The focus is not on the quantity of information gathered but rather on its quality and richness. All aspects of observation are considered to be worthwhile: the interpretive inquirer watches, listens, feels, asks, records and examines. In-depth interview, participant observation or document analysis are favoured tools in this approach. Interpretivism is often assimilated with naturalistic inquiry as it strives to understand naturally occurring phenomena in their naturally occurring states (Lincoln and Guba 1985). In contrast with positivism, which is based on the hypothetic-deductive framework, interpretivism relies on a holistic-inductive approach. The research phenomenon is investigated as a whole, and theoretical propositions are generated from the empirical field.

Denzin and Lincoln (1994) go further than this traditional trade-off between positivism and interpretivism by distinguishing five 'moments', presented as ways of thinking and carrying out qualitative inquiry. These moments are discussed in some detail in Chapter 1. In this context, the issue of trustworthiness can be linked to the second and third moments.

The second moment, i.e. the 'modernist' or 'post-positivist' phase, is characterised by an attempt to formalise and put more rigour into the qualitative research process. While creativity is enhanced, canons of 'good' qualitative research are suggested (Becker *et al.* 1961), and strong methodologies are developed to help the emergence of grounded theories (Glaser and Strauss 1967). The third moment describes 'blurred genres' (including constructivism, naturalistic inquiry, etc.). Here, qualitative research reaches maturity and multiplicity as researchers have a full set of paradigms, methods and strategies at their disposal. The focus is on 'thick descriptions' (Geertz 1973) in order to develop theory that makes sense out of a local situation. The issue of trustworthiness is highlighted, and criteria are proposed to evaluate qualitative studies, paralleling the positivist criteria of validity and reliability.

In contrast, the issue of trustworthiness is not relevant to researchers operating in the fourth and fifth moments, which are characterised by more reflective and critical approaches to qualitative inquiry. New ways of writing and developing theories are developed (narratives, 'tales from the field', more local/small-scale theories), whereas research is more participatory, activist and action-orientated. The representation of the Other becomes a key issue, and problems of silenced groups are (re)considered.

A paper by Ryan and Martin (2001) provides a good illustration of the irrelevance of trustworthiness to some qualitative tourism studies. Carrying out participant observation in a striptease club in Darwin, the authors examined the relationships between tourists and strippers. They acknowledge that the conclusions they reach are 'personal constructs from the respective meanings of the lives of both authors' (Ryan and Martin 2001: 147). Idiosyncrasies, uniquenesses and complex dynamics are integrated to the representation of a phenomenon which does not aim at trustworthiness but at possible (radical) changes in the participants' lives. Over time, friendships emerged between the researchers and some of the women. In the cases of three women, discussions led to statements about leaving the striptease business. From this example, one can conclude that criteria for assessing qualitative inquiry are relative and depend upon the paradigmatic stance each researcher takes. Smith and Deemer (2000) recognise this need for plurality and the value of multiplicity, but they caution that one should 'honor this need without giving over to excesses, to inquiry that is so fragmented that lines of connection have been lost and the social amelioration possibilities of our work have been rendered moot' (Smith and Deemer 2000: 894). In spite of that multiplicity of criteria, trustworthiness is very useful. A review by Riley and Love (2000) demonstrates that almost all qualitative tourism researchers are operating in the second and third moments. A more detailed examination of the criteria for establishing trustworthiness is made in the next section.

Criteria of trustworthiness

Qualitative approaches are often criticised by positivists because of the lack of objectivity and generalisability associated with them. Both reliability and validity are questioned, since Cronbach alphas and R^2 cannot be computed. It is not the usefulness of qualitative data that is at stake here, but rather the criteria by which the trustworthiness of a qualitative study can be judged. This issue goes beyond the quantitative/qualitative debate, as 'all research must respond to canons that stand as criteria against which the trustworthiness of the project can be evaluated' (Marshall and Rossman 1995: 143). There have been several attempts to rethink such terms as validity, generalisability and reliability in different qualitative research paradigms (Denzin 1997; Hammersley 1992; Kincheloe and McLaren 2000; Lather 1993), but Lincoln and Guba's (1985) typology is most often mentioned. They have developed four criteria for qualitative inquiry that parallel the quantitative terminology. Credibility (which relates to the quantitative criterion of internal validity) refers to how truthful particular findings are. Transferability (associated with external validity) is concerned with the extent to which the research findings are applicable to another setting or group. Dependability (related to reliability) consists of looking at whether the results are consistent and reproducible. Finally, confirmability (associated with objectivity) pertains to how neutral the findings are.

In qualitative research, *credibility* can be questioned owing to the 'subjective' nature of the data collected, since there is no rigid separation between the researcher and their subject but rather an interactive, co-operative and participative relationship. According to Henderson (1991), credibility in qualitative studies is mostly a question of personal and interpersonal skills (e.g. limiting biases due to the presence of the researcher, developing trust with informants, and avoiding reactive effects or selective perception). Moreover, research is credible when the suggested meanings are relevant to the informants and when the theoretical propositions conform with the interview and observation data.

The *generalisation* issue constitutes another frequent criticism raised against qualitative research: that findings are based on small and non-representative samples stemming from non-random sampling procedures. Addressing this point, it should be argued that a distinction must be made between statistical and analytical generalisation. While statistical generalisation of qualitative data is most often not possible or desirable, the analytical transfer of theoretical propositions to other objects (people, settings, phenomena, etc.) is conceivable provided the researcher knows and gives details about the context of the study, integrates findings with existent literature and describes how related objects are similar. That is why the term *transferability* is more appropriate than generalisability.

Dependability is a third criterion involved in appraising the trustworthiness of qualitative inquiry. In interpretive research, reality is not single and immutable but multiple and contextual. Therefore, knowledge generated is not absolute but bound by time, context, culture and value, and replicability is nonsense because of the ever-changing nature of the social world (Marshall and Rossman 1995). Dependability should, then, be considered as the correspondence between the data recorded by the researcher and what actually occurred in the setting.

Finally, objectivity or *confirmability* issues need to be addressed since they are often seen as essential foundations in social research. Guba and Lincoln (1994) argue that a researcher can never be totally objective. However, the data analysis process is made objective by looking for a variety of explanations about the phenomenon being studied, reporting theoretically meaningful variables, and giving others access to factual data in order to assess the way major interpretations emerged from the empirical material. To quote Henderson (1991: 138), 'to be objective means to see how the world would appear to an observer who has no prejudices about what s/he observes'.

In conclusion, Lincoln and Guba's typology is one of the most comprehensive efforts to date in establishing scientific canons for qualitative inquiry. The proposed criteria help bridge the gap between quantitative/positivist and qualitative/interpretive research. Moreover, Lincoln and Guba's trustworthiness criteria are not limited to contemplation but are connected with practical tools and techniques to implement them. These are described in the following sections.

Techniques to enhance trustworthiness

There are a wide range of techniques to augment the trustworthiness of qualitative findings (Erlandson *et al.* 1993). *Credibility* may be enhanced by techniques such as prolonged engagement, persistent observation and referential adequacy. Prolonged engagement consists of examining the research setting over an extended period; persistent observation involves actively seeking out sources of data identified by the researcher's own emergent design; while referential adequacy relates to the provision of contextual information in order to support data analysis and interpretation. These techniques were used by Forsyth (1995) in his study of tourism and agricultural development in Thailand. The project aimed to identify the factors affecting the adoption of tourism, and how tourism then impacted on agricultural practices in a way that affected soil fertility. The researcher immersed himself in the village of Pha Dua by living with three families in one large house belonging to an ex-headman and religious leader. Altogether, he spent about six months living in Pha Dua and studied the Thai language for a year before starting field research. The credibility of the researcher also affects the way findings are received.

Issues of training, experience, perspective, status and presentation of self in the research project need to be addressed (Marshall and Rossman 1995). Finally, member checks offer a good way of enhancing the credibility of analyses and interpretations. Informants are invited to read their transcripts and/or a summary of the analysis provided by the researcher, and to comment on it. Any remark, disagreement with interpretations or additional information should be reintroduced to the analytical process.

Transferability is enabled by using purposive or theoretical sampling, and by writing *thick descriptions*. Theoretical sampling consists of consciously adding new cases to be studied according to the potential for developing new insights or for expanding an emerging grounded theory (Glaser and Strauss 1967). The sample must be as varied as possible in order to provide the broadest range of information. Writing thick descriptions is concerned with describing the data extensively and compiling them in an orderly way so as to give other researchers the opportunity to appraise the findings and also the extent to which they could be transferred to other settings. By writing such thick descriptions of the effects of tourism on the host community in the pilgrimage town of Pushkar (India), Joseph and Kavoori (2001) provided a framework for understanding the impacts of Western tourism on religious communities.

There are a number of ways in which *dependability* in qualitative tourism studies can be increased. Having a research plan, while being able to be flexible with it, and documenting how changes in the plan occurred is the first thing to do. Furthermore, prolonged engagement and the use of an 'auditor' or a second opinion in data interpretation is useful. Finally, leaving an audit trail helps another person to redraw how the authors came to the conclusions they reached and to check whether he or she obtains the same conclusions. For example, in his paper about Jamaican children's representations of tourism, Gamradt (1995) asked children to complete an 'activity book' containing sentence completion items and a space for drawing 'a visitor who has come to Jamaica from far away'. By providing many of those completed sentences and drawings to support his propositions, the author gave access to the way he interpreted his data.

An external auditor may also be used for assessing the *confirmability* of a qualitative study. By reviewing the interviews and analytical procedures, the auditor may confirm the adherence to sound research practices (Riley 1996). Therefore, a reflexive journal should be kept in which the researcher records information about the interviews, the analyses/interpretations and about him- or herself on a daily basis. On the basis of that information, the auditor can act as a 'devil's advocate' to seek out and correct the researcher's prejudices.

Of course, the criteria and techniques for establishing trustworthiness are intertwined. The next section is devoted to still another technique, triangulation, which is probably the most comprehensive way of building trustworthiness into the research design.

Triangulation

Derived from topography and first used in the military and navigation sciences, the concept of triangulation has been fruitfully adapted to social science inquiry (Campbell and Fiske 1959; Jick 1979; Webb *et al.* 1966). Information coming from different angles or perspectives is used to corroborate, elaborate or illuminate the research problem. Triangulation limits personal and methodological biases and enhances a study's trustworthiness.

Denzin (1978) identifies four basic types of triangulation. By combining data sources, methods, investigators and theories, triangulation opens the way for richer and potentially more credible interpretations. The researcher can also 'guard against the accusation that a study's findings are simply the artefact of a single method, a single data source, or a single investigator's bias' (Henderson 1991: 11). The application of triangulation in qualitative tourism research has been discussed further by Decrop (1999a).

Data triangulation involves the use of a variety of data sources in a study. Data may be either primary (interview, observation, etc.) or secondary (textbooks, novels, promotional material, minutes of meetings, newspapers, letters, etc.). In addition to written material, pictures and films are valuable documentary sources. Data triangulation also encompasses the field notes written during and immediately after each interview or observation session. Field notes are not limited to verbal activities but include information on non-verbal behaviour, communicational aspects (audience reaction) and global elements (group behaviour, body gestures, combined verbal and non-verbal communication, etc.), which often give another point of view, if not direct insight to the research question.

Method triangulation pertains to the use of multiple methods to study a single problem – that is, different qualitative methods or a combination of qualitative and quantitative techniques (but not in a hierarchical order like qualitative exploration and quantitative inference; see Hall and Rist 1999). Since each method has its own limitations, and single methodologies may result in selective perception, the use of multiple methods paves the way for more credible and dependable information.

Investigator triangulation is concerned with using different researchers to look into the same body of data. Such triangulation can help reduce personal biases in analysis and interpretation. Biases may stem from the investigator's subjective understanding, gender, 'race' or culture. This type of triangulation takes a lot of extra effort and time that requires teamwork. Outsiders may also be asked to examine a part of the data and to confirm or invalidate prior interpretations. This method of triangulation helps to meet the dependability requirement – that is, that under the same circumstances, the same interpretation will occur.

Theoretical triangulation involves using multiple perspectives to interpret a single set of data. In the inductive process of building a theory, it is important to bring multiple sources of evidence together in order to define a construct or a causal relation. Glaser and Strauss (1967) suggest continuing to ask questions and make comparisons throughout the analysis process. Confronting emerging hypotheses with existing theories and looking for alternative interpretations further help in making more credible conclusions.

In addition to the four basic types of triangulation identified by Denzin (1978), more particular types may be useful for establishing trustworthiness. *Informant triangulation* simply involves considering a broad range of informants and comparing what they say. Both typical and atypical informants are worth investigating. This relates to the theoretical sampling issues discussed earlier in this chapter as a technique that assists with the transferability of the findings.

Multilevel triangulation may be considered as a particular kind of informant triangulation. It is inspired by Miles and Huberman (1984), who make a distinction between different levels of analytical interest in qualitative research. Indeed, qualitative research may involve people (social subjects) or phenomena (social objects) at different levels, which can be triangulated. For example, individual interviews can be supplemented by data involving the groups to which the interviewees belong (family, club, social class, subculture, etc.). In the same way, specific facts and behaviours (what people do or say) may be triangulated with broader activities (regular sets of behaviour) or processes (ongoing flows, changes over time) in which those facts occur.

Moreover, the same people or phenomena may be observed at different points in time, thereby leading to comparisons in a kind of *longitudinal triangulation*. Those comparisons increase the credibility and the dependability of the findings. For example, one can check whether informants' sayings or opinions show discrepancies over time. However, it should be noted that a longitudinal design alone cannot ensure trustworthiness. Leonard-Barton (1995) suggests a synergistic use of a longitudinal single site with replicated multiple sites because it enhances credibility, transferability and dependability at the same time.

Interdisciplinary triangulation (Janesick 1994) is the last type of triangulation to be discussed here. Interpretations become richer and more comprehensive when investigators, methods and theories from different disciplines (psychology, sociology, anthropology, marketing and geography) are considered for a particular research problem. This type of triangulation is especially relevant in tourism research, since in essence tourism is a multidisciplinary phenomenon. Tribe discusses the multidisciplinary nature of tourism in some detail in Chapter 3.

Implementing trustworthiness: an illustration

This last section considers how trustworthiness and triangulation are implemented in the field. The author's own PhD research (Decrop 1999b) serves as a case study. The research focuses on vacationers' judgement and decision-making processes. Existing models (e.g. Moutinho 1987; Um and Crompton 1990; Woodside and Lysonski 1989) postulate a very rational and sequential tourist decision-making process. While useful for measurement and prediction, those models fail to explain the complexity and the context of vacation decision-making. Moreover, the experiential aspects of vacation choices such as fun, feelings and fantasies (Holbrook and Hirschman 1982) or nostalgia and daydreaming, are not taken into account. Finally, the models lack relativism since all cases are reduced to one average decision-making process, which overlooks both the contingency of decision-making and the adaptability of the consumer (Payne *et al.* 1993). For these reasons, an interpretive qualitative approach was chosen in order to investigate vacationers' decision-making processes in more depth. This case is a typical example of the limitations of quantitative research and the potential of qualitative approaches for tourism-related issues.

Denzin and Lincoln's four criteria of trustworthiness have been considered as part of the research design and further implemented through different techniques used while collecting and analysing the data. Credibility was enhanced by the prolonged engagement in the setting, as informants were interviewed three or four times over a one-year period. There was a growing relationship of trust with informants, who became more and more spontaneous and talkative. Furthermore, analyses and interpretations were contextualised through extensive descriptions of the sample, of Belgian culture and of the vacation market. Finally, data, categories and concepts were continuously compared and checked against the empirical material in order to make findings and conclusions credible. Thick descriptions of the data, the informants and the context of the study were given in the form of introductory paragraphs and summary tables to promote the transferability to other products and cultures. In addition, interview quotations were used in order to support the theory generation process and to give readers the opportunity to interpret the data themselves. The widest possible range of information was included in these thick descriptions through purposive sampling. Four types of vacation decision-making units were interviewed: families (with different numbers of children, married as well as non-married), couples (young and older, married and non-married), singles (young and older males and females), parties of friends (small and large groups). The sample was also varied in respect of informants' socio-demographic backgrounds. Moreover, the members of the thesis's committee served as external auditors in assessing the dependability and the confirmability of the findings.

Finally, a range of triangulation techniques were used. *Informant triangulation* was ensured in many ways. First, many comparisons were made between the broad range of informants considered in the study. Moreover, since all members of a decision-making unit were invited to participate in the interviews, *multilevel triangulation* was possible – that is, seeing to what extent individual and group definitions, opinions and behaviour converged; within families, the relationship between parents and children was particularly interesting. The credibility and dependability of the findings were further enhanced by *longitudinal triangulation*. Informants were interviewed up to four times over the year, including up to three times before the summer vacation (February, April and June) and once after it (November). In doing so, the evolution of vacation plans and decisions could be tracked, and information on key variables could be compared at different moments of the decision-making process. *Data were triangulated* in a variety of ways. Pictures were collected from vacation photograph albums (at the end of the second series of interviews). Moreover, statistical data about the vacation behaviour of the Belgian population were gathered to shed additional light on the socio-cultural context of Belgian holidaymakers. Finally, field notes were written during and immediately after each interview describing the setting, the relationship with the informants, personal comments about what was done, said and not said during the interview, etc. Field notes also included critical remarks about the quality of the data and were compiled in a 'roadbook', which was used for analysis and interpretation together with the interview transcripts.

In addition to informant and data triangulation, *method triangulation* was attempted, by combining interview, observation and projective methods. On the one hand, two projective techniques (Haire 1950) were used to gather vacation information in a more indirect way than explicit questioning. After the first interview, one person from each decision-making unit was asked to close their eyes and to describe how they saw one typical day of their next summer vacation. After the second interview, a reflexive technique was used in which informants were asked to show some photographs or a photograph album of their vacations the previous summer and to comment on them. On the other hand, vacation behaviour was considered in settings other than decision-makers' homes. Sessions of non-participant observations were held at a vacation trade fair and in two travel agencies. The first task consisted of following 17 visitors to the Brussels vacation trade fair ('Salon des Vacances') over a four-day period. Their movements around the trade fair and their verbal and non-verbal behaviour (their stops at particular stands, social interaction processes, extent and nature of information gathered, etc.) were recorded. In addition, 50 other informants were observed in two local travel agencies during sessions that lasted three days. The author was sitting in the back of the agency and discreetly writing down everything he could see regarding the customer and the setting, and also all he could hear from the customer (waiting behaviour, personal interaction with the sales agent, post-conversation behaviour, etc.). The data generated by those observation sessions were critically compared with the interview transcripts.

Conclusion

Qualitative research is often considered as exploratory and criticised for overlooking canons of good science. In this chapter, such canons have been presented, as well as the suggestion that qualitative tourism researchers consider and implement the criteria that make a qualitative study trustworthy. It could be argued that 'good researchers, representing every paradigmatic stance, are similarly awed by the depth and complexity of the fields they are investigating' (Erlandson *et al.* 1993: 20). Several techniques have been proposed in this chapter to assist in achieving trust-worthiness in qualitative studies. Of course, this listing is not exhaustive: trustworthiness is above all a state of mind, which requires a balance between creativity and rigour and between art and science. The search for exceptions to both convergence and divergence is the key to making propositions both stronger and richer. Trustworthiness refers to methodo-logical adequacy but does not guarantee the overall quality, relevance or interest of a qualitative study. Authenticity is another important issue since interpretive research does not separate the investigator from their object of investigation but intrudes on the people and the social context being studied: 'the researcher establishes a partnership with the stake-holders in the study, a partnership that requires a fair and honest exchange of the separate constructions of all participants and in return offers opportunity for growth and empowerment' (Erlandson *et al.* 1993: 160). The implementation of that partnership brings authenticity to the research. In that sense, crystallisation (Richardson 1994; Janesick 2000) is said to complement triangulation by being another lens with which to view qualitative studies; 'crystallization recognizes the many facets of any given approach to the social world as a fact of life' (Janesick 2000: 392). The crystal metaphor replaces the triangle one, with the basic proposition of including various disciplines as part of multifaceted qualitative research designs. This is especially relevant when one is studying tourism problems, since tourism is a multifaceted and multidisciplinary phenomenon. In con-clusion, having some criteria by which to appraise the quality of qualitative tourism inquiry is important. However, such criteria depend on the para-digmatic stance each researcher takes. In reference to the framework pre-sented in the first chapter of this book, the establishment of trustworthiness is very relevant for authors operating in the second and third moments of qualitative research, while other criteria, such as authen-ticity, are more important for those working in the fourth and fifth moments.

Summary

Addressing the trustworthiness issue is important in helping to make qualitative and interpretive tourism studies more rigorous. However, the relevance of the issue of trustworthiness depends on the paradigmatic stance the qualitative researcher takes. Lincoln and Guba (1985) propose four basic criteria of trustworthiness (i.e. credibility, transferability, dependability and confirmability) which parallel positivists' reliability and validity constructs. Different practical techniques may be used to achieve those criteria: prolonged engagement, persistent observation, theoretical sampling, thick descriptions, reflexive journals, member checks, dependability and confirmability audits, etc. More than any other technique, triangulation offers a comprehensive means by which to apply the trustworthiness criteria. Triangulation consists in looking at the same phenomenon or research question from more than one source of evidence. Different types of triangulation (e.g. data, method, investigator and theoretical triangulation) may be used to make findings more trustworthy.

Questions

1   Trustworthiness is an important issue confronting qualitative tourism researchers. However, the issue is not relevant for all researchers. Explain why it is desirable to consider trustworthiness, and contrast two cases/research problems in which the first one should confront the trustworthiness issue whereas in the second there would not be the same need to do so.
2   Describe Lincoln and Guba's four trustworthiness criteria and explain how they may be paralleled with positivistic criteria.
3   For a tourism research problem of your choice, think about a qualitative research design in which a few techniques for enhancing the trustworthiness of your findings could be implemented.
4   Why is triangulation useful in qualitative tourism research? How do triangulation modes relate to Denzin and Lincoln's trustworthiness criteria?

References

Becker, H. S., Geer, B., Hughes, E. C. and Strauss, A. L. (1961) *Boys in White: Student Culture in Medical School.* Chicago, IL: University of Chicago Press.

Campbell, D. T. and Fiske, D. W. (1959) 'Convergent and discriminant validation by the multitrait-multimethod matrix', *Psychological Bulletin*, 30: 81–105.

Decrop, A. (1999a) 'Triangulation in qualitative tourism research', *Tourism Management*, 20: 157–162.

Decrop, A. (1999b) *Judgment and Decision Making by Vacationers.* Namur: Presses Universitaires de Namur.

Denzin, N. K. (1978) *The Research Act: A Theoretical Introduction to Sociological Methods*. New York, NY: McGraw-Hill.

Denzin, N. K. (1997) *Interpretive Ethnography*. Thousand Oaks, CA: Sage.

Denzin, N. K. and Lincoln, Y. S. (1994) 'Entering the field of qualitative research'. In N. K. Denzin and Y. S. Lincoln (eds) *Handbook of Qualitative Research*. Thousand Oaks, CA: Sage.

Erlandson, D. A., Harris, E. L., Skipper, B. L. and Allen, S. D. (1993) *Doing Naturalistic Inquiry: A Guide to Methods*. Newbury Park, CA: Sage.

Forsyth, T. J. (1995) 'Tourism and agricultural development in Thailand', *Annals of Tourism Research*, 22: 877–900.

Gamradt, J. (1995) 'Jamaican children's representations of tourism', *Annals of Tourism Research*, 22: 735–762.

Geertz, C. (1973) *The Interpretation of Cultures: Selected Essays*. New York, NY: Basic Books.

Glaser, B. and Strauss, A. (1967) *The Discovery of Grounded Theory*. Chicago, IL: Aldine.

Guba, E. G. and Lincoln, Y. S. (1994) 'Competing paradigms in qualitative research'. In N. K. Denzin and Y. S. Lincoln (eds) *Handbook of Qualitative Research*. Thousand Oaks, CA: Sage.

Haire, M. (1950) 'Projective techniques in marketing research', *Journal of Marketing*, 14: 649–656.

Hall, A. L. and Rist, R. C. (1999) 'Integrating multiple qualitative research methods (or avoiding the precariousness of a one-legged stool)', *Psychology and Marketing*, 16: 291–304.

Hammersley, M. (1992) *What's Wrong with Ethnography? Methodological Explorations*. London: Routledge.

Henderson, K. A. (1991) *Dimensions of Choice: A Qualitative Approach to Recreation, Parks, and Leisure Research*. State College, PA: Venture Publishing.

Holbrook, M. B. and Hirschman, E. C. (1982) 'The experiential aspects of consumption: consumer fantasies, feelings and fun', *Journal of Consumer Research*, 9: 132–140.

Janesick, V. J. (1994) 'The dance of qualitative research design: metaphor, methodolatry, and meaning'. In N. K. Denzin and Y. S. Lincoln (eds) *Handbook of Qualitative Research*. Thousand Oaks, CA: Sage.

Janesick, V. J. (2000) 'The choreography of qualitative research design: minuets, improvisations and crystallization'. In N. K. Denzin and Y. S. Lincoln (eds) *Handbook of Qualitative Research*, 2nd edn. Thousand Oaks, CA: Sage.

Jick, T. D. (1979) 'Mixing qualitative and quantitative methods: triangulation in action', *Administrative Science Quarterly*, 24: 602–611.

Joseph, C. A. and Kavoori, A. P. (2001) 'Mediated resistance: tourism and the host community', *Annals of Tourism Research*, 28: 998–1009.

Kincheloe, J. L. and McLaren, P. (2000) 'Rethinking critical theory and qualitative research'. In N. K. Denzin and Y. S. Lincoln (eds) *Handbook of Qualitative Research*, 2nd edn. Thousand Oaks, CA: Sage.

Lather, P. (1993) 'Fertile obsession: validity after poststructuralism', *Sociological Quarterly*, 35: 673–694.

Leonard-Barton, D. (1995) 'A dual methodology for case studies: synergistic use of a longitudinal single site with replicated multiple sites'. In G. P. Huber and A. H.

Van de Ven (eds) *Longitudinal Field Research Methods: Studying Processes of Organizational Change*. Thousand Oaks, CA: Sage.

Lincoln, Y. S. and Guba, E. (1985) *Naturalistic Inquiry*. Beverly Hills, CA: Sage.

Marshall, C. and Rossman, G. B. (1995) *Designing Qualitative Research*. Thousand Oaks, CA: Sage.

Miles, M. B. and Huberman, A. M. (1984) *Qualitative Data Analysis: A Sourcebook of New Methods*. Beverly Hills, CA: Sage.

Moutinho, L. (1987) 'Consumer behaviour in tourism', *European Journal of Marketing*, 21: 3–44.

Ozanne, J. L. and Hudson, L. A. (1989) 'Exploring diversity in consumer research'. In E. C. Hirschman (ed.) *Interpretive Consumer Research*. Provo, UT: Association for Consumer Research.

Payne, J. W., Bettman, J. R. and Johnson, E. J. (1993) *The Adaptive Decision-Maker*, Cambridge: Cambridge University Press.

Richardson, L. (1994) 'Writing: a method of inquiry', in N. K. Denzin and Y. S. Lincoln (eds) *Handbook of Qualitative Research*. Thousand Oaks, CA: Sage.

Riley, R. W. (1996) 'Revealing socially constructed knowledge through quasi-structured interviews and grounded theory analysis', *Journal of Travel and Tourism Marketing*, 5: 21–40.

Riley, R. W. and Love, L. L. (2000) 'The state of qualitative tourism research', *Annals of Tourism Research*, 27: 164–187.

Ryan, C. and Martin, A. (2000) 'Tourists and strippers: liminal theaters', *Annals of Tourism Research*, 28: 140–163.

Smith, J. K. and Deemer, D. K. (2000) 'The problem of criteria in the age of relativism'. In N. K. Denzin and Y. S. Lincoln (eds) *Handbook of Qualitative Research*, 2nd edn. Thousand Oaks, CA: Sage.

Um, S. and Crompton, J. L. (1990) 'Attitude determinants in tourism destination choice', *Annals of Tourism Research*, 17: 432–448.

Walle, A. H. (1997) 'Quantitative versus qualitative tourism research', *Annals of Tourism Research*, 24: 524–536.

Webb, E., Campbell, D. T., Schwartz, R. D. and Sechrest, L. (1966) *Unobtrusive Measures: Non-reactive Research in the Social Sciences*, Chicago, IL: Rand McNally.

Woodside, A. G. and Lysonski S. (1989) 'A general model of traveler destination choice', *Journal of Travel Research*, 27: 8–14.

10 New wine in old bottles

An adjustment of priorities in the anthropological study of tourism

Dennison Nash

Aim of the chapter

- In their studies of development, anthropologists and other social scientists have often used ethnographic methods to study changes induced by Western contact. This pattern continues in their studies of tourism. Here it is argued that this focus leaves out a good deal of what transpires in tourism development, especially that aspect which generates and directs it, an area in which the ethnographic method also could be applied. This chapter considers the problems attending such application and its potential for illuminating the developmental processes involved.

Not too long ago, anthropologists embarking on the study of tourism would have been surprised to encounter another social scientist with a serious interest in that subject; but times have changed enough for us now to have some confidence in the viability of this field of study and be able to attend more fully to questions about what we should be studying and how. Anthropology has always been concerned with the cultures of humankind, especially in those less developed societies on the Western periphery where anthropologists had become accustomed to doing ethnographic work. These societies, like societies everywhere, have always been in the process of change (see, for example, Bee 1974), which is something that anthropologists have not always stressed in their writing about them. In the early days of anthropology there were grand schemes that accounted for social change in terms of endogenous or exogenous factors. Remnants of the latter viewpoint are to be found in theories of development (see, for example, Novak and Lekachman 1964; Burns 1999: 137–160) and acculturation (see, for example, SSRC Seminar 1954) involving contact between more powerful Western societies and less powerful societies on their periphery. A diachronic picture of what was going on in this 'Third World', where ethnographic studies were almost *de rigeur*, came to characterise a good deal of anthropological writing.

Following their early encounters with tourism, anthropologically orientated scholars such as Forster (1964) in the Pacific, Nuñez (1963) in
Mexico and Kottak (1966) in Brazil have continued on a path of study that
continues still; and though Crick (1989) has stressed the diversity of touristic subject matter (see, for example, Cohen 1979; Graburn 1983), anthropological work on the subject has often followed this by now well-worn
path established in the perspectives of acculturation and development.
These days, an anthropological study of tourism is likely to involve an
ethnographic investigation of change supposedly fostered by Western
tourism in some society or sub-society on the Western periphery, as is the
case with most of the contributions to the classic compendiums edited by
Smith (1977, 1989), which have the subtitle *The Anthropology of Tourism*.
We should be reminded, however, that this centre–periphery tourism that
has tended to preoccupy anthropologists is only a small part of the tourism
going on in the world today, most of the international form of which
involves contact between peoples of developed societies – say, between
France and the United States (see Høivik and Heiberg 1980: 70); and this
is only one of the contemporary, as well as historic and prehistoric, exceptions to the touristic paradigm that has come to prevail among anthropologically orientated scholars.

Even though such studies (see, for example, Smith 1977, 1989), which
involve only a minuscule amount of the world's tourism in this and other
eras, can be valuable additions to our knowledge, they and other investigations of tourism's impact, which, according to Swain *et al.* (1998), have
constituted the most frequent kind of study of tourism by social scientists,
do not come close to exhausting the possible subject matter of tourism. In
this chapter, while not omitting this kind of study – that is, of tourism's
impact on less powerful peripheral societies – I will argue that if they are
really interested in the subject of *tourism*, anthropologists should begin to
concentrate on another, possibly more significant aspect of the phenomenon and study it by a method which has become their forte, namely ethnography. By ethnographically attending to this aspect, which involves the
production of tourists and tourism, we can be led into a consideration of
tourism's ultimate causes, the search for which is at least as important as, if
not more important than, tracing its consequences for destination areas on
metropolitan peripheries.

To broaden our view anthropologically, let us think of tourism as
involving the production and subsequent migration of leisured travellers
(Nash 1981; but see also Cohen 1974), as well as all attendant social activities, including those affecting peoples in tourism destination areas.
Looked at in this way, it is a widespread phenomenon, and indeed may be
universal (Nash 1979), which is a particularly inviting prospect for the
anthropologically orientated, many of whom claim an interest in all of
humankind. But even while perhaps paying lip service to such an orientation, anthropologists studying tourism have tended to follow customary

constraints that pull them towards the periphery of the contemporary world where developing, Third World peoples continue to exist under the influence of Western imperialism; and in their consideration of tourism's consequences there, they have frequently made it clear that they are on the side of these peoples, who have always claimed most of their attention. Their work has sometimes been called 'cautionary' in nature, the most negative of several possible 'platforms' which Jafari (1987) has identified for tourism study; and such work has sometimes been ideologically (rather than scientifically) driven, which is not unusual for those following the paradigms of development or acculturation.

The use of ethnography in tourism and other studies has, according to Dann *et al.* (1988: 4), often been too 'impressionistic'. Unfortunately for the development of a mature social science of tourism, that tendency continues as a new generation of tourism researchers begins to carry on (Boissevain 1996). As far as most of these ethnographic studies are concerned, Nash (2000: 33–34) says that 'though offering much to admire, none escape criticism in their application of theory and method, their conception of cultures and their contexts, and their consideration of the ethnographer as a factor in the research report'. This suggests that even though there has indeed been an increase in the number of ethnographic studies of tourism, the quality of such research, as well as its focus, remains questionable.

Fortunately, not all anthropological studies of tourism have been guided by the paradigm just discussed. There have also been some instances of anthropological concern with other areas of the touristic process, which Nash (1981) conceived of as extending from tourism-generating areas to recipient destination areas. Depending on historical circumstances, this process can include a variety of actors other than the so-called 'hosts' in destination areas (Nash 1981; 462–463). Some of these actors, mainly tourists, have been studied ethnographically (Graburn 1983; Foster 1986; Hartmann 1988; Bruner 1994), but such research is not yet well developed, and other actors in the process such as tour guides (Cohen 1985), travel agents and transport personnel have mostly escaped attention.

Perhaps the most lamentable omission, however, has been that area of the touristic process which has to do with the generation of tourism. Despite calls from Crick (1989) and Nash (1981) and some continuing attention from Graburn (1977, 1983, 1989, 1995), Moeren (1983), Reimer (1990), Selwyn (1990) and a few others, there is not much of anthropologically orientated substance to report on what could be an exceedingly fertile area for tourism research. It is here, after all, where the *sine qua non* of tourism, the tourist, is produced, and though we should keep in mind that there is give and take all along the acculturative-developmental line of the touristic process and that different actors in it contribute, to different extents, to its operation, there remains the nagging notion that some, particularly those in the productive area of this often neo-colonial relationship, are more important than others. By studying the activities of these people ethnographically, anthropologists and other social scientists may be able to do considerably more to open up the field of tourism study than they have so far accomplished.

Attention to this area has naturally been greater among sociologists interested in tourism such as MacCannell (1976) and Urry (1990), and they have made important theoretical contributions that have spawned a good deal of interest among anthropologists and others. Some of this has even resulted in empirically orientated studies (Waller and Lea 1999). But on the whole, one is disappointed by the lack of empirical grounding for sometimes stimulating generalisations about the tourism-'demanding' populations involved, who often come across as too homogeneous for what common sense and empirical inquiry have told us about them. In truth, they have hardly been studied adequately in a manner specified by Pearce (1982: 19) who says, 'Unless some point is worked out whereby evidence may be used to substantiate or reject a sociological perspective, then all accounts of tourists' experiences, motivations, and perceptions of tourist space remain equally valid and hence inadequate.' A proper use of ethnography, involving small-scale, intimate study of the social action of groups involved in this (or any other) touristic situation, could help to repair the deficiency suggested by Pearce (1982).

In sum, even with the aid of theoretically informed views about the production of tourism, anthropologists and others are not very far along in their quest for a scientific understanding of the process of tourism generation or production. It would seem, therefore, that we need more than calls for additional ethnographic studies (e.g. Crick 1989) of tourism. A comment from an administrator of an important granting agency concerning the quality of proposals received is not encouraging. She said that submitted tourism research projects were 'not very good'. Nash's review (2000) of ethnographic research on tourism tends to confirm this judgement.

A paradigm for studies of development

By looking at studies of development and acculturation generally, one can see where most of those studying tourism development have acquired their priorities. As mentioned earlier, they have tended to follow lines of inquiry that deal with the effect of forces emanating from more developed Western 'centres' on less developed peoples, as in the study of the changes elicited by what has been loosely termed colonialism, neo-colonialism or imperialism (Magubane 1996). On the applied side of this effort, some anthropologists who might have worked under government auspices have become involved in privately sponsored research concerning the 'human problem', which, as Lewis (1995: 99) puts it, has had the aim of 'limiting and ameliorating the ensuing dislocation' by such (Western) forces'.

In her discussion of the process of 'social impact assessment' involving (Western) transcultural oil companies and local communities in Latin America, Laura Rival (1997) points out that after a period of fruitless opposition and confrontation, some indigenous peoples there have begun to accept a new model of 'equal partnership' proposed by certain oil companies. She suggests further that applied anthropologists involved in this process can use the ethnographic method to study the social conditions in which the model of equal partnership is realised, a project which could profit by a comparison with more theoretically orientated work already done on more unequal imperialistic relationships (Asad 1973; Said 1978).

One could anticipate that in such study the (Western) outside forces would no longer be taken as simply given, but would be treated as facts to be investigated in terms of some theoretical scheme derived from the burgeoning literature on power (Mintz 1985; Wolf 1991, 1999). This would be especially the case with Rival's (1997) suggestion about anthropologists' investigations of consultancy firms and the processes by which they reach their decisions about development. In taking a position anticipated by Laura Nader (1972) some decades ago, Rival (1997) concludes that ultimate forces in the development process should become subject to the same kind of anthropological scrutiny (including, presumably, ethnographic study) that anthropologists are accustomed to use on dependent peoples around the world. So far, only beginning ethnographic steps in this direction have been taken, as in Miller's (1997) investigation of advertising agencies in Trinidad, or by Handler and Gable (1997) in their study of the realisation of a policy of 'New History' in colonial Williamsburg, which fall in line with Rival's (1997) argument.

An exemplary case for studies of tourism development

In order to apply Rival's (1997) views to the study of tourism development, consider as an example a plan (*rapport d'étude*) prepared by a metropolitan consulting agency for the tourism development of a small island dependency in the Caribbean, a plan which the author was privileged to see, but which appears not to have been realised. This plan, in the best neo-colonialist fashion, seems to have been prepared without grass-roots consultation with the local population. In pursuing a somewhat imaginative exercise, let us assume that some typical capitalist market research at home had been used as a basis for developing the proposal, which advocated three kinds of tourism development in different areas of the island dependency: tourism of sport and rest *en famille*, tourism for restoring fitness, and tourism for getting away from ordinary life – all of which appear to be typical 'compensatory' touristic practices for peoples living in modern society (associated features such as the development of self-sufficient agriculture and fishing for touristic use also were proposed, presumably without grass-roots consultation).

Continuing this imaginative exercise, let us assume that the author, having the relevant language in hand, had been able to obtain *carte blanche* to study the preparation and possible realisation of this proposal. Of course, he or she would have wanted to know about the nature of the consulting firm and the mission given it by the metropolitan government and its various agencies as well as the constraints it imposed on the market researchers. What were the directions given for the market research, by whom and how was it carried out, and what exactly did it reveal? By perusing various documents, including the *rapport d'étude*, participating in discussions, and questioning those involved – all well-established techniques in ethnography – one could begin to get at the perceived sources of the agency's proposal in terms that are routinely considered in more controlled studies of industrial and organisational decision-making (Stevenson and Naylor 1990), but which have not yet been applied significantly in the real world.

The ethnographic procedures used in this fieldwork would, of course, attend to both associative and dissociative social processes within the marketing firm and be extended in a study of give and take between it and significant others. An ethnographer doing this work could profit from studies of the production of touristic representations by Handler (1988), Bruner (1994), and especially Howell (1994), whose interactive analyses of the politics associated with the production of meaning in heritage sites comes closest to the requirements for such analysis. In brief, Howell (1994) sees the production of cultural conservation and heritage displays in such sites as involving contested interactions between different factions, each of which has a stake in the productive process. By participating in such interaction, she says, an applied anthropologist can help the parties involved to clarify the background of their positions and what is going on between them, and help them work out constructive conclusions for development. Whether applied or not, any research on the production of the tourism development project in the Caribbean dependency would have to deal with similar issues.

Given the lack of anthropological consideration of the tourism-generating situation, the presumed difficulty of access to relevant subjects, and the uncertain qualities of the cultures involved, ethnographic procedures would seem to be appropriate for exploring particular development scenarios. That such investigations also can have generalising relevance is pointed out by McCall and Bobko (1990: 390), who in their review of psychological studies of organisational behaviour say that 'we shouldn't neglect the fact that studies based on a single organism can also lead to [scientific] breakthroughs'.

To continue on the methodological side of our imaginative exercise, let us think that all along the way there would be problems concerning the researcher's acceptance by clients as well as subjects and adaptation, problems which invariably occur in ethnographic work but which would be given a special twist in this kind of study. Understandings with clients or subjects about the research and subsequent publication, which probably would not be easily realised, would have to be worked out with those responsible for drawing up the plan of tourism development, as well as their masters. People in charge probably would want to be convinced that the project would be viable, that it would do them no harm, and, indeed, that it could possibly help them, not unusual requirements for doing ethnographic work anywhere these days and particularly salient where issues of legitimacy come to the fore.

Finally, if researchers were fortunate enough to be able to follow through in the realisation of their proposal, a measure of the significance of the study could be found in its fate. Was it adopted and carried through or did it gather dust on some shelf somewhere, as one anthropologist speculated (in a letter to the author) about his report on projected tourism development to a Micronesian government? This question about implementation brings to mind the possibility of doing further ethnographically based investigations (Butler 1993: 143–146) about the realisation of tourism plans in destination areas, illuminated by Howell's (1994) work, mentioned earlier, which is a prime example of government-funded, ethnographically based tourism research.

Fortunately, an anthropologist could be helped in dealing with these and other research problems by attending to the existing literature on ethnographic fieldwork, especially in one's own culture (Messerschmidt 1981; Whyte 1997; Wright 1995; Nash 2000), and, perhaps, doing some preliminary studies near home, which, more than likely, would be in some Western society. Moreover, there are certain graduate programmes devoted to this sort of thing, but unfortunately they do not tell us much about ethnographic research among the powerful, which means that anthropologists would have to scramble a bit at the outset (but see Hoffman 1980).

These days, reflexive observations about the influence of subjects, their 'gatekeepers' and others, past and present, on the researcher during the course of a study would naturally be called for. (In Chapter 8, Hall considers the issue of 'gatekeepers' in tourism in more detail.) Though there is a growing literature on such matters (Davies 1999), we do not appear to be very far along in such reflexive operations where the ethnographic study of tourism is concerned (Nash 2000). Phillimore and Goodson discuss this concern in more detail in Chapter 1. Nevertheless, considering that the benefits of a tourism study of this kind could be considerable for anthropology and other social sciences, any efforts to improve this aspect of the ethnographical component would seem to be well worth the effort involved.

First, as suggested earlier, there would be an opportunity to fill out the anthropological picture of the touristic process, which so far has been concerned mostly with ethnographic studies of tourism's consequences for host peoples, particularly on the metropolitan periphery. Following this new line of attack, we would learn more about the production of tourism in its various forms in contemporary metropolitan areas. A host of research questions come to mind, such as, for example, the reasons for the (Japanese) government-inspired initiative of not too long ago to promote Japanese international tourism or certain corporate advertising campaigns to promote specific kinds of tourism (Reimer 1990; Selwyn 1990). In sum, by following this line of research we would learn more about why, according to Graburn (1983: 19), 'specific touristic modes are attached to particular social groups in the historical period in which they are found'.

Second, in pursuing such research, anthropologically orientated tourism researchers could take advantage of their alleged interdisciplinary sensitivity (see Nash 1996: vii) to tap a variety of more or less well-grounded theories about the productive aspects of the touristic process. By taking some tourism to be a superstructural manifestation of basal social elements, as does Nash (1995), they could begin to follow the processes involved along the power trail of its production and study it, perhaps by a 'multi-sited ethnography' as suggested by Marcus (1995) and elaborated by Bestor (2001). As a result, they would be able to tell whether the tourism in question is a simple *spillover* of basal or infrastructural activities in modern society, as with Boorstin (1964), a *compensation* for those activities, as MacCannell (1976) would have it, or (more likely), as Cohen (1993) suggests, both, depending on the social position of the actors involved. In working this through, such hypotheses and others suggesting that (modern) tourism reflects a need for alternation (Graburn 1989), a desire for nostalgic fixedness in a rapidly changing world (Graburn 1995), a search for authenticity (MacCannell 1976; see also Cohen 1988), a particular strategy in a consumer-oriented economy (Selwyn 1990; see also Cohen 1988), a means of gaining *distinction* in a certain system of social stratification (Urry 1990), or a response to oppressive industrial conditions (Krippendorf 1986) could be useful for framing the process of decision-making that is involved in the generation and shaping of contemporary tourism. Combined with solid ethnographic study of the power-imbued social processes to which these and other theoretical questions allude, they could provide a basis for the kind of tourism study that combines theoretical sophistication and methodological competence, which, as Dann *et al.* (1988) remind us, ought to be essential in all tourism social science. Such studies would replace airy theoretical speculations about the connections between basal conditions in tourism-generating situations – that is, the tourism-relevant aspects of 'demanding' home societies, and various aspects of tourism they propagate with theoretically informed and ethnographically based observations of the social actions involved.

Finally, a chronicle of all the social processes associated with the accomplishment of this research programme should be developed. Since such study would largely break new ground, anthropologists involved should leave as full a record as possible in the form of field notes and journal entries about self-involved methodological procedures, particularly with regard to gaining entry and subsequent adaptation. Considering all the factors weighing against such an enterprise, it would be helpful to have accounts of difficulties encountered, their effects on the researcher, and the ways in which they were dealt with. Even if largely absent in tourism research (see Nash 2000), these kinds of self-reflexive considerations have gained some ground in anthropology ever since Nash and Wintrob (1972) and some others first noted a trend in this direction, and are increasingly *de rigeur.*

There is more to be gained, though, than methodological advice from accounts of attempts to undertake and carry out ethnographic research on the productive aspect of the touristic processes that prevail in our world. In their stories of the responses of gatekeepers and subjects to research initiatives, as well as from occasionally revealed breaches of secrecy within productive organisations such as the World Bank (Kahn 2001), anthropologists (perhaps unknowingly) will be contributing to the picture of a culture that they know not nearly well enough. In this way, any inadvertent revelations during the course of a study may be turned into a modicum of research success.

Conclusion

The project proposed here, which involves the transfer of ethnographic procedures from the recipient end of the tourism development to its generation in Western metropolitan centres would broaden and deepen our understanding of the production of tourism in a particular historical context – one which is dominated by an increasingly global Western imperialism. But other strategies for analysing the production of tourists and tourisms in other historical situations will need to be developed also, if anthropologists are to maintain their pan-human credentials. For example, travellers at leisure among the aboriginal San, who, according to Silberbauer (1972), tended to visit relatives in allied bands from time to time with no other apparent purposes than diversion and pure sociability, could be examined. How would tourism researchers have studied the production of this tourism or proto-tourism if they were working with Silberbauer's data; and how would they have used the information from him and other specialists to get at this process when participant observation of aboriginal San is now nearly impossible? Can they rely on the memory of older informants, a not infrequent device in anthropological ethnographies? The possibility of grasping touristic actions in such situations without all the conventional ethnographic tools may be somewhat beyond them, but alternative techniques continue to provide anthropologists with information about the societies of humankind. Besides providing direct information, they can also sharpen up the cross-cultural sensitivity that should be in the background of all anthropological projects.

Considering the weight of tradition that has tended to pull anthropologists towards the periphery of what is increasingly recognised as a Western capitalist world system, it will not be easy to change priorities in such a way as to reorient the anthropological study of tourism. Researchers writing proposals to granting agencies or seeking work in tourism research will, of course, have to deal with any remaining questions about the legitimacy of tourism study itself (see Nash 1996: 1–3). They also will have to convince reviewers or employers that ethnographic inquiry by an anthropologist at or near the centre of what they may take to be world capitalist culture, not in some host society on its periphery, is acceptable and that they have the expertise to handle it. In all such persuasion, it might be helpful to remember that the gatekeepers and subjects involved may at times be operating under conditions of what amounts to institutionalised secrecy not unlike those obtaining in some of the classical studies of sorcery.

With only a little reflection, however, they will realise that any anxiety they experience about undertaking such projects could derive from powers at the heart of the Western, capitalist domain. Gaining access and functioning in the world of the powerful promises to be a different, perhaps more difficult chore than in the cultures of the less powerful on the periphery. So anthropologists' trepidation about the ethnographic study of worlds where tourism initiatives are generated these days would seem to be not entirely misplaced.

How should anthropologists prepare themselves to undertake research in this new venue? First, they should have the objectives of science and of anthropology well in hand. These involve researchers concerning themselves with the human condition wherever it exists, seeking as intimate an acquaintance with human subjects as possible, and being theoretically informed, empirically grounded, and self-reflexive and critical about themselves and their work. There should be no excuse for not being in command of the research product they have to sell. Second, they should be creative in their various sallies in the manner suggested by Smith (1992: 3–4) not too long ago for anthropologists contemplating the study of the comparatively new subject of tourism. Third, they ought to have some understanding of problems associated with research initiatives among the powerful such as those illuminated in studies of the cultures of policy professionals (Donnan and Macfarlane 1989; Wright 1995), as well as other studies in whatever discipline, which means that they should give up narrow parochialisms and take advantage of possible cross-disciplinary fertilisation that can help them on their way.

To bring this chapter to a close, it should be noted that there are, or ought to be, some (how many?) anthropologists who are not exactly ill prepared for a project involving ethnographic study in modern Western society. These anthropologists would be suitably adventurous, adequately supported and trained, and could acquire whatever expertise was necessary for carrying out a research agenda on the productive aspect of tourism in our world. Among these, there ought to be a number who have (to paraphrase Max Weber in his discussion of religious virtuosos) the requisite 'musicality' for gaining an intimate ethnographic understanding of the people involved in the production of tourism as well as the reflexivity for keeping themselves and their methods in the picture. If these anthropologists also were adequately supported and situated favourably enough for establishing requisite contacts and gaining co-operation in their ethnographic ventures, they might be able to carry through a project which would have significant consequences not only for an ethnographically based study of tourism, but for other anthropological projects as well.

Summary

- In their comparatively brief history of tourism study, anthropologists have tended to be 'fixated' on the consequences of Western-initiated tourism development in smaller, less developed societies, this despite the fact that, anthropologically viewed, tourism has other aspects to be studied.
- Anthropological investigations of tourism have usually made use of the traditional method of anthropological research, namely ethnography, which has a long and distinguished history of application in the less developed world.
- But this method can be applied to other aspects of tourism – in particular, those situations in which tourism is generated and directed.
- The application of ethnographic methods in such situations is discussed by noting some small steps taken already in this direction as well as the problems and possibilities involved in such research.

Questions

1   The title of this chapter refers to 'new wine' and 'old bottles'. What do these phrases refer to?
2   The author argues that certain disciplinary constraints have prevented anthropologists from broadening their consideration of tourism. What are they?
3   Give an example of an ethnographic research project that would explore some of the forces involved in the generation of tourism.
4   What are some of the new problems that anthropologists would encounter in doing the kind of ethnography proposed in this chapter and how could they be resolved?

Acknowledgement

An earlier version of this chapter was presented in a session (Anthropological Contributions to Travel and Tourism) of the 2001 annual (November) meeting of the American Anthropological Association in Washington, DC. The author wishes to thank Tim Wallace for organising the session, and other participants and members of the audience for their comments.

References

Asad, T. (1973) *Anthropology and the Colonial Encounter*. London: Ithaca.

Bee, R. (1974) *Patterns and Processes*. New York, NY: The Free Press.

Bestor, T. (2001) 'Supply-side sushi: commodity, market and the global city', *American Anthropologist*, 103: 76–95.

Boissevain, J. (ed.) (1996) *Coping with Tourists: European Reactions to Mass Tourism*. Providence, RI: Berghahn Books.

Boorstin, D. (1964) *The Image: A Guide to Pseudo Events in America*. New York, NY: Harper and Row.

Bruner, E. (1994) 'Abraham Lincoln as authentic reproduction: a critique of postmodernism', *American Anthropologist*, 96: 397–415.

Burns, P. (1999) *An Introduction to Tourism and Anthropology*. London and New York, NY: Routledge.

Butler, R. (1993) 'Pre- and post-impact assessment in tourism development', In D. Pearce and R. Butler (eds) *Tourism Research: Critiques and Challenges*. London: Routledge.

Cohen, E. (1974) 'Who is a tourist?: A conceptual clarification', *Sociological Review*, 22: 527–555.

Cohen, E. (1979) 'A phenomenology of tourist experiences', *Sociology*, 13: 179–202.

Cohen, E. (ed.) (1985) 'Tourist guides: pathfinders, mediators and animators', *Annals of Tourism Research*, 12: 5–29.

Cohen, E. (1988) 'Authenticity and commoditization in tourism', *Annals of Tourism Research*, 15: 371–386.

Cohen, E. (1993) 'The study of touristic images of native people: mitigating the stereotype of a stereotype', In R. Butler and D. Pearce (eds) *Change in Tourism: People, Places, Processes*. London: Routledge.

Crick, M. (1989) 'Representations of international tourism in the social sciences: sun, sex, savings and servility', *Annual Review of Anthropology*, 18: 307–344.

Dann, G., Nash, D. and Pearce, P. (1988) 'Methodology in tourism research', *Annals of Tourism Research*, 15: 1–28.

Davies, C. (1999) *Reflexive Ethnography: A Guide to Researching Self and Others*. London: Routledge.

Donnan, H. and Macfarlane, G. (eds) (1989) *Social Anthropology and Public Policy in Northern Ireland*. Aldershot: Avebury.

Forster, J. (1964) 'The sociological consequences of tourism', *International Journal of Comparative Sociology*, 5: 217–227.

Foster, G. (1986) 'South Seas cruise: a case study of a short-lived society', *Annals of Tourism Research*, 13: 215–238.

Graburn, N. (1977) 'Tourism: the sacred journey'. In V. Smith (ed.) *Hosts and Guests: The Anthropology of Tourism.* Philadelphia, PA: University of Pennsylvania Press.

Graburn, N. (1983) 'The anthropology of tourism', *Annals of Tourism Research,* 10: 9–34.

Graburn, N. (1989) 'Tourism: the sacred journey'. In V. Smith (ed.) *Hosts and Guests: The Anthropology of Tourism,* 2nd edn. Philadelphia, PA: University of Pennsylvania Press.

Graburn, N. (1995) 'Tourism, modernity and nostalgia'. In A. Akbar and C. Shore (eds) *The Future of Tourism: Its Relevance to the Contemporary World.* London: Athlone.

Handler, R. (1988) *Nationalism and the Politics of Culture in Quebec.* Madison, WI: University of Wisconsin Press.

Handler, R. and Gable, E. (1997) *The New History in an Old Museum.* Durham, NC: Duke University Press.

Hartmann, R. (1988) 'Combining methods in tourism research', *Annals of Tourism Research,* 15: 88–105.

Hoffman, J. (1980) 'Problems of access in the study of elites and boards of directors'. In W. Shafir, R. Stebbins and A. Turnowitz (eds) *Fieldwork Experience: Qualitative Approaches in Social Research.* New York, NY: St Martin's Press.

Høivik, T. and Heiberg, T. (1980) 'Centre–periphery tourism and self-reliance', *International Social Science Journal,* 32: 69–98.

Howell, B. (1994) 'Weighing the risks and rewards of involvement in cultural conservation and heritage tourism', *Human Organization,* 53: 150–159.

Jafari, J. (1987) 'Tourism models: the sociocultural aspects', *Tourism Management,* 8: 151–159.

Kahn, J. (2001) 'World Bank weighing steps against one of its economists', *New York Times,* 7 September, p. C1.

Kottak, C. (1966) 'The structure of equality in a Brazilian fishing community', PhD Thesis, Columbia University, New York, NY.

Krippendorf, J. (1986) 'Tourism in the system of industrial society', *Annals of Tourism Research,* 13: 517–532.

Lewis, I. M. (1995) 'Anthropologists for sale?'. In A. Akbar and C. Shore (eds) *The Future of Anthropology: Its Relevance to the Contemporary World.* London: Athlone.

McCall, M. and Bobko, P. (1990) 'Research methods in the service of discovery'. In M. Dunnette and L. Hough (eds) *Handbook of Industrial and Organizational Psychology,* vol. 1, Palo Alto, CA: Consulting Psychologists.

MacCannell, D. (1976) *The Tourist: A New Theory of the Leisure Class.* New York, NY: Shocken.

Magubane, B. (1996) *The Making of a Racist State: British Imperialism and the Union of South Africa.* Trenton, NJ: Africa World Press.

Marcus, G. (1995) 'Ethnography in/of the world system: the emergence of multi-sited ethnography', *Annual Review of Anthropology,* 24: 96–117.

Messerschmidt, D. (ed.) (1981) *Anthropologists at Home in North America: Methods and Issues in the Study of One's Own Society.* Cambridge: Cambridge University Press.

Miller, D. (1997) *Capitalism: An Ethnographic Approach.* Oxford and New York, NY: Berg.

Mintz, S. (1985) *Sweetness and Power: The Place of Sugar in Modern History*. New York, NY: Viking Penguin.

Moeren, B. (1983) 'The language of Japanese tourism', *Annals of Tourism Research*, 10: 93–108.

Nader, L. (1972) 'Up the anthropologist: perspectives gained from studying up'. In D. Hymes (ed.) *Reinventing Anthropology*. New York, NY: Random House.

Nash, D. (1979) *Tourism in Pre-industrial Societies*. Aix-en-Provence: Centre des Hautes Études Touristiques.

Nash, D. (1981) 'Tourism as an anthropological subject', *Current Anthropology*, 22: 461–481.

Nash, D. (1995) 'Prospects for tourism study in anthropology'. In A. Ahmed and C. Shore (eds) *The Future of Anthropology: Its Relevance to the Contemporary World*. London: Athlone.

Nash, D. (1996) *Anthropology of Tourism*. Oxford: Pergamon.

Nash, D. (2000) 'Ethnographic windows on tourism', *Tourism Recreation Research*, 25: 29–36.

Nash, D. and Wintrob, R. (1972) 'The emergence of self-consciousness in ethnography', *Current Antropology*, 13: 527–541.

Novak, D. and Lekachman, R. (eds) (1964) *Development and Society: The Dynamics of Economic Change*. New York, NY: St Martin's Press.

Nuñez, T. (1963) 'Tourism, tradition and acculturation: *weekendismo* in a Mexican village', *Ethnology*, 2: 347–352.

Pearce, P. (1982) *The Social Psychology of Tourist Behavior*. New York, NY: Pergamon Press.

Reimer, G. (1990) 'Packaging dreams', *Annals of Tourism Research*, 17: 501–512.

Rival, L. (1997) 'Oil and sustainable development in the Latin American humid tropics', *Anthropology Today*, 13: 1–3.

Said, E. (1978) *Orientalism*. New York, NY: Vintage Books.

Selwyn, T. (1990) 'Tourist brochures as post-modern myths', *Problemy Turystyki*, 8: 13–26.

Silberbauer, G. (1972) 'The G/Wi bushmen'. In M. Bicchieri (ed.), *Hunters and Gatherers Today*. New York, NY: Holt, Rinehart and Winston.

Smith, V. (ed.) (1977) *Hosts and Guests: The Anthropology of Tourism*. Philadelphia, PA: University of Pennsylvania Press.

Smith, V. (ed.) (1989) *Hosts and Guests: The Anthropology of Tourism*, 2nd edn. Philadelphia, PA: University of Pennsylvania Press.

Smith, V. (ed.) (1992) 'Anthropology and tourism', Special issue of *Practicing Anthropology*, 14 (2), Spring.

SRRC (Social Science Research Council) Seminar (1954) 'Acculturation: an exploratory formulation, *American Anthropologist*, 56: 973–1002.

Stevenson, M. and Naylor, J. (1990) 'Judgment and decision-making theory'. In M. Dunnette and L. Hough (eds) *Handbook of Industrial and Organizational Psychology*, vol. 1. Palo Alto, CA: Consulting Psychologists.

Swain, M., Brent, M. and Long, V. (1998) 'Annals and tourism evolving: indexing 25 years of publication', *Annals of Tourism Research*, 25: 991–1014.

Urry, J. (1990) *The Tourist Gaze*. London: Sage

Waller, J. and Lea, S. (1999) 'Seeking the real Spain? Authenticity in motivation', *Annals of Tourism Research*, 26: 110–129.

Whyte, W. (1997) *Creative Problem Solving in the Field: Reflections on a Career.* Walnut Creek, CA: Altamira.

Wolf, K. (1991) Distinguished lecture: 'facing power', *American Anthropologist*, 92: 586–596.

Wolf, K. (1999) *Envisaging Power.* Berkeley, CA: University of California Press.

Wright, S. (1995) 'Anthropology: still the uncomfortable discipline'. In A. Akbar and C. Shore (eds) *The Future of Anthropology: Its Relevance to the Contemporary World.* London: Athlone.

11 From ontology, epistemology and methodology to the field

Jenny Phillimore and Lisa Goodson

Creativity, power, reflexivity and the need for transparency

Part I of this book has focused on some of the issues that influence the research process, ranging from philosophical debates to practical concerns. In this first part of the book we have taken some of the concepts now frequently discussed in social science research and considered them in some detail in relation to the study of tourism. Together, the chapters in Part I make a strong argument for tourism researchers to consider emergent thinking emanating from research in other disciplines and fields, and to progress towards taking on board a much broader range of approaches to research. A more embracing approach calls for greater consideration of epistemological, ontological and methodological issues in relation to the researcher as an individual, to the research problem and the research setting, and to how the different elements of the research process can fit together in a complementary fashion. In addition, contributors are calling for much greater levels of transparency in tourism research, with researchers being more open about their personal biography and their experiences in the field, and how these two interact and evolve over time to impact on the different phases of the research process.

Jenny Phillimore and Lisa Goodson showed in their review of post-1996 tourism research that research in the field is still largely based in Denzin and Lincoln's (1998) first to third 'moments' of qualitative research and underpinned by positivist and postpositivist inquiry paradigms. They identified overemphasis of the use of the third person to write depersonalised research accounts and, in relation to this, the tendency for many authors to write using an 'expert' voice that offered a single unifying interpretation of events, places, perceptions and identity without considering the wide range of different perspectives that Keith Hollinshead argues are very much a feature of tourism research. With some notable exceptions, they found that tourism research almost completely lacked any critical reflection of methodological issues, and indeed on occasions lacked even a basic insight into the methodological approach adopted. While there was emphasis on gathering the views of host communities using qualitative techniques, there was a tendency for this Other to be portrayed as one monolithic group, with very little consideration of difference either within or between groups. This sanitising of real life and oversimplification of complex issues and places often extended as far as generalising findings from relatively small-scale studies to national or regional or even international level. In addition, while the majority of work was based on data collected in relatively short periods of research, it was represented as 'the way it is' rather than as a snapshot of the situation in a particular place at the time the data were collected.

Hollinshead argued that in using quantitative methods, researchers have had difficulties envisaging the real cultural world of local groups in tourism, but that doing so is critical, given that tourism has become an increasingly important force in the de-making and making of held individual and societal realities. He argued that choice of research instrument depends on critical skills of applied philosophical awareness rather than methods-level decision-making. Hence, developing an understanding of epistemological and ontological issues and how they relate to tourism is critical. In particular, he highlighted a lack of understanding about the ontological 'hereness' of places, namely how sites are selected and projected and how such places are locally mediated by interest groups. Although many of these gaps in understanding are stark, as for example that between hosts and guests, others are more nuanced and lie between different sections of the same populations. He outlined the ways in which qualitative approaches are the most suitable tools with which to pick up on these subtle issues. The chapters of Part II focus on some of these subtle issues and show how research at a micro level can help inform understanding at both mezzo and macro levels.

Where Phillimore and Goodson were able to observe influences from other paradigms or the various 'moments' outlined by Denzin and Lincoln (1998), there was evidence that researchers were adopting an eclectic approach, mixing and matching elements from different research perspectives so that the methods adopted, the way the data were analysed, the extent to which respondents' voices were accounted for and the claims made for the data may all relate to different underlying paradigmatic principles. Phillimore and Goodson argue that this fracturing of approach relates in some part to the artificial nature of Denzin and Lincoln's typology and any analysis based on such a simplistic tool. However, it is also a reflection of the relative newness of tourism as a field and more so to its multi- and interdisciplinary nature. In adopting this fractured approach to research, some researchers are introducing new and more creative ways of doing qualitative tourism research and, as a result, are moving the study of tourism forward. The experimentation with different approaches and the way in which more experienced researchers have demonstrated and applied their research to a wide spectrum of research principles and practice have helped, albeit incrementally, to push the boundaries of tourism research.

A fundamental part of any research process is the decisions that under-pin it. Inquiry paradigms are fundamental to decision-making; our beliefs about what tourism is, what can be known about tourism and how this knowledge can be uncovered have a massive impact on the kinds of research we do. Evidence of this is presented in Part II of the book, where chapters by Adele Ladkin and Jennie Small consider the related tech-niques of life history and memory-work but approach their research prob-lems in entirely different ways to totally different ends (in Chapters 14 and 15 respectively). Without knowledge of their different beliefs and details of the ways in which they applied their techniques, it is difficult to judge their findings. We argue that only through this reflexivity can the indexi-cality of knowledge and, therefore, validity be located. Part II aims to bring the reflexivity called for in relation to the research process to the surface so that the indexical nature of knowledge can be highlighted in the context of the research process.

One of the key issues to emerge from Part I is that of power in research. For Tribe, this relates to the way in which tourism falls into two main fields, with the most coherent/consolidated being business aspects. He maintains (in Chapter 3) that in tourism, technical interests have sought control and management through scientific positivism: 'tourism studies depends usually on what we have gone looking for and how we have gone looking for it', so that we are creating rather than discovering tourism. Certainly this links with Phillimore and Goodson's concern that the dis-covery element of tourism research is currently lacking from the field. Tribe identified a direct link between the positivistic approach to know-ledge creation and an overemphasis on studies that suit the needs of busi-ness which he fears may lead to gaps in knowledge in areas that are not codified. To deal with this situation we need to look at tourism from other perspectives and be prepared to undertake research in ways that do not fit with the codes prescribed by positivism. Alain Decrop's chapter (Chapter 9) offered some middle ground in showing how 'trustworthiness' could replace the positivist criterion of reliability as a way of judging qualitative research. Others, such as Humberstone and Small in this book (Chapters 7 and 15), argue that any kind of codification results in constraints to cre-ativity. The offerings in Part II focus on research from a wide range of per-spectives and look at ways in which the power inherent in the research process can be dissipated so that research participants have a greater role in setting research agendas and interpreting findings.

Following on from Tribe's discussion of power in shaping the nature of tourism knowledge, Margaret Swain in Chapter 6 focuses upon intersect-ing power dynamics and how they shape research agendas and associated results. She suggests that embodiment theory, which to date has largely been absent from tourism, can help to inform research in tourism. Swain argues that as bodies physically take up space in tourism, we need to explore how space is created in order to understand how embodied participants in tourism interact on the basis of their embodied selves. In talking about embodiment, Swain brought together mind and body rather than approaching them with the traditional Cartesian dualisms and favouring of the mind over the body. This means considering the impact of an individual's physicality, sexuality, gender, ethnicity and class as well as their ideology, religion, etc., and refers to embodiment from the perspectives of both the researcher and the researched. She gave examples of how the researcher's body impacts on the way they are able to undertake research in particular settings. This is a theme picked up in several of the chapters in Part II as authors explore the impact of their body on their research, the individuals involved in their research and on their interpretation and representation of findings. For Swain, the combination of embodiment with ethics can help researchers strive to understand differences and promote more equitable human conditions in the course of tourism research. The need for an ethical approach to tourism research is a theme which runs through the book and an important guiding principle in several of the research projects discussed in Part II.

Barbara Humberstone's chapter (Chapter 7) builds upon the issues of power and ethics that Swain raises in her consideration of embodied research. She argues that all of us undertake research and live our lives according to a particular ideological standpoint, and thus no researcher is capable of objective, value-free research. Standpoint is very much about considering issues from the perspectives of others rather than the perspective of the so-called objective researcher. From this stance, she suggested that standpoint research offers new ways of thinking about the effects of tourism on both the environment and marginalised peoples, and offers an opportunity to rethink the supposedly neutral values of the traditional/natural scientific approach. Looking at issues from the perspectives of marginalised groups helps us to recognise their needs and environmental context, and helps us to find new ways of encouraging sustainability. However, in order to take these matters into account and develop a much more ethical approach to research in sensitive environments it is suggested that it is necessary to collaborate with marginalised groups. While Humberstone argues strongly for such an approach, she raises concerns about the extent to which such collaboration might be viewed as valid by the academy, a problem faced by Small in her chapter on memory-work with women in New Zealand.

Michael Hall in Chapter 8 picks up on the theme of reflexivity and also sets out in some detail just how constrained researchers are by the academy, publishers and sponsors. He expresses concerns that researchers are expected to research and write in a particular way in order to meet the requirements of the editorial boards attached to learned journals. In this context, while researchers are judged by their published output, their ability to take risks and try to publish ground-breaking qualitative research in the public domain will be severely limited. Overcoming this problem requires two approaches. First, the academy and publishers need to be more open to new developments and perhaps consider the issues of trustworthiness raised by Decrop. Criteria for acceptance of papers and sponsoring of research projects might be adapted to include measures such as transparency in the research process, involvement of the researched in the process, and consideration of the authenticity of data. In addition, we need to see the introduction of a series of edited books focusing on relating transparent accounts of qualitative research perhaps based on particular themes. Being more open and letting readers behind the veneer of polished articles will help researchers to build on existing knowledge and approaches.

Dennison Nash's unease about the lack of reflexivity in tourism research adds weight to that of Hall. Nash is concerned about the quality and focus of ethnographic work in tourism because so much of it appears to be ideologically driven, but few authors take account of the ideologies that underpin their research. In addition, Nash in Chapter 10 highlights the tendency for tourism ethnography to focus upon the lives of the powerless, and outlines in particular how there has been an overemphasis on acculturation and development in tourism research. He suggests that in order to build a more holistic picture of the tourism industry, we need to undertake more small-scale work to find out what makes a tourist. This research, he argues, should be based on the lives of tourists within the societies of tourist-generating countries, and greater emphasis should be placed on ethnographic research on the powerful. Researchers may find the change in focus uncomfortable because it would involve, where undertaking ethnography in global corporations, a massive shift in power relations. However, if tourism researchers are genuine about undertaking research for the benefit of the least powerful stakeholders in tourism, they must shift the focus from descriptive accounts of how the powerless are exploited to the unsavoury issue of how generating nations exploit them. In doing this we break with conventional neo-colonialist power relations and reframe power relations in research. Returning to the key issues of reflexivity and transparency in research, Nash again stresses the importance, when breaking new ground, of chronicling 'all the social processes associated with the accomplishment of this research programme'. We need to know how anthropologists build relationships with the powerful so that we can learn from their experiences.

From theory to practice

The chapters in Part II are written with various degrees of reflexivity, but all give some indication of how, whether using the first or the third person, it is possible to take account of self in the research process. The various contributions have been selected because they provide a wide variety of examples of the research process in relation to a range of different variables, including techniques from interviews to focus groups, different geographical locations from the urban metropolis of London to the remote Indonesian village of Wogo, and different perspectives from explicitly positivist to highly interpretive. The themes that emerge in Part I – decision-making, power and creativity – are present to varying degrees in each chapter of Part II, depending on the ideological stance of the contributor. These chapters are included not as idealised representations of field research to show putative qualitative researchers how it *should* be done, but as examples of some of the myriad ways in which qualitative research actually happens in the field. They raise some of the problems that researchers come across when in the thick of their research, and show how they deal with those problems. Rather than representing research, as many of the brief methodological accounts reviewed post-1996 have, as a seamless, linear process from research question to research paper, these contributions seek to show the reality of research as a messy, non-linear business.

The second part of the book begins with a chapter by Karen Thomas that demonstrates well the messiness and multidimensionality of research. Thomas, in her account of her work on the motivations of adventure tourists living in the United Kingdom writes of the process as a complex interplay of choices and decisions. She demonstrates how, at every step of the process, there are many different decisions that can be taken, each leading to its own distinct research route and potentially different outcome. While her research began with a solid positivist underpinning which fits well with the modes of research associated with much of the work on tourist motivations, she outlines how, over time, she moderated her stance as she began to realise the benefits of a more interpretive approach. This chapter demonstrates well the points made by Hollinshead, and Phillimore and Goodson, about research being a creative process based on the need to develop tools specifically for the job in hand regardless of 'convention' in a particular research area. Thomas's account of the development of a thematic framework to analyse her data also gives useful insight into an aspect of the research process not often considered.

Fiona Jordan and Heather Gibson in Chapter 13 discuss the way in which their project and joint working relationship developed, and, like Thomas, they consider the range of decisions that underpin the research process. The power relationship between the researched and researchers is one of the issues under consideration as Jordan and Gibson explore the ways in which their interviews with English and American solo women travellers sought to build trust and rapport in an attempt to reduce power differentials. The contributors also reflect on the inductive nature of their research and how they were able to allow research questions to emerge from the data as the research progressed, while retaining a framework that enabled comparison between interviews and across continents. Their discussion of the use of the 'constant comparison' method for data analysis shows some of the benefits of working collaboratively, in that greater discussion around the meaning of data is required when there are two interpretations rather than the one associated with sole authorship. Reflecting some of the issues raised by both Hall, on the way in which data is written, and Humberstone, in the way in which the researched are represented, these contributors address the problem of giving voice to respondents in an academic environment and show how difficult it is to operationalise 'fifth-moment' research, given the constraints of publishing. This chapter shows how the crisis of representation can inform the research process by demonstrating the way in which tourism researchers shape findings through the process of interpretation, and how unpacking and elucidating this process helps the reader assess more accurately the authenticity of the data according to the criteria outlined by Decrop.

Ladkin's approach is one that is underpinned by positivism and a traditional stance, but it is also informed by 'blurred genres'. She demonstrates well how multiple methodological approaches can combine in the triangulation which, Decrop argues, is one of the most powerful ways of using qualitative research, to create a useful and rigorous technique. Life history work is a good example of an approach that has evolved with traditional and changing beliefs about the nature of knowledge. Ladkin considers the potential for the technique, especially in filling one of the key gaps around longitudinal perspectives, moving research from the snapshot approach which, Phillimore and Goodson argue, is one of the key weaknesses in tourism research, to one that is actually grounded in a wider range of temporal moments. Such an approach in this context would help us to gather more reliable knowledge about destinations, decision-making processes and career choices, and would help to inform policy more reliably than snapshot research.

Small's chapter on memory-work is underpinned by a social constructivist paradigm and very much based in the fourth and fifth 'moments'. Her research is highly individualistic and focuses on set moments in time, making no attempt to generalise from findings but seeking to understand how women experience holidays at different points in their life. She introduces the notion of the researched as co-researchers and seeks to reduce power relationships in research by encouraging women to run their own research sessions and to interpret the 'rules' as they see best. In so doing, Small demonstrates well the approaches recommended by Swain and Humberstone. She gives women the opportunity to voice their thoughts and feelings about holidays from a range of different points in the life cycle, ensuring that their embodied selves are represented and that the research is shaped from the standpoint of participants rather than the researcher. Small decided that flexibility was important to maximise rapport, and thus empowered the 'co-researchers' to run their own research events as they pleased. Despite her best efforts to reduce power differentials and give her participants a role in interpretation of the findings, Small was also constrained by the rules of the academy.

Jill Belsky introduces the first ethnographic study included in this book. Her work, similar to Small's, focuses on a collaborative project, but in this example it is undertaken with her students over a period of years. Belsky looks at how different residents take advantage of opportunities differentially available to them through a community ecotourism project. She discusses in some detail the notion of critical reflection, a structured, analytical and emotional process that helps researchers to examine the ways in which they make meaning out of circumstances, events and situations, and demonstrates how through adopting this approach with her student co-researchers she has been able to facilitate discovery. Belsky stresses the importance of learning from everyday life in Belize in a naturalistic way and then using what is gleaned from the field to shape further inquiry in an iterative fashion. In addition, the relationships built up with local people through both longitudinal research and living in their houses enabled the research participants to become more comfortable with the situation and feel confident to raise issues as they wished, thus reducing power differentials. She stresses, as do Jordan and Gibson, how research in a team can bring significant added value. In Belsky's experience, the multiple interpretations that result from working on data analysis in diverse groups constituted of indigenous, as well as outside, researchers brought benefits from the range of different insights offered. She also looks at the role of research in post-colonialism and argues that qualitative methods have a vital key role to play because they allow the voices of the least powerful to be heard. Belsky stresses, as Nash has previously argued, the highly political nature of tourism research, and shows how participatory research with a wide range of players in the tourism industry can give local people the opportunity to take some control of the research topic, process and product.

Stroma Cole's chapter outlines her experiences of ethnographic research in the remote Indonesian village of Wogo. Whereas, like that of Belsky, Cole's research took place over a long period of time, Cole was a lone researcher in the field, and her research was very much impacted by the range of different personal relationships she develops with local people. Unable to share her interpretations with co-researchers, Cole seeks to validate her interpretations by checking her findings with local informants. This approach, and the close personal relationships developed, reduced power differentials in the research relationship and gave local people the opportunity to shape the research, thus giving voice to the standpoint of marginalised peoples, as proposed by Humberstone. Cole considers how the changing shape of her embodied self affected the research in different ways over time as she moved from young female tour operator to academic researcher and mother. For Cole, her relationship with the villagers leads to a range of benefits from having an ability to identify conformist responses to gaining access to discussions that would not be open to outsiders. However, she also discusses the flip side to familiarity and equitable power relationships in the range of expectations she has to live up to, as villagers viewed her as a member of the family. In keeping with Hollinshead's discussion of the need for fine-grained details about the 'hereness' of place, she also notes the importance of gleaning information about everyday life outside of formal research and shows how being on-site for long periods of time can enable the researcher to gain an understanding of the social, economic and political fundamentals of a society which provides qualitative research with the wider and more complex, but informed, context within which to interpret their emergent findings.

In the final chapter of this book, Guy Jobbins discusses power relationships between himself, his respondents and the translators he employed in Morocco and Tunisia. Jobbins entered the field somewhat unnerved by the notion of qualitative research and concerned that he would not be able to trust his respondents. Through the process of undertaking research, he learned the importance of building social relationships with respondents and his translators so that they might build trust in him. This proved critical, given that much of his research was undertaken with government officials unaccustomed to researchers and reluctant to speak openly. Perhaps the most elucidating moment for Jobbins was when he became aware of the extent to which he was imposing his interpretation upon respondents. Forced to review power relations completely, Jobbins not only reconsidered the way he interpreted data in North Africa but began to rethink the way he communicated in everyday life. Jobbins's account demonstrates very clearly the importance of critical reflection in research at every level, as contributors in Part I have argued, from the point of setting research questions, to interactions in the field, to interpreting data and disseminating findings.

Taking tourism research forward

There has been much criticism about the state of tourism research from outside the field, particularly from leisure theorists. This book has argued that the time has come for tourism researchers to be more self-critical and more adventurous. This may mean adopting more of a qualitative approach to research, even though it does not necessarily require the wholesale adoption of qualitative methods. In tourism we need to acknowledge, as other researchers have, that research is a messy business. We need to give a great deal more consideration to the creative process of knowledge production. Currently we know little about how the research process happens in tourism, because the focus is upon the outcomes of research. It is important for all those working in the field to develop and share their knowledge about the kinds of struggles that arise when they are developing research problems and working in the field, and the ways in which problems can be dealt with. Part II offers the reader some insight into the real world of tourism research and some indication of how the issues discussed in Part I actually surface in the field. Reading accounts of research should be interesting and inspiring. We hope that Part II will encourage researchers to go out into the field, try new techniques, seek different ways of working with research participants, and develop more reflexive approaches to writing up their work.

Reference

Denzin, N. K. and Lincoln, Y. S. (1998) *The Landscape of Qualitative Research: Theories and Issues.* London: Sage.

Part II

From research theory to practice

12 The research process as a journey

From positivist traditions into the realms of qualitative inquiry

Karen Thomas

Aims of the chapter

- To outline how research can be seen as a product of a wide range of influences.
- To explore the use of qualitative focus groups within a multi-method study.
- To examine how the philosophical underpinnings of positivism have impacted upon the design and implementation of a focus group study.
- To explore the research process as a journey shaped by key phases of decision-making.

Introduction

At the heart of the research process lies a complex interplay of choices and decisions which mould the nature and direction of research. Thus, many of the fundamental challenges facing the social science researcher relate to the core activity of decision-making and the justification of the strategy and method(s) adopted (Crotty 1998: 2). Studying the theory and philosophical foundations of social science research reveals that 'different research ... methods are not just responses to different research needs but also embody quite different ontological and epistemological perspectives' (Arksey and Knight 1999: 15). Denzin and Lincoln (2000: 19–22) articulate this, outlining how the final choice of research strategy and method should be seen as a culmination of issues at the level of the researcher and the research paradigm, the latter involving the interconnected issues of 'ontology (... What is the nature of reality?), epistemology (What is the relationship between the inquirer and the known?), and methodology (How can we know the world, or gain knowledge of it?)'.

An understanding of the theory and philosophical foundations of social science research not only provides a knowledge base from which to evaluate current practice, but helps to ensure that researchers are 'better able to set forth the research process in ways that render it transparent and accountable' (Crotty 1998: 216). Although the need for reflexive accounts of the research process is accepted, a large divide can often be seen between the texts outlining the more philosophical issues related to research and those focusing upon research in practice. To help address this divide, the chapter draws from the author's doctoral research, which evaluates the application and utility of the International Tourist Role (ITR) scale (Mo *et al.* 1993) as a foundation for the development of theory and practice of tourism marketing (King 1997). The qualitative research component, explored here, was developed to evaluate how far the dimensionality of the ITR scale could be used as a conceptual framework to help marketers to communicate more effectively. The main body of the chapter sets out the development of a methodology designed to explore consumer reaction to visual imagery within the context of the novelty-seeking preference structures identified by the ITR scale.

A key influence behind this chapter has been the need to examine the research process in a reflexive manner, rendering explicit the process by which the research was designed and the key influences acting upon the decisions made. This will enable the discussion to outline how the research can be seen as a product of a wide range of influences. These influences arise from the assumptions of the researcher, and the research paradigm, the conceptual foundations and research objectives, together with the influences acting upon the wider research setting, which are more pragmatic in nature.

At a more abstract level the chapter draws an analogy between the research process and a journey. The different stages of the journey are shaped by key phases of decision-making and reflect the evolving nature of the research. The beginning of the journey was dominated by the positivist traditions of quantitative motivation research, but as the study evolved and expanded into the realms of qualitative inquiry, possibilities for research of a more interpretive nature opened up. While the underpinnings of positivism did remain quite strong during the design and implementation of the research, the concluding section will reflect upon the final stages of the journey and show how an appreciation of alternative paradigms, and their critiques of positivism, can help the critical evaluation of the whole research process.

Philosophical and conceptual underpinnings to the research

Before the qualitative research component can be explored, the chapter must first set out the context of the whole study, in order that its philosophical and conceptual underpinnings can be understood (the beginning of the journey). The early stages of the research were heavily influenced by the positivist traditions dominating tourism research and, notably, the research in tourism motivation striving to produce standardised and robust role scaling systems. A review of the motivation literature revealed that previous research had been piecemeal, highlighting that future research should be based upon testing existing techniques (King 1997: 122). Following a review of the various role scaling systems, the International Tourist Role (ITR) scale (Mo *et al.* 1993) was selected as a basis for the research. This 20-item scale, based upon Cohen's (1972) conceptualisation of novelty, proposes that within the context of international travel, the novelty construct is three-dimensional (Destination-Orientated Dimension, Travel Services Dimension and Social Contact Dimension), and a subsequent review of the literature added a fourth Arousal-Seeking Dimension to form the modified 26-item ITR scale (King 1997).

From a positivist perspective very much based in the 'traditional moment' (Denzin and Lincoln 1998), the quantitative component set out to test the dimensionality of the modified ITR scale. The research was based throughout on a purposive sample taken from two market niches, with respondents randomly sampled from the mailing lists of an adventure travel operator and a long-haul Asia specialist. The research gave further support to the ITR scale (Mo *et al.* 1993) as a valid and robust role scaling technique, and supported the Arousal-Seeking Dimension as a complementary scale (King 1997). The study went further to suggest that the potential of the scale could only be realised only if it is set within a broader research framework, providing the context for a multi-method approach incorporating both quantitative and qualitative research.

The quantitative component was highly etic, using a survey methodology to develop a typology combining ITR scale preference structures with pre-travel search behavioural characteristics. The focus of the qualitative component, however, was to evaluate the modified ITR scale as a conceptual framework. This set out to explore whether an understanding of how participants define and value concepts related to novelty-seeking can be used to enhance targeting through the creation of more salient imagery. To achieve this, the ITR scale needed to provide a tool through which it is possible to explore consumer preference structures and provide themes that are easily transferable to marketing communication messages. Based upon an understanding of the limitations of quantitative methods, in that they 'cannot fully address questions of understanding and meaning' (Riley and Love 2000: 166), this component moved beyond using the ITR scale as a measurement scale, to allow further exploration of the meanings attributed to the dimensions of novelty through the eyes of the tourist, taking on a more emic perspective (a turning point in the journey). Given, however, the philosophical foundations underpinned by positivism and the conceptual foundations set around the dimensionality of novelty-seeking, the following section demonstrates how the qualitative method developed needed to be more 'rigorously defined' (Denzin and Lincoln 2000: 21) and to allow the exploration of emic categories of knowledge to be structured around the dimensions of the modified ITR scale.

Method

The objective of the qualitative component of the research was to evaluate whether an enhanced knowledge of the preference structures, identified by the modified ITR scale, can be useful in helping marketers position products and communicate more effectively with consumers via salient text and visual images. This required the development of a technique to explore consumer reaction to visual imagery and written text, and to examine how consumers value and define concepts related to the modified ITR scale and how such knowledge can aid the synthesis of product promotion with the perceived needs and expectations of the target market. The remainder of the chapter evaluates the process by which the methodology was designed and the key influences acting upon the decisions made.

Choice of research technique

It has become widely accepted that focus groups are a 'rich source of qualitative data for the social science researcher' (Oates 2000: 195). Their evolution has seen them used in a variety of contexts and within a whole range of research paradigms, from their use within the largely positivistic framework of market research to their alliance with more interpretivist paradigms (Cunningham-Burley *et al.* 1999: 188). The philosophical foundations of this research, underpinned by positivism and the conceptual framework established around the dimensionality of novelty-seeking, required that the technique should maintain some structure, but be 'sensitive to the multiple levels of meaning and the multiple codes' (Leiss *et al.* 1990: 225) utilised in motivation and image reception.

Focus groups are regarded as being particularly effective in capturing the complexities of motivation (Carey 1994: 226; Krueger 1994: 45) and hence the links between motivation and image reception. The key qualities which led to the choice of the technique related to the synergistic effect of group interaction and their potential to break down the researcher–researched power relationship. While these qualities could be regarded as undermining some of the foundations of positivism, the decision to include some element of qualitative research was based on a questioning of the detached nature of positivistic inquiry. These qualities were felt to be essential in order that the research could move towards more emic categories of knowledge and capture the naturally emerging language of the participants.

The synergy which can be generated between participants is a key differentiating characteristic between group and individual interview techniques (Berg 1995: 69; Oates 2000: 186). 'As participants answer questions, the responses spark new ideas or connections from other participants. Answers provide mental cues that unlock perceptions of other participants, cues that are necessary in order to explore the range of perceptions'

(Krueger 1994: 54). This synergistic effect, achieved through the potential for participants to draw from one other, to react to the responses of others, and compare experiences and perceptions (Berg 1995: 69; Gaskell 2000: 46; Stewart and Shamdasani 1998: 509), is particularly important where the research requires people to explore the complexities of motivation and imagery.

Building upon the quality of synergism is the advantage the focus group provides for breaking down the researcher–researched power relationship. As group interaction partly replaces the interviewer–interviewee relationship found in individual interviewing, a greater emphasis can be given to participants' viewpoints (Berg 1995: 72), giving more weight to the 'participants' ways of understanding, their language and what they feel is important' (Oates 2000: 188). Thus, while it must be appreciated that focus groups are artificial (Oates 2000: 187) and that this research was being developed within a positivistic framework, the technique was still seen to be useful in addressing the need to explore how the participants value and define concepts related to novelty-seeking and expose the reasoning behind their perceptions. Furthermore, in a peer group setting, 'participants are more likely to describe their experiences in locally relevant terms, rather than attempt to impress or please the researcher, or use language and concepts that they believe to be the researchers'' (Goss and Leinbach 1996: 117). For a summary of the advantages and disadvantages of the focus group method, see Box 12.1.

Box 12.1 Advantages and limitations of the focus group method

Advantages

- The synergistic effect of group interaction: participants can react to and build upon the responses of other group members.
- The potential to break down the researcher–researched power relationship: to empower participants and encourage a more collaborative process of knowledge production.
- Flexibility.
- The ability to explore how participants value and define key concepts, in their own words.
- The ability to allow participants to rationalise views and experiences/ expose reasoning behind perceptions.
- The potential to use visual stimuli.
- Time efficiency.

Limitations

- The artificial nature of the research setting.
- The fact that the influence of the peer group and/or dominant individuals may bias the results, increasing the potential for social desirability bias.

- The fact that the influence of the researcher as the moderator may influence participants.
- The need for a skilled moderator.
- The fact that the small numbers of participants limit the ability to generalise to a wider population.

Research design

One of the key issues in the design of qualitative research is the extent to which methods should be pre-structured (Maxwell 1998: 85). Central to this decision were the philosophical and conceptual underpinnings to the study discussed previously. While structured approaches can help to ensure comparability, unstructured approaches focus more on drawing out emic data, trading 'generalizability and comparability for internal validity and contextual understanding' (Maxwell 1998: 85). Though a technique was required to capture participants' reactions to imagery, drawing out the reasoning behind their perceptions and how they define concepts relating to novelty (emic data), the research objectives required that this should take place within the context of the dimensions of novelty-seeking.

Following the consideration of a range of techniques, it was felt that some form of instrument was required to measure consumer reaction to the visual stimuli, to impose some degree of standardisation with respect to the ITR scale's conceptual foundations and thus enhance the validity of the results within the whole research. Despite this need to structure the discussion around the novelty construct, it was hoped that through using visual stimuli within the focus group setting, the discussion could 'determine deeper levels of meaning, make important connections, and identify subtle nuances in expression and meaning' (Stewart and Shamdasani 1998: 509). The following section will discuss the associated aspects of research design and instrument development necessary to capitalise upon the use of visual stimuli to generate rich qualitative data.

Instrument development

Consumer reaction was defined to incorporate both reception, characterised by the internalisation and interpretation of stimuli material, and the resulting affective response (Russell and Pratt 1980: 311–312) to the constructed meaning. The research instrument developed incorporated a set of four self-completion grids, modified from Russell *et al.*'s (1989: 494) affect grid. Although the original grid operationalised affect to incorporate the two orthogonal continuums of hedonic tone (unpleasant feelings–pleasant feelings) and arousal (high arousal–sleepiness), the nature of the grid seemed applicable for adaptation. Thus, four modified grids were developed to measure respondents' reactions to stimuli within the context of the four ITR scale dimensions. Figure 12.1a presents an example of one of the modified grids and Figure 12.1b the change in labelling for the subsequent three grids used in the research.

These new response grids adapted the wording of the hedonic tone (horizontal) continuum to very unappealing–very appealing. The most distinct changes, however, were made to the vertical continuum, whereby the arousal continuum was replaced by continuums to reflect the four dimensions of the modified ITR scale (see Figure 12.1). To enhance the validity and reliability of responses, attention was paid to the need to make the labelling meaningful and understandable to the respondents, and therefore an appropriate operationalisation of the individual dimensions. The final modification made was to reduce the size of the grid from a 9×9 to a 5×5 grid. As it was intended to incorporate the grids within a focus group setting, to stimulate qualitative discussion, a quantification of the results was not required.

To aid the administration of the instrument and improve reliability and validity, it was important to give concise instructions to the participants. Each photograph or text extract was given an identifying letter, and completion of the grid incorporated a two-step procedure to place the letter in the most appropriate grid square. In the first dimension (Figure 12.1), the participants were required to assess, first, how far the type of destination provided them with either unfamiliar or familiar travel experiences (i.e. the most appropriate placing for the letter on the vertical scale), and second, how far this would appeal to their holiday preferences, and thus record their judgement from very unappealing to very appealing.

A fundamental operational quality of the instrument is that, as 'a single-item scale', it is 'short and easy to fill out and . . . can, therefore, be used rapidly and repeatedly' (Russell *et al.* 1989: 493), enabling the reactions to the stimuli within each dimension to be recorded on a single grid. The most valuable feature of the instrument is its role in encouraging qualitative discussion to explore the rationale for placement of judgements and, therefore, evaluate how the participants value and define concepts related to the modified ITR scale. This operationalised the original objective to explore how an understanding of the preference structures based around the dimensionality of the ITR scale could be used to create more salient text and visual images.

Selection of resource stimuli and their incorporation within the focus group

The selection of resource stimuli was pivotal to the success of the research, given the aim of developing participants' thoughts around the four dimensions, in order to explore how they define and value concepts related to novelty-seeking. As the time available to test consumer reaction to the

(a)

Example of a Modified Grid for the Type of Destination/Holiday Environment Dimension

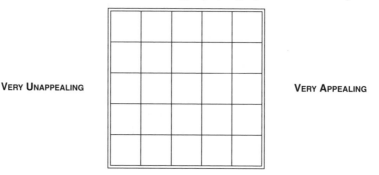

VERY UNFAMILIAR
(Destination: provide very new and different travel experiences,
i.e. unfamiliar environment; culture; people; language or travel facilities.)

VERY UNAPPEALING   VERY APPEALING

VERY FAMILIAR
(Destination: familiar environment; culture; people; language or travel facilities.)

(b)   The labelling used for the ITR dimensions represented by the other three grids.

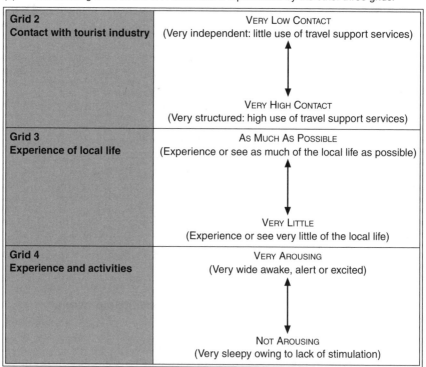

Grid 2 Contact with tourist industry	VERY LOW CONTACT (Very independent: little use of travel support services) ↑ ↓ VERY HIGH CONTACT (Very structured: high use of travel support services)
Grid 3 Experience of local life	AS MUCH AS POSSIBLE (Experience or see as much of the local life as possible) ↑ ↓ VERY LITTLE (Experience or see very little of the local life)
Grid 4 Experience and activities	VERY AROUSING (Very wide awake, alert or excited) ↑ ↓ NOT AROUSING (Very sleepy owing to lack of stimulation)

Figure 12.1 Response grids: example of a modified grid for the type of destination/
holiday environment dimension.

stimuli was limited, the session was planned to explore reaction to the visual stimuli, with extracts of text and associated response grids given to the participants at the end of the session to return in pre-paid envelopes. Furthermore, the decision to allow only 10 seconds for each photograph viewing was supported by piloting, as this was deemed to measure initial reactions and mirror more closely the way in which consumers react to imagery in marketing communications.

To maximise the discussion, photographs and short textual extracts were selected from four brochures that were judged a priori to be likely to produce a range of reactions, both positive and negative, across all four dimensions. As the whole research was based upon a case study approach and purposive sampling of two tour operators from different market niches (a long-haul Asian specialist and an adventure travel operator), their brochures were used as a source for two-thirds of the resource stimuli. The remaining two brochures were chosen to represent other, contrasting market niches, namely a mass-market long-haul brochure and a sun, sand and sea short-haul brochure. Furthermore, to ensure that participants reacted to the image presented and not to the image they had of a particular country, where possible, photographs were selected whose destinations are not easily recognisable. The only exception to this was an image of the Taj Mahal, which was incorporated to gauge reactions towards a recognised tourist sight in an exotic country.

Qualitative discussion

While the response grids allowed the participants to record their reactions to the resource stimuli, the most valuable part of the focus group was the qualitative discussion designed to investigate their reactions to the visual stimuli, exploring their rationale behind the placement of photographs within the response grids. The planning and implementation of this part of the focus group incorporated the consideration of many issues related to the nature and characteristics of appropriate questions, the role of the moderator, and group dynamics. As the latter parts of the chapter show, this also represented a key stage in the journey, in which inspiration could be drawn from other paradigms and their critiques of positivism.

For each dimension, a series of questions and prompts were used to discuss, first, individual photographs, and second, a series of photograph pairs specifically chosen to capture contrasting elements of each dimension. For example, within the 'type of destination/holiday environment' photo set, one slide depicted a deserted town square in Nepal and another the Taj Mahal. While both are clearly Third World environments, the former was devoid of any tourist infrastructure and the latter depicted a highly recognised tourist sight.

The questions were phrased to explore consumer reaction in greater detail as the discussion developed, and to explore the connections between the reactions and consumer preference structures. Although the questions were structured around the dimensions, flexibility was vital to enable the discussion to probe additional concepts identified by the participants, such as security and risk. This helped the research to move closer towards a more collaborative process of knowledge production, within the constraints of the study. Throughout the discussion, the role of the moderator was pivotal: the moderator managed group dynamics and provided an environment in which common and contrasting themes and issues could be explored. This was useful to capitalise on the synergy required to empower the participants (Madriz 2000: 838), thus accessing deeper levels of meaning and aiding the exploration of the ways in which the participants rationalise their own views and experiences (Oates 2000: 188).

Analytical strategy

While the main body of this chapter has justified the decisions made regarding the research design, it is important to expand upon the development of the analytical strategy. As Riley and Love highlighted in their examination of the state of qualitative research, if the analytical procedures and interpretation are not made transparent it is difficult for 'non-qualitative researchers to understand and accept findings as reliable and valid' (2000: 182). As such, the process of analysis needs to be seen as part of the research design.

A crucial stage in the analysis was the familiarisation and immersion within the data. Only once this had occurred could the fracturing of the data begin through the coding and classifying of source material into theoretically defined categories. These categories were devised both inductively and deductively (Berg 1995: 177), requiring immersion in the transcripts to identify meaningful themes (inductive approach) and for the testing of previously devised categories (deductive approach). This was felt to satisfy the requirements of the research that the results should move towards a more emic perspective but be based upon the dimensional structure provided by the modified ITR scale.

Through the overlapping and often re-evaluative procedures of coding, summarising responses to each image and the comparative analysis of image pairs, a gradual move was made from the descriptive to the interpretive phase of analysis. For each dimension, a series of themes were generated into which the material was coded, and further scrutiny of these themes revealed interrelationships and hierarchies among the themes within the individual dimensions. This reinforces the view of Ryan and Bernard that:

> No matter how the researcher actually *does* inductive coding, by the time he or she has identified the themes and refined them to the point where they can be applied to an entire corpus of texts, a lot of interpretive analysis has already been done.
>
> (2000: 781)

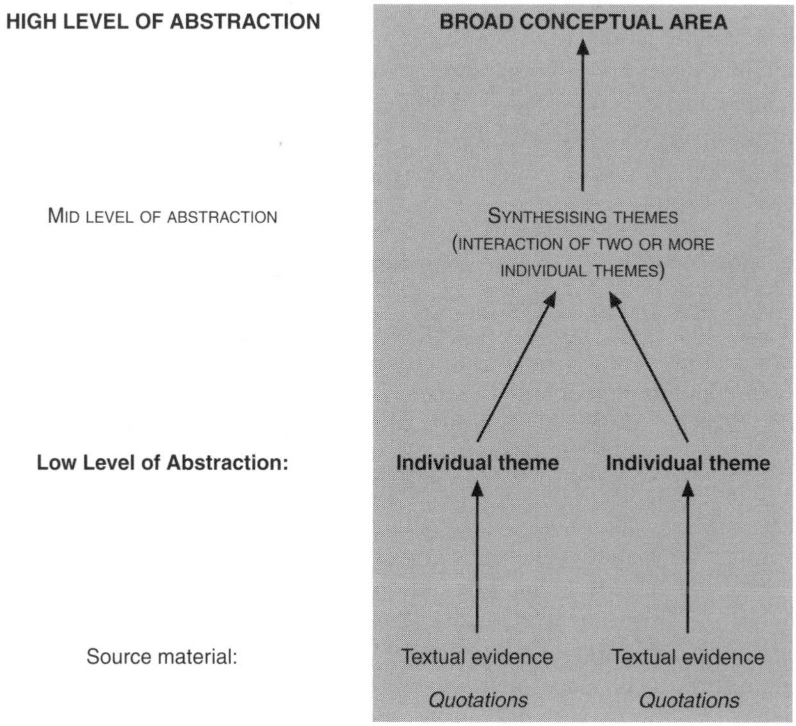

HIGH LEVEL OF ABSTRACTION · BROAD CONCEPTUAL AREA

MID LEVEL OF ABSTRACTION · SYNTHESISING THEMES (INTERACTION OF TWO OR MORE INDIVIDUAL THEMES)

Low Level of Abstraction: · Individual theme · Individual theme

Source material: · Textual evidence · Textual evidence

Quotations · *Quotations*

Figure 12.2 Hierarchical structure of the thematic framework.

'Visual displays are an important part of qualitative analysis' (Ryan and Bernard 2000: 784), and to illustrate the synthesis and abstraction of these themes, a series of thematic frameworks for each dimension were produced to provide an enhanced and more meaningful understanding of the data. The thematic framework (Figure 12.2) illustrates how themes have been developed through higher degrees of abstraction and synthesis from individual themes (low level of abstraction, supported by textual evidence) to the synthesising of themes (interaction of two or more individual themes at the middle level of abstraction) to broad conceptual areas (high level of abstraction). An important point to note is that the arrows in the thematic framework move up the hierarchy to depict how the hierarchical structure has been developed; the discussion of the analysis used a top-down approach to focus on the outcomes, in order to use the data to show an understanding of the broad conceptual areas. Each level of the hierarchy, however, was considered in the interpretation, to avoid the criticism that abstraction can lead to an unnecessary loss of richness and context.

Critique and evaluation

The flexibility associated with the focus group method was a key quality behind its selection. This chapter has so far shown how the final technique was a product of a wide range of influences, many of which arose from the nature of the inquiry paradigm. This section critically evaluates the methodology, focusing on some of the key themes which the chapter has sought to explore.

The value of a multi-method approach and the benefits gained from the incorporation of the qualitative research component within the study: The fact that this focus group research was only one part of a larger multi-method study had a significant influence on the way in which it was designed and implemented. The earlier discussion of the philosophical and conceptual underpinnings briefly outlined the influence of positivism on the research design. The statistical analysis which informed the testing of the dimensionality of the scale and the resulting segmentation analyses (King 1997) can be understood within this paradigm, as the research was developed to embody the 'conventional benchmarks of "rigor": internal and external validity, reliability and objectivity' (Lincoln and Guba 2000: 166, but see also Chapter 9, where Decrop discusses the way in these conventional benchmarks of 'rigor' can be translated for interpretive inquiry) in its efforts to produce measurement tools deemed to be standardised, reliable and robust within a positivistic framework. Within this context, the use of qualitative research was still appropriate, but when one is working from within such a realist ontology and objective epistemology, qualitative methodologies have a tendency to be more structured (Denzin and Lincoln 2000: 21). What follows will demonstrate that despite this, the qualitative nature of the focus group research did provide the study with considerable added value.

In attempting to assess what was gained from incorporating the qualitative research component within the study, it must first be noted that the design and implementation of this research component is distinctive in that it represents a significant move away from the typical application of such role scaling techniques. In particular, it moved beyond using the ITR scale as purely a measurement tool, towards focusing on its dimensionality as a conceptual framework. Although the study still required research which was focused around the dimensionality of the modified ITR scale, exploring the participants' reactions to the visual stimuli meant that the focus groups were able to develop more understanding of how the participants value and define concepts related to novelty-seeking. Furthermore, this provided the opportunity to develop more meaningful insights into the beliefs and preferences of those researched.

The interactive nature of the focus group setting also enabled the research to benefit from being able to 'follow up immediately people's responses and to explore the contradictions and inconsistencies that are part of everyday life' (Stroh 2000: 197), which could not have been achieved within a purely quantitative study. This is particularly important given that tourism is a phenomenon heavily influenced by status, and that negative stereotyping of mass tourism is rife. Research into motivation can, therefore, benefit from qualitative methods, given that instruments such as attitude scales find it difficult to avoid value-laden terminology. In an evaluation of the ITR scale, Jiang *et al.* raise the issue of social desirability bias and the problems of using attitude scales where particular words and phrases may 'trigger socially desired responses' (2000: 979). Although the influence of social desirability bias cannot be omitted in a group context, the requirement for respondents to make initial judgements via the response grids and the subsequent breaking down of power relationships during the discussion session helped to explore some of the more complex issues associated with motivation. The use of visual stimuli was also pivotal here, providing a stimuli for the discussion.

The above evaluation should not diminish the role played by the ITR scale in the whole study, but using only quantitative, etic methods of data collection could have resulted in what Walle describes as the 'dehumanisation of research in order to reduce bias and increase "rigor"' (1997: 525). The inability to let participants present their thoughts in their own words is often cited as a limitation of such quantitative tools as attitude scales, and thus the development of the qualitative research component aimed to move towards more emic categories of knowledge. In this context, the technique not only enabled the research to explore just how people responded but required them to articulate and rationalise their reactions. In particular, the qualitative nature of the research enabled the technique to move some way towards accessing the naturally emerging language, meanings and interpretations of the participants, which is vital for communications and image research. As is seen in what follows, the breaking down of the researcher–researched relationship was instrumental in achieving this.

Despite the degree of structure discussed above, the quality of the focus group to encourage group interaction and thus create a certain amount of synergy proved invaluable in starting to break down the researcher–researched power relationship. Through encouraging the participants to share a wide variety of opinions and respond to each other in order to explore both common and contrasting themes, the research could gradually move towards being more sensitive to emic categories of knowledge. The group context provided an environment in which the research was able to benefit from discovering not just what people think but why they think in that way. The focus groups highlighted a complex interplay of novelty/familiarity preferences across and within the multidimensional space. The qualitative nature of the research, and in particular the use of visual stimuli within the focus group setting, helped to unravel some of those contradictions. This was particularly important given that consumer reaction encompasses reception (internalisation and interpretation) and the affective response to the constructed meaning.

Limitations: Research into tourism motivation and imagery requires sophisticated methodologies, with the ability to delve into 'deeper levels of meaning ... and identify subtle nuances in expression and meaning' (Stewart and Shamdasani 1998: 509) in order to understand 'consumers' ways of seeing' (Morgan and Pritchard 1998: 243). While the incorporation of a qualitative method did attempt to move forward the parameters of this type of research, the positivistic underpinnings and the structure inherent in this qualitative research design do have many implications for the nature of the knowledge produced through the research.

The implications of using qualitative methods in such a structured manner are that in striving to ensure rigour and validity in the design of research instruments, both highly contested qualities, the researcher can impose constraints on the development of knowledge. Jiang *et al.* 2000 raise a note of caution that while the ITR scale captures the essence of Cohen's original typology, the significant changes which have occurred within tourism and consumer behaviour mean that Cohen's original work perhaps cannot be seen to 'address all latent aspects of the novelty concept in inter-national tourism context today' (Jiang *et al.* 2000). It is only through moving into the use of a more open and unstructured approach to qualitative research design that the research could have begun to address this issue, thus drawing more closely on alternative, more interpretivist paradigms. Set within this framework, debates surrounding the use of the response grids, the influence of the researcher through the choice of the images and textual extracts, and the focus on the dimensionality of the ITR scale would all need to be re-evaluated as the researcher could move towards a more interpretive approach to the use of focus groups. This might involve more of a leaning towards research based in the crisis of representation and fifth moments whereby the research participants themselves are given more opportunity to frame the research questions and have a wider role in interpreting the data, as Goodson and Phillimore suggest in Chapter 2.

As attitudes about what constitutes knowledge are continually chal-lenged, and the limitations of positivism are debated within tourism, alternative avenues for research are opened up. A more collaborative and flexible process of knowledge exploration and production may be able to move us closer towards understanding the social construction of imagery. Such an approach may also produce research which is able to mirror more closely the contention that the consumer is a pivotal component of meaning transfer, and that 'how places are viewed or interpreted is dependent on the perspective of the viewer' (Young 1999: 388). As the final section now shows, this would represent the start of a new journey, thus reinforcing the circularity of the research process.

Reflecting upon the 'journey'

The final reflections upon this research draw from the analogy made throughout the chapter between the research process and a journey. The researcher should see the process as a journey, not only one in which a diversity of decisions need to be made to orientate oneself along a labyrinth of different paths, but also as a journey in which assumptions should be continually scrutinised and re-evaluated.

The beginning of this journey was heavily influenced by the positivist traditions influencing tourism research and, in particular, the highly quantitative, etic research being conducted into tourism motivation. The development of a multi-method approach, however, was based upon the recognition that quantitative methods alone would not allow the research to 'fully address questions of understanding and meaning' (Riley and Love 2000: 166). It was believed that a qualitative method would allow the conceptual foundations of the ITR scale to be explored and to move us further towards more emic categories of knowledge concerning how participants define and value concepts related to novelty-seeking, and that this would provide themes which were meaningful in a marketing context.

In hindsight, the development of the qualitative focus group study can be seen as a turning point in that journey. From the positivist traditions of the quantitative motivation research, the exploration into the realms of qualitative inquiry, which involved engaging with the participants rather than communicating through the standardised instrument of the questionnaire survey, helped to raise questions about the dominance of positivism, discussed by a number of authors (e.g. Decrop 1999: 157; Riley and Love 2000: 180). 'The central idea that there are objective facts "out there" to be discovered by rigorous enquiry' has its limits (Arksey and Knight 1999: 10), and many authors have 'begun to explore paradigms beyond positivism' (Riley and Love 2000: 166). Thus, as the 'limitations' discussion shows, future research may benefit from more interpretive methodologies drawing inspiration from alternative paradigms such as the multiple meanings of constructionism. While this should not undermine the findings or the methodological process through which this original research was conducted, the analogy has been drawn to show the benefit of regularly questioning the basis of the philosophical foundations and assumptions which underpin the choice of research methodologies and methods – 'assumptions about the social world and the nature of human understanding of it' (Arksey and Knight 1999: 4).

The chapter began with the premise that studying the theory and philosophical foundations of social science research not only provides a knowledge base from which to evaluate current practice, but helps to ensure that researchers are 'better able to set forth the research process in ways that render it transparent and accountable' (Crotty 1998: 216). In the light of this conclusion, one further point to add is that writing research in a reflexive manner also helps the researcher to reflect upon the research process and the influences which have shaped its development (the start of a new journey). This is vital not only in the creation of sound research, but also to help ensure that research does not unquestioningly become too entrenched in established 'ways of doing things', as it is this that can lead to parochialism, and constrain both conceptual and methodological development within the relatively immature field of tourism research. As Walle states, 'tourism needs to forcefully articulate ... that it is a broad and distinct field and that it embraces a variety of appropriate research strategies' (1997: 535).

Summary

- Studying the theory and philosophical foundations of social science research not only provides a knowledge base from which to evaluate current practice, but helps to ensure that researchers are 'better able to set forth the research process in ways that render it transparent and accountable' (Crotty 1998: 216).
- The qualitative nature of the focus group technique provided considerable added value to the study. Exploring the participants' reactions to the visual stimuli meant that the research was able to develop more understanding of how participants value and define concepts related to novelty-seeking, gradually moving towards being more sensitive to emic categories of knowledge.
- Although the research was originally influenced by the positivist traditions of quantitative motivation research, as the study evolved and expanded into the realms of qualitative inquiry, possibilities for research of a more interpretive nature opened up. A more collaborative and flexible process of knowledge exploration and production may be able to move us further towards understanding the social construction of imagery and to understand 'consumers' ways of seeing' (Morgan and Pritchard 1998: 243).
- At a more abstract level, an analogy can be drawn between the research process and a journey. The different stages of the journey are shaped by key phases of decision-making and reflect the evolving nature of research. The journey demonstrates the benefit of regularly questioning the basis of the philosophical foundations and assumptions which underpin the choice of research methodologies and methods. This process is vital not only in the creation of sound research but for continued methodological developments within the field of tourism research.

Questions

1   What are the benefits of a multi-method approach and what specific qualities has the focus group technique brought to this study?
2   What implications did the positivistic underpinnings and the structure inherent in this qualitative research design have on the nature of the knowledge produced through the research?
3   Writing research in a reflexive manner helps the researcher to reflect upon the research process and the influences which have shaped its development. What are the key influences which have impacted upon this research design?
4   In what ways could future research use a less structured and less positivistic approach, and how might this impact upon the research process?

References

Arksey, H. and Knight, P. (1999) *Interviewing for Social Scientists*. London: Sage.

Berg, B. L. (1995) *Qualitative Research Methods for the Social Sciences*, 2nd edn. London: Allyn and Bacon.

Carey, M. A. (1994) 'The group effect in focus groups: planning, implementing and interpreting focus group research'. In J. M. Morse (ed.) *Critical Issues in Qualitative Research Methods*. London: Sage.

Cohen, E. (1972) 'Towards a sociology of international tourism', *Social Research*, 39: 164–182.

Crotty, M. (1998) *The Foundations of Social Research: Meaning and Perspective in the Research Process*. London: Sage.

Cunningham-Burley, S., Kerr, A. and Pavis, S. (1999) 'Theorizing subjects and subject matter in focus group research'. In R. S. Barbour and J. Kitzinger (eds) *Developing Focus Group Research: Politics, Theory and Practice*. London: Sage.

Decrop, A. (1999) 'Triangulation in qualitative tourism research', *Tourism Management*, 20 (1): 157–161.

Denzin, N. K. and Lincoln, Y. S. (1998) *The Landscape of Qualitative Research: Theories and Issues*. Thousand Oaks, CA: Sage.

Denzin, N. K. and Lincoln, Y. S. (2000) 'Introduction: the discipline and practice of qualitative research'. In N. K. Denzin and Y. S. Lincoln (eds) *Handbook of Qualitative Research*, 2nd edn. London: Sage.

Gaskell, G. (2000) 'Individual and group interviewing'. In M. W. Bauer and G. Gaskell (eds) *Qualitative Researching with Text, Image and Sound: A Practical Handbook*. London: Sage.

Goss, J. D. and Leinbach, T. R. (1996) 'Focus groups as alternative research practice: experience with transmigrants in Indonesia', *Area*, 28 (2): 115–123.

Jiang, J., Havitz, M. E. and O'Brien, R. M. (2000) 'Validating the international tourist role scale', *Annals of Tourism Research*, 27 (4): 964–981.

King, K. D. (1997) 'An Evaluation of a tourist role scale technique as a foundation for the development of theory and practice of tourism marketing strategies', unpublished PhD thesis, University of Birmingham.

Krueger, R. A. (1994) *Focus Groups: A Practical Guide for Applied Research*. London: Sage.

Leiss, W., Kline, S. and Jhally, S. (1990) *Social Communication in Advertising: Persons, Products and Images of Well-Being*, 2nd edn. London: Routledge.

Lincoln, Y. S. and Guba, E. G. (2000) 'Paradigmatic controversies, contradictions, and emerging confluences'. In N. K. Denzin and Y. S. Lincoln (eds) *Handbook of Qualitative Research*, 2nd edn. Thousand Oaks, CA: Sage.

Madriz, E. (2000) 'Focus groups in feminist research'. In N. K. Denzin and Y. S. Lincoln (eds) *Handbook of Qualitative Research*, 2nd edn. Thousand Oaks, CA: Sage.

Maxwell, J. A. (1998) 'Designing a qualitative study'. In L. Bickman and D. J. Rog (eds) *Handbook of Applied Social Research Methods*. London: Sage.

Mo, C.-M., Howard, D. R. and Havitz, M. F. (1993) 'Testing an international tourist role typology', *Annals of Tourism Research*, 20 (2): 319–335.

Morgan, N. and Pritchard, A. (1998) *Tourism Promotion and Power: Creating Images, Creating Identities*. Chichester: Wiley.

Oates, C. (2000) 'The use of focus groups in social science research'. In D. Burton (ed.) *Research Training for Social Scientists*. London: Sage.

Riley, R. W. and Love, L. L. (2000) 'The state of qualitative tourism research', *Annals of Tourism Research*, 27 (1): 164–187.

Russell, J. A. and Pratt, G. (1980) 'A description of the affective quality attributed to environments', *Journal of Personality and Social Psychology*, 38 (2): 311–322.

Russell, J. A., Weiss, A. and Mendelsohn, G. A. (1989) 'Affect grid: a single-item scale of pleasure and arousal', *Journal of Personality and Social Psychology*, 57 (3): 493–502.

Ryan, G. W. and Bernard, H. R. (2000) 'Data management and analysis methods'. In N. K. Denzin and Y. S. Lincoln (eds) *Handbook of Qualitative Research*, 2nd edn. London: Sage.

Stewart, D. W. and Shamdasani, P. N. (1998) 'Focus group research: exploration and discovery'. In L. Bickman and D. J. Rog (eds) *Handbook of Applied Social Research Methods*. London: Sage.

Stroh, M. (2000) 'Qualitative interviewing'. In D. Burton (ed.) *Research Training for Social Scientists*. London: Sage.

Walle, A. H. (1997) 'Quantitative versus qualitative tourism research', *Annals of Tourism Research*, 24 (3): 524–536.

Young, M. (1999) 'The relationship between tourist motivations and the interpretation of place meanings', *Tourism Geographies*, 1 (4): 387–405.

13 Let your data do the talking

Researching the solo travel experiences of British and American women

Fiona Jordan and Heather Gibson

Aims of the chapter

- To critically analyse the value of in-depth interviewing and grounded theory analysis in generating qualitative data on the experiences of tourists in general and marginalised groups in particular.
- To discuss our own experiences of working collaboratively to research issues in tourism.
- To evaluate the challenges of researching the experiences of tourists who are sometimes marginalised from 'mainstream' tourism research.

Introduction

Over the past five years we have interviewed more than 50 British and American women who take solo holidays. The idea for the project was originally conceived following Fiona's LSA (UK Leisure Studies Association) 1997 presentation of a study in which she investigated the lack of provision among British travel companies for women wishing to take a solo holiday (Jordan 1998). Heather had just finished interviewing both men and women in their retirement years about their travel experiences (Gibson 2002) and was intrigued by one particular traveller's tale, that of a female artist who spoke of both the joys and the tribulations of travelling solo. We decided that it would be interesting to compare the experiences of solo women travellers from both the United Kingdom and the United States, and embarked on a research project that has evolved through various phases both methodologically and theoretically. It is the story of this study that is told in this chapter. While our study specifically explores the experiences of women travelling solo, we would suggest that a number of the issues of research design, data collection, analysis and presentation of findings discussed here have wider relevance for tourism researchers. In particular, people investigating the experiences of tourists whose voices have often been marginalised in larger-scale quantitative studies may be interested in our reflective account of working with research participants and working collaboratively with each other.

The chapter examines the use of in-depth interviewing as a technique for gathering qualitative data on the experiences of tourists. Using the case study of our research with women travelling alone, we will discuss the principles underlying the design of this research project, the value and the difficulties of working collaboratively and internationally, and the challenges of 'giving voice' to those often marginalised in the quantitatively orientated tradition of tourism research. The intention of this chapter is to combine consideration of the challenges associated with this type of research with ideas for researchers who might undertake similar studies. In order to contextualise our review of the method (specific research technique) used in this study, we will begin by outlining briefly the methodological stance (research perspective) that guided the selection of interviews as our preferred means of data collection.

The choice of research method for any project should be made only following consideration of what Finn *et al.* (2000: 5) describe as the 'dark, murky waters of ontology, the form and nature of reality, and epistemology i.e. what counts as knowledge'. All research is influenced by the philosophical position of the researchers, the nature of the project and its intended audience. The research philosophy underpinning the design of a study therefore impacts on both the way data are gathered and how they are used to create knowledge. Thus, the conceptual foundations of our study and our methodological approach 'set out a vision for what [our] research is and how it should be conducted' (Potter 1996: 50). In any research project it is imperative that the method chosen is appropriate for the goal of the study rather than choosing a method because it is conventional or because it is the one most familiar to the researcher. The goal of our study, was initially to gain an in-depth understanding of the nature of solo travel as experienced by women in the 30 to 50 age group (this age range was expanded in later phases of the study).

To achieve this, we determined that empirical research designed from a qualitative perspective would be best. Such an approach to data collection would enable us to gather rich descriptive accounts of women's travel stories by providing them with an opportunity to respond to open-ended questions rather than having them respond by choosing a number from a scale on a questionnaire. This latter method would have been appropriate if we were simply interested in investigating broad patterns and trends in solo travel (Babbie 1995; Henderson and Bialeschki 2002). However, as little in-depth experiential research on this topic existed at the time, we decided to adopt a qualitative approach as a starting point for the generation of research themes (Bryman 1988; de Vaus 1993; Veal 1992, 1997).

In contrast to much of the existing work within tourism research, our project was therefore conceived within an interpretive paradigm, corresponding to a phenomenological research approach (Henderson 1991). In an interpretive paradigm, 'the central endeavour is ... to understand the subjective world of human experience' (Cohen and Manion 1994: 36). The increasing adoption of such qualitative approaches by tourism researchers differs from the previously positivist orientation of much research in this subject field (Walle 1997). According to Botterill, 'the "normalization" of a positivist epistemology' has, 'unduly limited the development of tourism research as social science' (2001: 212). While not denying the value of quantitative research *per se*, or of combining quantitative and qualitative approaches to researching tourism issues, we concur with Johnson's view that 'too often tourism research is presented as methodologically precise and statistically impeccable but otherwise disembodied' (2001: 181). In our research we were concerned with theorising experiences of tourism and, in doing so, endeavouring to 'get inside the heads and hearts of the tourists themselves to ask the questions that seem of the utmost importance in moving tourism to a more person centred phenomenon' (Wearing and Wearing 2001: 155). We believed that research designed within an interpretive paradigm would enable us to do this effectively. We thus chose to gather our data using in-depth, one-to-one, face-to-face interviews in which we encouraged women to tell us about their experiences of travelling solo within a semi-structured framework. In so doing, we explored topics that we might not have thought about when designing the research project but which the women themselves identified as being significant. Thus, instead of imposing our preconceived notions of solo travel on the women, we used the semi-structured questions as a guide to stimulate discussion about their actual experiences while travelling solo.

Where are we coming from? Theoretical perspectives underpinning our research

It is important to understand the assumptions underlying a study, as they have a direct influence on the choices researchers make in relation to method, questions asked, data analysis and interpretation. So, before discussing the design and execution of our research, we will briefly outline its theoretical underpinnings. Our theoretical framework was initially derived from feminist analyses of women's everyday leisure (Deem 1986; Green *et al.* 1990; Wimbush and Talbot 1988; Henderson *et al.* 1996; Wearing 1996) and our general backgrounds in tourism studies (Gibson 1994, 1996; Jordan 1997). Given that we conceptualised solo travel as a specialised form of women's leisure, we felt that it was important to draw upon relevant theories from both leisure and tourism studies to inform our research design. As we were particularly interested in exploring barriers and constraints affecting women's participation in solo travel, and at the time there was little research into the subject of women as tourists (Kinnaird and Hall 1994, 1996; Squire 1994), we looked to the leisure studies literature, within which feminist frameworks for analysis of constraints were more evident than in tourism studies. One of our intentions was to render visible women's experiences as tourists and to contribute to a feminist critique of the lack of gender research within tourism studies (Enloe 1989; Richter 1994, 1995; Swain 1995). During the course of this project we have seen interest developing in this subject area, as reflected in a growing body of work on women and tourism (Craik 1997; Davidson 1996; Deem 1996a,b; Herold *et al.* 2001; Marshment 1997; Sanchez Taylor 2001; Wickens 2002). It is however, still argued by some that a feminist critique of tourism remains underdeveloped (Black 2001; Pritchard 2001), and thus we believe that our work has an ongoing contribution to make within this field. In addition, from a methodological point of view, our study has suggested future directions for the exploration of the travel experiences of other marginalised groups such as gay and lesbian tourists and tourists from black and ethnic minority communities.

Since 1998 we have been collecting data through in-depth interviews with women in the United States and the United Kingdom. The flexibility afforded by the use of interviewing as a research method has been very valuable in our study. The themes generated by ongoing analysis of our interview data have been incorporated into further interviews through a technique known as constant comparison, and our theoretical orientation has changed as a consequence of this ongoing data analysis combined with continuing secondary research. These developments have been charted in papers presented at conferences (see, for instance, Gibson and Jordan 1998a,b; Jordan and Gibson 2000, 2002). During the collection and analysis of initial interview data it became apparent that the diversity and empowering nature of the participants' travel experiences presented a challenge to our original, constraints-based theoretical framework (Gibson and Jordan 1998b). In order to try to make sense of what we were being told, we reinterpreted these data in relation to a revised theoretical position that took greater account of the individuality of these women's travel narratives. We began to draw on the work of Foucault (1977, 1979, 1984), post-structuralist analyses of leisure and tourism (see, for instance, Aitchison 2000, 2001; Fullagar 2002; Wearing 1996; Wearing and Wearing 2001), and the emerging leisure and tourism geographies (see, for instance, Aitchison *et al.* 2000). These concepts allowed us to theorise women's solo holiday experiences in relation to identification of problematic holiday spaces, women's strategies of resistance in reclaiming holiday space, and the empowering nature of solo travel. In the following section we discuss the feminist research perspectives that underpinned the design of this study.

Which way? Designing the research project

It is not our intention in this chapter to rehearse in detail the ongoing debate concerning the overall categorisation and labelling of feminist perspectives or what a 'feminist' research project should look like. There are already a number of texts and articles that cover this subject comprehensively (Charles and Hughes-Freeland 1996; Dupuis 1999; Harding 1987, 1991; Clarke and Humberstone 1997; McCarl Nielsen 1990; Roberts 1981; Stanley 1990; Stanley and Wise 1993). We will, however, refer to key feminist perspectives that constituted the guiding framework for our research design. We believe that the principles often associated with feminist projects have potential for wider application in the design of research within tourism more generally.

Key tenets of feminist research design centre on the relationship we, as researchers, establish with those who participate in our research and on acknowledging the part that we play in shaping both the research process and the findings of our studies. Therefore, we took heed of Morris *et al.*'s (1998) advice for designing a feminist research project, and so the guiding principles for our study can be summarised as follows:

> a concern to make the formerly invisible visible by focusing on women's lives; a commitment to doing research *for* women and not just *on* them [authors' original emphasis]; a rejection of hierarchical relationships within the research process by making those being researched into partners and collaborators; a commitment to reflexivity based on notions of openness and intellectual honesty.
>
> (Morris *et al.* 1998: 220–222)

The use of interviews and their attendant power relationships between the researcher and the researched has been a topic of much discussion in feminist literature (Cotterill and Letherby 1993; Eichler 1988; McCarl Neilson 1990; Reissman 1991; Stanley 1990). It can be argued that despite valiant attempts at maintaining neutrality, researchers naturally tend to see the world through the lens of their own experiences, which influences all aspects of the research process. The simple fact that a researcher has control over the way in which a study is designed, the selection of respondents, the choice of interview questions and the way in which interviewees' responses are interpreted has the potential to create an inequitable relationship. Such unequal distribution of power in a research situation can lead to problems of bias in data analysis and selectivity in the presentation of research findings. In order to redress the balance of power, many feminist researchers have argued for respondents to be viewed as more equal participants in research rather than being regarded as subjects or objects of study. We would contend that the relationship between any qualitatively orientated researcher and those taking part in the research is a critical one, and consideration of ethical and power relations should not be confined to the realms of women researching women. Increasingly there is interest (Carter and Clift 2000; Pritchard *et al.* 1998a,b) in researching the motivations, experiences and representation of marginalised groups such as gay tourists, tourists with disabilities and tourists from black and ethnic minority groups, often overlooked by the tourism industry. Such research requires particular attention to be paid to the researcher–participant relationship and its attendant power relations, especially if the researcher is part of a majority group in society, as he or she may not be able to fully understand the experiences of being part of a minority group (for further discussion of such issues in the context of leisure research, see Henderson 1998, and Henderson and Ainsworth 2001). Our intention throughout the design and execution of this research has been to provide a forum through which the voices of the women participating in the study could be heard, albeit mediated by our interpretation of their stories. In fact, their stories have actually shaped our theoretical journey, a facet of the research that we regard as a strength.

Analysing, interpreting and theorising other people's personal experiences in the way that our research has done necessitates adding a researcher's perspective to the stories told, a factor that may initially seem at odds with the feminist precepts of working with women collaboratively in research such as this. The very fact of having a 'researcher' interpret the stories of the 'researched' implies a power relationship that can make the researcher the more powerful through their control over the data. As Jackson and Jones (1998: 8) point out, on the other hand, even in everyday life we constantly interpret and endeavour to make sense of our own experiences and those of people around us. Our responsibility as researchers, then, is to make this process visible in the way we write up our research. As Phillimore and Goodson show in Chapter 1, however, the majority of tourism researchers write from the perspective of an 'expert' interpreting the data for their audience but giving no account of the interpretation process or how that process and the resultant findings are influenced by the personal biography of the researcher(s).

In accordance with the qualitative, feminist and person-centred orientation of our research, we have always been conscious of the importance of reflecting on, and rendering visible, our roles as researchers. Dupuis (1999: 59) advocates the adoption of such reflexive methodologies in qualitative leisure research, stating that 'strong, rigorous qualitative research and good science' rely on the inclusion of personal experience, emotions and collaboration within the research process. Reflexive research, Dupuis claims, involves self-introspection throughout the research process; methodological recognition of the way in which the researcher is connected to the world in which the research is taking place and the impact of this on the research; acknowledgement of the interrelationship between the research and research participants and the ways in which all parties shape the research; and realistic writing up that encompasses both the 'commonalities' and the 'contradictions' found in the research (Dupuis 1999: 59–60). In our case, both of us are experienced travellers as well as being knowledgeable about the workings of the tourism industry in the

United Kingdom and the United States, and well versed in a broad range of literature in leisure and tourism studies, an asset Strauss and Corbin (1998) identify as the 'theoretical sensitivity' of the researcher. Thus, using a mixture of real-world experience and theoretical knowledge, we were able to converse knowledgeably with our participants about travel and also make sense of their stories by using relevant theoretical insights.

Talking travel: using in-depth interviews as a data gathering technique

As mentioned in our introduction, when we embarked upon this study there was very little by way of empirical data pertaining to the experiences of female tourists, and no data on the experiences of contemporary solo women travellers (Gibson 2001). Undertaking quantitative, survey research could have provided us with some descriptive data about these travellers and enabled us to map general patterns of travel amongst women in the United States and the United Kingdom. However, these types of basic market data are already published in sector reports (e.g. Bureau of Transportation Statistics *American Travel Survey* 1999 or Mintel's *Singles on Holiday* report 2001) and would not have provided the depth of knowledge of women's actual experiences while travelling solo that we were seeking. Understanding the experiences and meanings of travel for solo women travellers, and tourists in general, is important, as tourism can be viewed as a special form of leisure with the potential to provide people with the 'freedom to be' (Kelly 1987), or freedom to resist everyday oppressions (Wearing 1998). Unless tourism researchers focus their investigations on the meanings that travel has for individuals, we will only ever have superficial knowledge of travel patterns without really understanding why and how different people travel.

Interviews were chosen as an appropriate method to accomplish our aim of gathering qualitative experiential data, in that, as Seidman states, 'At the root of in-depth interviewing is an interest in understanding the experience of other people and the meaning they make of that experience' (1991: 3). Using interviews also enabled us to adopt an inductive approach whereby we were not testing part of a theoretical model (deductive reasoning), but identifying patterns in our data that might lead to the development of grounded theory based on the actual travel experiences of the women themselves. Like all research methods, each has strengths and weaknesses. For semi-structured in-depth interviews, some of the pros and cons are outlined in Box 13.1 (adapted from Finn *et al.* 2000 and Henderson 1991).

Box 13.1 Advantages and limitations of semi-structured in-depth interviewing

Advantage

- An adaptable technique enabling probing of specific themes taking account of each participant's particular experiences. This in turn makes it a flexible technique because as data collection progresses and new ideas relevant to understanding the research topic emerge, interview schedules can be refined to reflect these insights.
- The face-to-face nature of the encounter enables the researcher to read body language and other non-verbal forms of communication in addition to speech, which can elicit valuable insights.
- The personal nature of interviews may enable the researcher to develop empathy with participants thereby creating a more comfortable environment for both parties.
- Little equipment is required, so the location can be varied easily to accommodate preferences of the researcher and the participant.
- Possible interconnections between experiences and views can be explored.
- The technique is useful for generating experiential data that can then be theorised.
- It can provide contextual background for studies using multiple methods.
- Semi-structured interviews provide inexperienced researchers with some structure while also allowing them to develop their own approach to interviewing.
- Semi-structured interviews also allow for comparability across interviews, as the same questions are asked of each respondent.
- They can provide rich, descriptive data with many colourful and illustrative examples of different tourist experiences.
- The technique has high validity, as interviewers can ensure that questions are understood by the interviewees by adapting the wording, or probe to elicit more in-depth responses.

Limitations

- Misinterpretation of views by the researcher and/or the participant is possible.
- The method requires training and confidence to be fully effective in data-gathering, and some theoretical insight to be able to probe for more detail on valuable ideas as an interview is conducted.
- It can be difficult to replicate, i.e. it has lower reliability (although we would argue that this is not necessarily a disadvantage in qualitative research, as what you sacrifice in reliability you gain in validity).
- The researcher may consciously or unconsciously steer the interviewee towards expressing views that agree with the research themes sought.

- Interviewing can be an awkward and uncomfortable experience, especially for inexperienced researchers.
- The value of the data is dependent on the honesty of the interviewee (or their desire to say what they think the interviewer wants to hear).
- The interviewer may be reactive to responses from the interviewee rather than structuring the interview themself.
- Finding a location to suit both the researcher and the interviewee can be problematic (i.e. not next to the coffee bean grinder in a café, as Fiona found out!).
- The technique relies on interviewees to volunteer to participate in what can be a time-intensive process. Interviews can last anywhere from one to two or more hours.
- Recording of interviews can be problematic if the interviewee does not want to be taped or is conscious of being recorded.
- Interviews, transcription and analysis of interview data are all time-consuming activities.
- Interviews can generate a large amount of data extraneous to the topic, and it may be problematic to generate comparable themes (this may be particularly difficult where more than one researcher is gathering interview data for the study).

Calling solo women travellers: finding women to interview

One of the issues related to researching 'hard-to-find' groups is that very often there is no sampling frame from which to draw a random sample. For example, for this study there are no comprehensive lists of solo women travellers. Thus, rather than selecting a random sample of women in the 30 to 50 age range (which was the original age group defined for this project) and asking about their general attitudes to solo travel, we decided to call for women who travelled alone to volunteer for interview. Before commencing the interviews, the researchers discussed (via email) key issues such as how to go about finding participants to interview and what topic areas should form the basis for the interview structure. Having tested our ideas through a small pilot study of six interviews, we faced the challenge of how to find other suitable interviewees. Initially we used snowball sampling, whereby women who had already participated in the study were asked to recommend other solo women travellers who might wish to participate. However, snowball sampling can mean that your sample is homogeneous and reflects only the experiences of like-minded solo travellers who happen to know one another. Thus, in order to encourage women with diverse solo travel experiences to volunteer for interview, we designed and distributed simple A4 paper fliers throughout the areas where we live and work. No financial incentives were offered (we did not have any budget for this project), but we discovered that the topic of solo travel was one that interested women enough to persuade them to participate in the study.

In all, over 50 women volunteered for interview. Thus, the sample of women interviewed in this study can be categorised as purposive (defined in terms of age range and solo travel experience) because of the way it was selected. The women we interviewed were aged between 25 and 66 and had very different travel styles and travel experiences. As the data collection progressed, through constant comparison we identified themes in the data which led to the use of theoretical sampling in the selection of the later participants. The diversity of the women who volunteered for interview has resulted in the data collected providing fascinating insights into the varied ways in which women define and experience solo travel and has given us many intriguing travellers' tales to illustrate the findings. Of course, a self-selected sample of this sort can always be critiqued as possibly biased towards those people who have the time and inclination to volunteer to take part in the project. However, as with all 'hard-to-reach' populations, sometimes researchers have to work within the constraints of available research participants and take account of the possibility that they have spoken to only a selection of the possible research participants in their discussion of the generalisability of their data.

Talking travel: interviews with solo travellers

Prior to the interviews we wrote to, or emailed, each woman, with a list of topics that would be covered and details of why we were gathering these data. In order to ensure that there was a record of consent for the interview, we asked every woman to sign an informed consent form (Berg 1989). When using interviews as a means of data collection it is always important to ensure that the status and use of data collected are clearly outlined to the participants. In this particular study, the vast majority of the women who volunteered for interview said that they were not concerned about issues of confidentiality as they were happy for their travel stories to be told. We have, however, elected to anonymise their comments when presenting the data in conference papers, in accordance with ethical guidelines underpinning social science research.

Hoinville and Jowell (1987: 71) identify what they call a 'continuum of formality' in interviewing, which ranges from totally structured to completely informal. This adaptability of interviews is widely recognised as one of the advantages of this research technique. In this study, interviews carried out with solo female travellers were semi-structured, using a predetermined interview guide consisting of 13 questions with various different probes to elicit more detail on a particular topic. Given that our intention was to interview women about their experiences of solo travel and then undertake cross-cultural comparison of the themes emerging from the interview data, we felt that it was important to ensure that our interview plans followed a similar, topic-based structure to allow for comparative analysis of data collected. Henderson (1991) recommends semi-structured interviews for novice interviewers or where a topic is new to the interviewer. As interviewers become more experienced and more knowledgeable about the topic, an interview can lose much of its structure and become more conversational in style. In our case, the list of topics was more of a guideline to ensure that we covered roughly the same issues in each interview, than a set interview template. Our interview guide covered such topics as reasons why the women chose to travel solo, what they liked and disliked about solo travel, and their encounters with other people while on the road. Because our data collection has continued over a number of years, and consistent with constant comparison methods, we have had the chance to investigate emerging themes by integrating them into later interviews and exploring them in more depth than would otherwise have been the case.

As discussed earlier, there are ethical issues associated with the power relations of interviews that invest researchers with a responsibility towards participants. The personal and subjective nature of interviews, which allows researchers to develop trust and empathy with participants, is a key benefit of the use of this technique for exploring travel experiences. Allowing people to tell their own stories generates data that are rich with description, and the real value of qualitative data lies in this richness. A number of the interviewees said that they welcomed the opportunity to chat about travel experiences and they hoped that doing so would help raise the profile of solo travel and overcome its perception as something strange for women to do. Similarly, during, or at the end of, most interviews we were asked about our own travel experiences and how those had led us to undertake this research. The process of sharing experience rather than just asking questions is an important aspect of the 'social interaction' of interviewers (Easterby-Smith *et al.* 1991). As researchers, if we want other people to tell us about their experiences, then surely we should be prepared to share our own. The researcher sharing their own experiences with research participants in this way can help to establish a slightly more equal rapport, as each knows more about the other. This can assist in addressing some of the issues of unequal power relations between the researcher and the research participant discussed earlier.

Letting the data speak: analysing and presenting interview data

As the interviews were conducted, they were transcribed and verified for accuracy in an ongoing process. Likewise, constant comparison of the data was also a continuous process (Glaser and Strauss 1967). The constant comparison method is used in conjunction with grounded theory. It is a process that starts during the interview stage, where patterns emerging in the data are incorporated into the later interviews to substantiate their relevance to other individuals. The idea is to work towards theoretical saturation whereby the patterns in the data have all been explored so that any theory generated from the data is as comprehensive as possible. When saturation is achieved, constant comparison techniques are used to analyse the data, first in the form of open coding, whereby behaviours are classified into broad categories, and then, as the data analysis continues, the categories are refined by means of axial coding and selective coding.

We used manual coding to develop themes (open, axial and selective coding; Strauss and Corbin 1998), which were compared and contrasted, often by email, between the United Kingdom and United States. We did find it extremely helpful to brainstorm the initial ideas with each other and to compare and contrast the themes. The value of such collaborative work lies in data analysis being simultaneously undertaken by two people rather than the findings resulting from just one person's interpretation of data. This allows for comparison or contrast of themes that seem to be emerging from data gathered, and acceptance or rejection of these ideas as reflecting the experiences of different women. Thus, two sets of themes are being identified and explored, allowing for a breadth of ideas to be generated. In this way it became apparent that, for instance, there were some cross-cultural differences in the ways that women in the United Kingdom and the United States chose to travel. Women from the United Kingdom were more likely to seek out organised small group travel and package holidays than women from the United States. But it was also evident that there were many ways in which women from the two countries experienced similarities in their solo travel. Certain aspects, such as the empowering nature of travelling solo, and the difficulties of dining alone, were common across cultures. The collaborative nature of this project has thus added an extra dimension to our research; in particular, it enabled us to open up our data to multiple interpretations, which is one of the key principles in Denzin and Lincoln's (1998) 'crisis in representation' moment. When we have presented our findings at leisure and tourism conferences over the years, the cross-cultural nature of the project has additionally resulted in interest from women in various countries to expand the project even further.

As researchers trained in positivistic methods, and who exist in an academic environment whereby the gatekeepers to publication continue to demand scientific rigour in research, which is often based on the positivistic values rather than the criteria for trustworthiness that Decrop discusses earlier in the book, we often find ourselves balancing between emphasising the value of the individual stories the women have shared and ensuring the 'credibility' of our data. Using grounded theory methods has helped us deal with this somewhat, as an inductive approach, whereby theory is generated from the data, helped us to work from the women's experiences rather than imposing our ideas and biases on the data, as far as this is possible.

Making sense of our data

Generally in grounded theory projects, a literature review is undertaken in parallel with data analysis, and where existing theory is drawn upon, this should be re-examined in relation to the ongoing analysis of data. There is some debate as to whether 'true' grounded theory analysis can actually be undertaken. O'Brien (1993: 11), for instance, argues that 'Whether theory is acknowledged or not, pure "empirical" research is impossible. At the simplest level theory may involve assumptions about how we know the world.' As we outlined earlier, we did have in mind theoretical positions when designing this research, and thus the data collection was to some extent underpinned by theory. Indeed, Strauss and Corbin (1998) suggest that the researchers are not without knowledge and that the theoretical sensitivity of the researcher is useful in that it plays a role in identifying and interpreting the patterns within the data. However, given that our intention was to explore the extent to which the theory could usefully help us to make sense of the experiences of women as solo tourists, our theoretical starting point was not fixed and we were happy to let the themes emerging from our initial interviews direct the topics to be further explored in later interviews. This has led us to explore areas of theory that we might otherwise not have considered relevant prior to our empirical work and, for us, has demonstrated the value of adopting a grounded theory approach.

Having collected and analysed data from women on both sides of the Atlantic, we then faced the challenge of presenting these data. Wolcott (1994: 17) suggests that 'qualitative researchers need to be storytellers' if they are to do justice to descriptive research. Our main concern was how to theorise the accounts of women's solo travel without appropriating their stories and making them ours; in other words, how to 'give voice' to these women. Again this relates back to our earlier discussion about the inequities in the power relationships between the researcher and the research participants. A question that should be consistently present in the researchers' minds at this point should be 'Is this what my interviewees were actually telling me?' The writing up and presentation of experiential data are potentially problematic areas for qualitative researchers and an area which Phillimore and Goodson have demonstrated has barely been addressed in a sophisticated manner by the majority of tourism researchers. Callaway (1992), for instance, warns that a disjunction can occur when personal interview data are written up in an impersonal or scientific way: '[W]omen and men in the field conduct their work in personal, face-to-face encounters through the medium of dialogue. Later, back home, these multiple levels of personal discourse become transmuted into impersonal and distant printed words' (1992: 30). Similarly, Sparkes (1996: 52) states that researchers cannot go back 'to the cozy self deluding days when texts were seen as neutral and innocent representations of the realities of others'. He calls for researchers to explore alternative ways of presenting data that reflect the role of researchers within the process.

One means of addressing these issues within tourism research may be for more researchers to adopt a personal and reflexive style of writing, resisting the more traditional scientific modes of data presentation that characterise the positivist research tradition (as discussed at the start of this chapter). This move to a more sophisticated approach as seen in recent sociological research, a development Denzin and Lincoln (1994) associate with the crisis in representation, would add a richer dimension to qualitative tourism research and enable readers to better assess its 'trust-worthiness'. However, one of the potential hurdles to pursuing this style of writing is that some journal editors and reviewers are wary of manuscripts written in the first person, an issue that Hall discusses in some detail in Chapter 8 of this book.

Another way of endeavouring to overcome the inherent difficulties in presenting the views of research participants, while also acknowledging the interpretive role of the researcher, may be to adopt a variety of narrative styles within one text (Sparkes 1996). Cotterill and Letherby highlight the flexibility of narrative accounts in presenting data: 'The narrative technique allows respondents to "tell the story" in whatever way they choose and, importantly, validates individual experience and provides a vehicle through which this experience can be expressed to a wider audience' (1993: 73).

In presenting our work we have sought to address the difficulties of 'giving voice' to the women who have participated in our study by incorporating various narrative styles ranging from the use of traditional short quotation supporting our points to the inclusion of longer individual stories or focusing on the travel careers of just a few of the participants. Again, this is an area where the collaborative nature of the research has been of great value. Having two researchers brainstorm various ways in which to present data has allowed each of us to determine the extent to which we think that particular modes of representation are reflective of the views of the women we interviewed. This process has also contributed both to the creativity of presentation and to enhancing our confidence in trying out new ideas. We are, however, still wary of the seemingly glib claim to 'give voice' to those marginalised by tourism research – something that we have found is easier to say than to do.

Conclusion

So, what have we as academics learned from the 'doing' of this research? To us, as full-time lecturers and researchers, working collaboratively has been both motivating and beneficial in terms of time management. It seems unlikely that either of us would have made as much progress as we have, particularly with data analysis and writing, had we not been doing so in relation to conference deadlines. Working with each other has helped us to prioritise this work where otherwise the minutiae of daily lecturing and other research work would have distracted us.

Discussing themes emerging from the data and comparing and contrasting these across countries and cultures has, we believe, added an extra dimension to our understanding of women's solo travel experiences. For example, the very structure of the tourism industries in the two countries has an obvious influence on the women's travel choices. In the United Kingdom, where travel has been traditionally more packaged than in the United States, more women spoke of their experiences of joining educational or activity tours to alleviate some of the awkwardness of travelling solo. In contrast, the US women tended to be much more independent travel planners and travellers. Thus, contextualising our data in a cross-cultural understanding of tourism industry practices was crucial, as was an understanding of the differing value systems at work in each of the two countries as they relate to annual vacations and tourism in general.

As Cotterill and Letherby (1993) suggest, collaborative working in this way involves the interweaving of personal and intellectual biographies. Having researched, presented and written together over so many years, we have evolved a way of working that allows our writing to be truly a product of both of our ideas. We each contribute our individual views and the data we have gathered within our particular cultural contexts. These ideas are then blended together to form a cross-cultural insight into the topic that neither of us could achieve as successfully alone. Sadly, research by Aitchison demonstrates that this type of co-authorship by female academics is unusual in the context of tourism, 'raising questions of women's apparent isolation within the research environment' (2001: 16). From our own experiences we would certainly encourage tourism researchers to consider collaborative investigation in relation to its potential to broaden our understanding of the cross-cultural nature of tourism and its value in personal terms.

We would not want to give the impression that it is always easy to write collaboratively and internationally. Practical issues such as time zone differences and varying pressures from workloads can constitute problems when deadlines loom. While it is not always easy to work across the Atlantic in this way, technology has played a key role in enabling us to progress this project. We have made extensive use of email in analysing the data and sharing ideas, supported with annual face-to-face meetings at conferences, summer visits home by Heather to the United Kingdom, and the occasional telephone call. We have also found that when writing together from a distance, the 'Track Changes' tool in Microsoft Word has been very helpful. In working this way we are, however, very dependent on technology, and if that fails (one of our conference papers was completely wiped from our computers and had to be scanned in from hard copy), then it creates major problems.

For both of us, this has been a fascinating project to be involved in. It has evolved over the years and moved on in directions that we could never have anticipated when we began. We have listened to many amazing tales of solo travel and, for two women with a love of travel, this has been an enjoyable experience. Our relationships with the women who participated in the study have been not just about collecting objective data for analysis, but about conversations with people. As Cotterill and Letherby (1993: 77) point out, 'the research process may make the participants of the research think about things they have never thought about before or indeed think about things in a different way'. A number of the women interviewed said that they felt they had learned from the experience of talking about their travel. A few women who had not travelled for a while, or who had little access to travel, said that talking to us had reminded them of the value of travel in their lives. We have even been contacted by women who have heard about our study and have thanked us for focusing attention on solo travel. Our project has in some ways helped them legitimate their travel behaviour to the outside world. It is particularly nice to reflect on research that seems to have been a positive experience for both the researchers and those who have participated in the research. For us this has certainly been a very rewarding journey.

Summary

In reflecting upon the value of our study for future tourism research in general, we would like to emphasise a few lessons we have learned:

- This project shows the value of reflexive research. Working inductively from the women's travel stories and thinking about them within the context of our own travel experiences has led us in directions that we would never have imagined at the outset.
- This project has resulted in an in-depth understanding of the experiences of a group of British and American women who travel solo. The meanings and sense of empowerment they derive from travelling would not have been evident with more traditional methods.
- Adopting an inductive approach to researching tourism experiences not only provides an insight into the meanings that tourism has for different tourists but also provides a basis for theorising those experiences in a meaningful way.
- The value of cross-cultural research in tourism cannot be overestimated. Not only is the cultural context of tourists influential in shaping their attitudes and behaviours, but the structure of the tourism industry in particular countries is also instrumental in these processes.

Despite the influence of globalisation and fears of homogenisation in destinations, there are still subtle differences among tourists around the world, and it should not be assumed that all tourists are alike.

- Tenets of feminist research aimed at redressing the balance of power between researchers and research participants have the potential to be applied more widely in the context of tourism research, particularly in researching the experiences of marginalised groups.

Questions

1 There are many different types of tourist, from the traditional package tour to the backpacker on a round-the-world tour. Choose a type of tourism that you are interested in finding out more about. Develop five or six semi-structured interview questions that you might use if you were to talk to a person who participates in your chosen form of tourism. Hint: structure your questions in such a way that the interviewee is encouraged to talk rather than providing one-word answers.

2 Consider the ethical issues raised within a research project where the experiences of disabled tourists are to be researched by non-disabled researchers and how these challenges might be dealt with in the design of the research project.

3 Choose a tourism topic that you consider could be most usefully researched collaboratively and plan a collaborative project, incorporating decisions such as who to collaborate with, what practical research design issues need consideration and what extra dimension will be added to the research by its collaborative nature.

4 In pairs, plan and carry out a semi-structured interview about a previous holiday that each of you has taken. In doing so, think through the key issues that you would like to explore within the interview.

References

Aitchison, C. (2000) 'Poststructural feminist theories of representing Others: a response to the "crisis" in leisure studies' discourse', *Leisure Studies*, 19: 127–144.

Aitchison, C. (2001) 'Theorizing Other discourses of tourism, gender and culture: can the subaltern speak (in tourism)?', *Tourist Studies*, 1 (2): 133–147.

Aitchison, C., Macleod, N. E. and Shaw, S. (2000) *Leisure and Tourism Landscapes: Social and Cultural Geographies*. London: Routledge.

Babbie, E. (1995) *The Practice of Social Research*, 7th edn. New York, NY: Wadsworth.

Berg, B. L. (1989) *Qualitative Research Methods for the Social Sciences*. Boston, MA: Allen and Bacon.

Black, P. (2001) 'Walking on the beaches looking at the ... bodies'. In K. Backett-Milburn and L. McKie (eds) *Constructing Gendered Bodies*. Basingstoke: Palgrave.

Botterill, D. (2001) 'The epistemology of a set of tourism studies', *Leisure Studies*, 20: 199–214.

Bryman, A. (1988) *Quantity and Quality in Social Research*. Contemporary Social Research 18. London: Routledge.

Bureau of Transportation Statistics (1999) *American Travel Survey*. Washington, DC: BTS.

Callaway, H. (1992) 'Ethnography and experience: gender implications in fieldwork and texts'. In J. Okely and H. Callaway (eds) *Anthropology and Autobiography*. ASA Monographs 29. London: Routledge.

Carter, S. and Clift, S. (eds) (2000) *Tourism and Sex: Culture, Commerce and Coercion*. London: Pinter.

Charles, N. and Hughes-Freeland, F. (eds) (1996) *Practising Feminism: Identity, Difference, Power*. London: Routledge.

Clarke, G. and Humberstone B. (eds) (1997) *Researching Women and Sport*. Basingstoke: Macmillan.

Cohen, L. and Manion, L. (1994) *Research Methods in Education*, 4th edn. London: Routledge.

Cotterill, P. and Letherby, G. (1993) 'Weaving stories: personal auto/biographies in feminist research', *Sociology*, 27 (1): 67–79.

Craik, J. (1997) 'The culture of tourism'. In C. Rojek and J. Urry (eds) *Touring Cultures: Transformations of Travel and Theory*. London: Routledge.

Davidson, P. (1996) 'The holiday and work experiences of women with young children', *Leisure Studies*, 15: 89–103.

de Vaus, D. A. (1993) *Surveys in Social Research*, 3rd edn. London: UCL Press.

Deem, R. (1986) *All Work and No Play: The Sociology of Women and Leisure*. Milton Keynes: Open University Press.

Deem R. (1996a) 'No time for a rest? An exploration of women's work, engendered leisure and holidays', *Time and Society*, 5 (1): 5–25.

Deem, R. (1996b) 'Women, the city and holidays', *Leisure Studies*, 15: 105–119.

Denzin, N. K. and Lincoln, Y. S. (eds) (1994) *Handbook of Qualitative Research*. London: Sage.

Denzin, N. K. and Lincoln, Y. S. (eds) (1998) *The Landscape of Qualitative Research: Theories and Issues*. Thousand Oaks, CA: Sage.

Dupuis, S. L. (1999) 'Naked truths: towards a reflexive methodology in leisure research', *Leisure Sciences*, 21: 43–64.

Easterby-Smith, M., Thorpe, R. and Lowe, A. (1991) *Management Research: An Introduction*. London: Sage.

Eichler, M. (1988) *Nonsexist Research Methods: A Practical Guide*. Boston, MA: Allen and Unwin.

Enloe, C. (1989) 'On the beach: sexism and tourism'. In C. Enloe (ed.) *Bananas, Beaches and Bases: Making Feminist Sense of International Politics*. London: Pandora.

Finn, M., Elliott-White, M. and Walton, M. (2000) *Tourism and Leisure Research Methods: Data Collection, Analysis and Interpretation*. Harlow: Longman.

Foucault, M. (1977) *Discipline and Punish: The Birth of the Prison*. London: A. Lang.

Foucault, M. (1979) *The History of Sexuality*, vol. 1, *An introduction*. New York, NY: Pantheon.

Foucault, M. (1984) *The History of Sexuality* vol. 2, *The Use of Pleasure*. London: Penguin.

Fullager, S. (2002) 'Narratives of travel: desire and the movement of feminine subjectivity', *Leisure Studies*, 21: 57–74.

Gibson, H. (1994) 'Some predictors of tourist role preference for men and women over the adult life course', unpublished doctoral dissertation, University of Connecticut, Storrs, CT.

Gibson, H. (1996). 'Thrill seeking vacations: a lifespan perspective', *Loisir et Société/Society and Leisure*, 19 (2): 439–458.

Gibson, H. (2001) 'Theoretical perspectives on gender and tourism'. In Y. Apostolopoulos, S. Sonmez and D. Timothy (eds) *Women as Producers and Consumers of Tourism in Developing Regions*. Westport, CT: Praeger.

Gibson, H. (2002) 'Busy travellers: leisure-travel patterns and meanings in later life', *World Leisure*, 44 (2): 11–20.

Gibson, H. and Jordan, F. (1998a) 'Travelling solo: a cross-cultural study of British and American women aged 30–50 years', paper presented at the Fourth International Conference of the Leisure Studies Association, Leeds Metropolitan University, Leeds, July 1998.

Gibson, H. and Jordan, F. (1998b) 'Shirley Valentine lives! the experiences of solo women travellers', paper presented at the World Leisure and Recreation Association Fifth World Congress, Leisure in a Globalised Society – Inclusion or Exclusion?, São Paulo, October 1998.

Glaser, B. and Strauss, A. (1967) *The Discovery of Grounded Theory: Strategies for Qualitative Research*. Chicago, IL: Aldine.

Green, E., Hebron, S. and Woodward, D. (1990) *Women's Leisure, What Leisure?*: Basingstoke: Macmillan.

Harding, S. (ed.) (1987) *Feminism and Methodology*, Bloomington, IN: Indiana Press; Milton Keynes: Open University Press.

Harding, S. (1991) *Whose Science? Whose Knowledge? Thinking from Women's Lives*. Milton Keynes: Open University Press.

Henderson, K. (1991) *Dimensions of Choice: A Qualitative Approach to Recreation, Parks and Leisure Research*. State College, PA: Venture Publishing.

Henderson, K. A. (1998) 'Researching diverse populations', *Journal of Leisure Research*, 30 (1): 157–170.

Henderson, K. A. and Ainsworth, B. (2001) 'Researching leisure and physical activity with women of color: issues and emerging questions', *Leisure Sciences*, 23: 21–34.

Henderson, K. A. and Bialeschki, M. D. (2002) *Evaluating Leisure Services: Making Enlightened Decisions*, 2nd edn. State College, PA: Venture Publishing.

Henderson, K. A., Bialeschki, M. D., Shaw, S. M. and Freysinger, V. J. (1996) *Both Gains and Gaps: Feminist Perspectives on Women's Leisure*. State College, PA: Venture Publishing.

Herold, E., Garcia, R. and DeMoya, T. (2001) 'Female tourists and beach boys: romance or sex tourism?', *Annals of Tourism Research*, 28 (4): 978–997.

Hoinville, G. and Jowell, R. in association with Airey, C., Brook, L., Courtenay, C. *et al.* (1987) *Survey Research Practice*. London: Heinemann Education.

Jackson, S. and Jones, S. (1998) 'Thinking for ourselves: an introduction to feminist theorising', in S. Jackson and S. Jones (eds) *Contemporary Feminist Theories*. Edinburgh: Edinburgh University Press.

Jordan, F. (1997) 'An occupational hazard? Sex segregation in tourism employment', *Tourism Management*, 18 (8): 525–534.

Jordan, F. (1998) 'Shirley Valentine, where are you? Tourism provision for mid-life women travelling alone'. In C. Aitchison and F. Jordan (eds) *Gender, Space and Identity: Leisure, Culture and Commerce*. Eastbourne: Leisure Studies Association Publications.

Jordan, F. and Gibson, H. (2000) 'Whose space is this anyway? The experiences of solo women travellers', paper presented at the Institute of British Geographers Annual Conference, University of Sussex, Brighton, 4–7 January 2000.

Jordan, F. and Gibson, H. (2002) 'Sun, sea, sand, but most of all sex: discourses of the productive consumption of tourism as sexualised space(s)', paper presented at the Leisure Studies Association International Conference, Leisure: Our Common Wealth?, University of Central Lancashire, Preston, July 2002.

Kelly, J. R. (1987) *Freedom to Be: A New Sociology of Leisure*. New York, NY: Macmillan.

Kinnaird, V. and Hall, D. (eds) (1994) *Tourism: A Gender Analysis*. New York, NY: Wiley.

Kinnaird, V. and Hall, D. (1996) 'Understanding tourism processes: a gender-aware framework', *Tourism Management*, 17 (2): 95–102.

McCarl Nielsen, J. (ed.) 1990, *Feminist Research Methods: Exemplary Readings in the Social Sciences*. London: Westview.

Marshment, M. (1997) 'Gender takes a holiday: representation in holiday brochures'. In M. Sinclair (ed.) *Gender, Work and Tourism*. London: Routledge.

Mintel (2001) *Singles on Holiday, Leisure Intelligence – UK Report December 2001*. London: Mintel International Group.

Morris, K., Woodward, D. and Peters, E. (1998) 'Whose side are you on? Dilemmas in conducting feminist ethnographic research with young women', *International Journal of Social Research Methodology*, 1 (3): 217–230.

O'Brien, M. (1993) 'Social research and sociology'. In N. Gilbert (ed.) *Researching Social Life*. London: Sage.

Potter, W. J. (1996) *An Analysis of Thinking and Research about Qualitative Method*. Hillsdale, NJ: Lawrence Erlbaum.

Pritchard, A. (2001) 'Tourism and representation: a scale for measuring gendered portrayals', *Leisure Studies*, 20: 79–94.

Pritchard, A., Morgan, N., Sedgley, D. and Jenkins, A. (1998a) 'Reaching out to the gay tourist: opportunities and threats in an emerging market segment', *Tourism Management*, 19: 273–282.

Pritchard, A., Morgan, N., Sedgley, D. and Jenkins, A. (1998b) 'Gay tourism destinations: identity, sponsorship and degaying'. In C. Aitchison and F. Jordan (eds) *Gender, Space and Identity: Leisure, Culture and Commerce*. Eastbourne: LSA Publications.

Reissman, C. K. (1991) 'When gender is not enough: women interviewing women'. In J. Lorber and S. A. Farrell (eds) *The Social Construction of Gender*. London: Sage.

Richter, L. K. (1994) 'Exploring the political role of gender in tourism research'. In W. Theobold (ed.) *Global Tourism: The Next Decade*. Oxford: Butterworth-Heinemann.

Richter, L. K. (1995) 'Gender and race: neglected variables in tourism research'. In R. Butler and D. Pearce (eds) *Change in Tourism: People, Places, Processes*, London: Routledge.

Roberts, H. (ed.) (1981) *Doing Feminist Research*. London: Routledge.

Sanchez Taylor, J. (2001) 'Dollars are a girl's best friend? Female tourists' sexual behaviour in the Caribbean', *Sociology*, 35 (3): 749–764.

Seidman, I. E. (1991) *Interviewing as Qualitative Research: A Guide for Researchers in Education and the Social Sciences.* London: Teachers College Press.

Sparkes, A. C. (1996) *Writing the Social in Qualitative Enquiry.* Educational Research Monograph Series, Research Support Unit, School of Education, University of Exeter.

Squire, S. J. (1994) 'Gender and tourist experiences: assessing women's shared meanings of Beatrix Potter', *Leisure Studies*, 13: 195–209.

Stanley, L. (ed.) (1990) *Feminist Praxis: Research, Theory and Epistemology in Feminist Sociology.* London: Routledge.

Stanley, L. and Wise, S. (1993) *Breaking Out Again: Feminist Ontology and Epistemology*, 2nd edn. London: Routledge and Kegan Paul.

Strauss, A. and Corbin, J. (1998) *Basics of Qualitative Research: Techniques for Developing Grounded Theory*, 2nd edn. Thousand Oaks, CA: Sage.

Swain, M. B. (1995) 'Gender in tourism', *Annals of Tourism Research*, 22: 247–266.

Veal, A. J. (1992) *Research Methods for Leisure and Tourism: A Practical Guide.* Harlow: Longman/ILAM Leisure Management Series.

Veal, A. J. (1997) *Research Methods for Leisure and Tourism: A Practical Guide*, 2nd edn. Harlow: Longman/ILAM Leisure Management Series.

Walle, A. H. (1997) 'Quantitative verses qualitative tourism research', *Annals of Tourism Research*, 24: 524–536.

Wearing, B. (1996) 'Grandmotherhood as Leisure?', *World Leisure and Recreation*, 38 (4): 15–19.

Wearing, B. (1998) *Leisure and Feminist Theory.* London: Sage.

Wearing, S. and Wearing, B. (2001) 'Conceptualizing the selves of tourism', *Leisure Studies*, 20: 143–159.

Wickens, E. (2002) 'The sacred and the profane: a tourist typology', *Annals of Tourism Research*, 29 (3): 834–851.

Wimbush, E. and Talbot, M. (eds) (1988) *Relative Freedoms: Women and Leisure.* Milton Keynes: Open University Press.

Wolcott, H. F. (1994) *Transforming Qualitative Data: Description, Analysis and Interpretation.* London: Sage.

14 The life and work history methodology

A discussion of its potential use for tourism and hospitality research

Adele Ladkin

Aims of the chapter

- To provide an overview of life history analysis.
- To examine the historical development of the life history method.
- To outline the appeal of and the criticisms of the life history method.
- To explore work history analysis and its relevance to the labour market.
- To examine the life and work history methodology applied to tourism and hospitality research.
- To draw conclusions regarding the implications for tourism research and knowledge.

Introduction

The purpose of this chapter is to explore the life and work history methodology, and to discuss its potential use in tourism and hospitality research. The intention is both to discuss the historical developments of the methodology and, using a selection of qualitative findings, to illustrate both the current application and the potential use of this type of research. Life and work histories are a distinct methodology in the sense that they are concerned with a particular branch of activity – in this case, all historical aspects of an individual's life. The methods used to collect life and work history information are varied, and comprise both quantitative and qualitative techniques.

Despite the well-documented usage of both the life and the work history methodology for a range of different applications (Dex 1991: 1), as with many of the less well established research methods, the use of the methodology is at the infancy stage in relation to tourism and hospitality research. As a consequence, this chapter contains a degree of speculation on the future applicability of the methodology for tourism research. Furthermore, while recognising the diversity of the life and work history methodology, this personal reflection focuses largely on career analysis, which is where the significant applications are found in relation to tourism and hospitality research (Ladkin 2002: 379; McCabe and Ladkin 2002: 159).

The collection of life and work history data has been used in social science research for a number of years, and has made significant developments at both a conceptual and a methodological level. Life histories exist within the umbrella term of 'life writing' along with life stories, autobiographies, journals, diaries, portraits, profiles, memoirs and case studies (Smith 1994: 287). Each of these types of life writing has a slightly different perspective, but all are concerned with biographical data, with a focus on the past. Work histories are concerned with information relating to an individual's working life, and work history data provide the information that is needed for career analysis (Ladkin 2002: 381). Collecting life and work history data provides the methodological approach to obtaining detailed work histories and career analysis.

Essentially, life and work histories are concerned with two separate aspects of an individual's past life, and consequently have experienced slightly different developments. For this reason, the developments of life and work histories are considered separately here. Life history analysis stems from research carried out by the Chicago School of Sociology that had a qualitative bias, and was opposed to the quantitative survey methods. Deviancy sociologists, occupational sociologists and oral historians have all advocated the use of the life history technique (Faraday and Plummer 1979: 773). In contrast to this, work history analysis grew out of large-scale quantitative studies that focused on employment. Specifically, its roots lie in the study of women's employment, but now it extends to both genders.

While they may relate to different research traditions, it can be argued that life and work histories of individuals are emphatically linked and, importantly, there are methodological issues which apply equally to both types of historical data (Ladkin 1999: 40). Furthermore, it has been argued that making a distinction between life and work methodologies acts as a barrier to the development of the methodology. Baker and Elias (1991: 215) argue that through small-scale and qualitative analysis of life and family histories, and the large-scale quantitative analysis of employment, life and work experiences are linked, and it is impossible to consider one without the other. In addition, Dex (1991: 10) argues that the distinction between quantitative and qualitative research methods continues to hamper the development of the life and work history methodology, as studies with a primarily qualitative approach have failed to acknowledge even the most basic quantitative techniques, and vice versa. In response to these criticisms, the work history analysis applied to tourism and hospitality research argues for the use of both quantitative and qualitative techniques. As such, the methodology blurs the definitions between methods and focuses on multiple approaches, thus slotting into Denzin and Lincoln's third moment in qualitative research, blurred genres (Denzin and Lincoln 1998; see Chapter 1 for a discussion of the five moments of qualitative research).

Life history analysis

Historical development

Life histories, like the many variants of biography, are contained within the biographical method (Smith 1994: 287). The biographical method contains a wealth of disciplinary strands, notably literature, history, social science, education, and feminist and minority perspectives (Smith 1994: 294). The life history approach as a research methodology for the social sciences has been in existence since at least the early twentieth century, traditionally in the area of sociology. Faraday and Plummer (1979: 773) argue that the first major sociological life history, *The Polish Peasant in Europe and America* by Thomas and Znaniecki 1918–1920 (cited in Faraday and Plummer 1979: 773), initiated the idea that the life history approach was the perfect type of sociological material. This was because it was not confined to the study of human personality, but could relate to general classes of data or facts, which in turn could be used for the determination of laws of society, a key outcome in research during the traditional period. Given the prevailing social climate of the time, the human condition of individuals in society was at the forefront of social research. From a sociology of knowledge perspective, the growing popularity of, and scholarly interest in, using life history analysis in interpreting human behaviour is a product of the well-documented forces that have increased awareness of the presence of the 'human element' in the formation of institutions and socio-political control mechanisms (Stanfield 1987: 429). Elder (1978: 19) agrees that this work by Thomas and Znaniecki opened up new ideas in relation to the temporal study of individuals and groups in situations of drastic change, thereby providing a research method that was capable of examining many aspects of society. Consequently, the life history approach became established as a way to study a whole range of issues relating to societies at any point in time. Further support for the life history approach continued through the Chicago tradition of sociological analysis (Elder 1978: 19), with a more modernist stance on the approach. The early Chicago School of Sociology provided a pragmatic approach to method and theory, and its researchers were essentially concerned with the study of individuals, groups and social organisations in a historical context, with an interest in processes of social change (Ladkin 1999: 38). According to Smith (1994: 298), life histories are part of the mosaic of community and institutional investigations that are capable of considering any abstract theory of person and community, and testing implicit assumptions about human beings in the larger sociological studies.

Conceptually, Hareven (1978: 5) and Elder (1978: 22) made advances in the 1970s with the introduction of the concept of life course analysis. This was the first time that life histories were given a framework for analysis, which resulted in a broadening of the types of data collected and recorded. Typical of the Chicago School, the conceptual framework of life course analysis has its roots in the study of individual histories and careers, and focuses on the overlaps between individuals' experiences and those coincidental in chronological (or historical) time, family members, structural or policy changes, or other individuals (Ladkin 1999: 38). This approach examines the interrelationships of an individual's life with other events and lives. It also focuses on transitions over time between various states, and is therefore particularly suited to longitudinal data. Owing to its universal appeal, many studies now use this life course analysis framework (Dex 1991: 4).

In terms of data collection, over the years a whole range of methods have been used to collect information regarding life histories. The most commonly used methods are autobiography, oral history, in-depth interviewing, survey interviewing, biography, biographical reference tools and fictitious biography. Although they are slightly different in focus, a common element is that all the techniques collect information over time. Furthermore, in each of these approaches there exists an overriding chronology of birth and death, and each is concerned with what happens between these two defining events. The wide choice of techniques and range of applications has meant that life history analysis continues to be widely used among researchers and has evolved with research traditions slotting into all the major research traditions outlined by Denzin and Lincoln (1998). In addition to sociologists, psychologists, political scientists and historians have all used life history data to research aspects of human behaviour and economic status (Stanfield 1987: 430). The recent uses of work history data for career analysis has introduced labour market analysis as a developing use for life and work history data (Ladkin and Riley 1996: 444; Ladkin 1999: 38; Ladkin 2002: 380; Ladkin and Juwheer 2000: 121).

Because of the number of different developmental strands of the life history method, a detailed discussion of each of these is beyond the scope of this chapter, but they have been considered in some detail by a number of authors, for example Stanfield (1987: 429) and Smith (1994: 286). However, the essential elements that the life history methodology contains are that it often has a focus on social history and people in society (Elliott 1990: 65); it uses autobiographical and biographical materials to explain motivations, attitudes and values that shape human behaviour and events (Stanfield 1987: 429); and it attempts to uncover the causes behind observed relationships in the lives of human beings (Baker and Elias 1991:

214). The key epistemological and methodological issues that surround the technique are considered here in relation to the attraction and the criticisms of the methodology.

The appeal of the life history methodology

The popularity of the life history method as a research tool in the social sciences is largely due to the qualities and strengths associated with this approach. A fundamental quality is that by exploring the past, life history analysis provides valuable information about the present. For example, an exploration into the earnings of the young in terms of their past employment history is one approach offered by Baker and Elias (1991: 214). The value of the past with regard to careers can provide detailed labour market analysis (Riley 1990: 38; Burchell 1992: 346, Ladkin and Riley 1996: 443; Ladkin 1999: 38; Ladkin 2002: 380). This has relevance for tourism research in terms of developing an understanding of tourism labour markets. Related to the general appeal of qualitative research, two key methodological issues provide an additional attraction of the life history method. The first is the detail and intimacy of the material gathered. Life histories provide an opportunity for the researcher to get close to the experiences and feelings of the respondent. The second takes a wider perspective and is that the life history method provides a unique position to explore how people relate to wider societal events by adopting a subjective approach (Gittens 1985: 95). Both of these have a value in tourist motivation research, as motivations to travel are influenced by both individual attitudes and external societal influences. Life and work history data also offer a valuable insight into attempts to unravel the relationships in social life and to gain a better understanding of social relationships. It has been suggested that it is possible to test economic, historical, psychological and sociological theories using life histories, as individuals' lives are the stage on which societal changes are played out (Ladkin 1999: 38). In essence, the wide application of what essentially is a reliable and accessible method of data collection is one of the main attractions of the technique.

A further reason for advocating the popularity of the life history method has arisen as a result of the methodological problems inherent in the collection of longitudinal data. While there is now acceptance and use of the life course concept, it is difficult to carry out large-scale longitudinal studies. Although longitudinal surveys have the unique ability to illustrate causal direction and processes of change, they are used infrequently, for two main reasons. The first relates to the problem of attrition in repeat-interview longitudinal research, whereby respondents are lost in the survey over time (Dex 1991: 5). In other words, any survey that spans a long period of time is likely to lose respondents, as, for a variety of reasons, not all the initial respondents will be in contact for the whole time period. The second relates to the high cost of collecting longitudinal data.

Even when the research is small scale, considerable cost is incurred when individuals or organisations have to be revisited. This economic factor tends to prohibit the use of longitudinal survey research methods (Dex 1991: 5). Both these factors rule out the use of longitudinal surveys in tourism research, as in any other branch of social science research. In response to these difficulties, support has grown for collecting longitudinal data by the recall or memory method (Dex 1991: 5). The recall method has the advantages that it is cheaper to carry out, it is unaffected by attrition, and a systematic record can be obtained and coded at one point in time (Ladkin 1999: 39). Support for the recall method of data collection is provided by a number of studies – for example, the study of the family and the life course in historical perspective (Hareven 1978: 10). Given the costs and inherent problems of longitudinal data sources, the recall method of data collection is likely to be the one most commonly used for the collection of life history information. Small's chapter in this volume (Chapter 15) provides an excellent example of how recall can be used in tourism research.

The final main attraction of the life history method is that life histories also have a range of important practical applications. First, it is a method that can be used for exploratory purposes in research areas where little is known, or the concepts have been poorly identified. The life history acts as a sensitising mechanism to issues and problems in a particular area (Faraday and Plummer 1979: 778). Second, life histories can also be used as a complement to other research methods. For example, life history data can be used to complement data derived from secondary sources, survey research, observation and in-depth interviews (Armstrong 1987: 14). The life history methodology has been used extensively in race relations research (Dex 1991: 13), and for sociological research into economic status and human behaviour (Stanfield 1987: 430). Although this list is not exhaustive, the practical applications in relation to many types of societal research have firmly established life histories as a valid research technique for discovering a range of individual and societal issues. Although the technique is presently underutilised in tourism research, there is no reason why these practical applications could not be used for research in tourism. The UK Economic and Social Research Council's endorsement of the value of life history analysis as a methodological tool has assisted in the justification of its validity.

Criticisms of the life history method

Despite the wide use and appeal of the life history approach among social scientists, the technique is not without criticism. The methods used to collect life history data and the validity of such data are the main epistemological issues surrounding the technique. For the most part, the criticisms are the same as the wider issues related to qualitative research – for example, bias, the credibility of the data, and doubts of scientific validity (Babbie 2001: 298). These criticisms of qualitative research are therefore also applied to tourism research. Specific to the life history technique, according to Faraday and Plummer (1979: 775), there are four types of methodological problems associated with its use: social science problems, technical problems, ethical and political problems and personal problems. The social science problem refers to the sociological justification for using the life history technique in terms of what it can contribute to sociological understanding, or, put another way, the 'why' questions of social research which deal with the purpose and justification of the study. The technical problems are the 'how' questions of social research which deal with the execution of the study – for example, given certain problems, how can research meet these requirements and how can the most appropriate techniques be employed? The ethical and political problems are the 'commitment' questions of social research, which concern the morality and politics of the research. Finally, the personal problems refer to the effectively ignored questions which recognise that the researcher is not merely an automation data processor but is a human being who absorbs the very research process into his or her daily experiences (Faraday and Plummer 1979: 793). These issues have relevance to tourism research where the life history methodology has been used. Specifically with regard to epistemology and the life history technique, the key question is to define the status of the findings. This is not the same as the validity, but is a matter of the 'truth' that is arrived at through the method (Faraday and Plummer 1979: 779). The life history technique is grounded in a pragmatist approach to knowledge in which the ultimate test of truth is experience. Like symbolic interactionism, life history is concerned not so much with grasping the totality, either of structures or of personality, as with depicting the immediate lived experiences as actual members in everyday society grasp them. For an extensive discussion on epistemology and the life history technique, the reader is directed to Faraday and Plummer (1979: 779). There needs to be some consideration of how the criteria for trustworthiness outlined by Decrop in this book (Chapter 9) can be applied to life history technique in order to demonstrate the rigour of the technique to those unfamiliar with it.

A further main criticism of the life history method relates to methodological problems associated with collecting life history data by the memory recall method. The fallibility of human memory and the issue of selective recall inevitably mean that the life history methodology is subjective (Walby 1991: 171). It is subjective in the sense that it is based on personal emotions and prejudices, and over time, true meanings are likely to have been lost or altered. Therefore, any research into the various aspects of tourism would have to acknowledge that memory may be subject to inaccuracy. Given this criticism, life histories become little more than a source of rich ethnographic data; they are not reliable enough to produce hard data that can be appropriately analysed using statistics. This fallibility of human memory cannot be considered in isolation from issues of interpretation, in the sense that memory distorts facts and feelings over time as events are recalled and processed by an individual. Methodological developments that improve the ability to correctly interpret memory recall data will greatly improve our understanding of this issue (Dex 1991: 15).

Dex (1991: 6) identifies a number of additional reasons why the memory recall method is problematic. The first relates to the age of the memory. For example, the error of recall is likely to be greater the further back in history the respondent is required to remember. It cannot be assumed that people will remember all experiences clearly over time, and some bias or lack of clarity is likely to occur in the recall of certain events. Second, there is a proven link between the strength of a particular memory in relation to reported or experienced events, with experienced events more likely to be remembered accurately (Larsen 1988: 334; Howes and Katz 1992: 112). This results in a likely bias in certain events or experiences, as there often exists a tendency for people to remember and recall those instances and impressions which have been of significant importance or interest to them (Dex 1991: 6). The question remains as to how well people can remember facts and attitudes towards travel experiences.

Finally, a further criticism concerns the problem of bias: bias arising from the subject being interviewed, bias from the researcher, and that arising from the subject–researcher interaction (Plummer 1983: 102). The bias in the data collected is linked to how the events are interpreted both as they are experienced, and again how they may be changed over time. This represents a significant methodological problem in the collection of life histories for any research purpose, including research into tourism.

However, to a certain extent these problems of recall can be countered by careful survey techniques and data collection methods. For example, Gittens (1979: 92) argues that the way a questionnaire is structured has been found to influence the quality of recalled data, and specific questions to a respondent are more likely to produce accurate answers, rather than vague or general questions (Ladkin 1999: 40). Riley *et al.* (2002: 79) identify a number of reasons why the recall method is appropriate for the collection of work- and career-related information. These include the point that recall of autobiographical events is superior to the recall of public events. Moreover, the recall of autobiographical events improves if events are important to the self-schema and are linked to long-term goals, event memorability, and event pleasantness. Therefore, although certain events exist in relation to the recall method, there are ways in which the errors can be minimised. The benefits associated with collecting life history data rather than through longitudinal research will ensure continued use of the technique.

Work history analysis

Although it can be argued that work history information is essentially part of life histories, the idea of collecting specific data related to people's working lives has received less attention. However, Dex (1991: 13) argues that work history analysis can be used to investigate occupational mobility and to analyse social class, unemployment, stratification and social mobility, the links between education and occupation, the immobility of labour, ageing and early retirement, and ongoing labour market issues. Further suggestions for utilising work history data include sociologists' interest in changes in the nature of work and conceptualisation of labour markets, and economists' interest in labour turnover and mobility, the analysis of learning differentials and pay discrimination, and the effects of social policies on labour market experiences and behaviour (Dex 1991: 13). Each of these could be applied to investigating work and labour markets in tourism research. Given the previous criticism that the development of both the life and work history methodology has been hampered by not considering both elements, further research in this area should aim to address this issue. Therefore, both life and work history are considered in relation to potential use in tourism and hospitality research.

As with life history analysis, a number of developments have taken place over time in the use of work history data. One such development is the analysis of women's work histories. These include examining the sequence order of work, the duration of jobs, an overview of the profile of one dimension of work history, and the effect of disruptive events on work activity (Dex 1984: 640). The role of women in the workforce and female employment in the hospitality industry is a research area that to date has not used the work history approach. Greater detail can then be added through further sets of relationship – for example, between structural aspects of the economy and the individual's work experiences, or between the individual's present status and previous work experience. This reinforces the value of the past as a dimension to career or work histories, as past work histories can provide information on both economic and social conditions over a period of time. Therefore, an exploration of work histories in tourism occupations could lead towards an understanding of changing labour market conditions. This is explored later in this chapter.

According to Dex (1984: 640), the examination of interrelationships of an individual's life to other events and lives to life course analysis suggests that an appropriate term is the work course analysis for the focus of working lives (Ladkin 1999: 40). As with life course analysis, work course analysis has become an acceptable conceptual framework on which to base work history research. Furthermore, life course analysis identifies the career line (Elder 1978: 22). In its most general sense, the career of an individual refers to a sequence of activities or roles through social networks and settings. From this perspective, a career line is equivalent to an individual's life history in a variety of roles – for example, parenthood, marriage and work life. Whatever a person's work life of jobs and employers, this represents their occupational career (Elder 1978: 24). Taken from a structural perspective, a career entails a succession of related jobs arranged in a hierarchy of prestige that a person moves through in an ordered sequence. From a subjective standpoint, a career can be seen as a projection of future events and their anticipated significance, or a vantage point based on interpretations of experience as a person moves through life (Elder 1978: 22). Elder (1978: 24) argues that expectations in a career projection are influenced by childhood experience and have some bearing on subsequent interpretations of accomplishments. Whatever viewpoint is taken, it is clear that an individual's work history is linked to many other events within their life, and that a person's life history is a product of multiple histories defined by time and event sequence. Occupations in tourism are no exception to this rule.

The collection of work histories for use in conjunction with labour market analysis is an important development in the use of work history analysis. The work history approach allows research to focus on work processes and labour market changes, rather than just exploring relationships between employers and employees within the workplace, and on the involvement of workers in trade unions and other political organisations, which was a feature of previous work history research. Using a study of youth employment, Baker and Elias (1991: 214) tested whether the experience of unemployment leads to reduced labour earnings late on in an individual's economic life, or whether a spell of unemployment increases the chances of experiencing a further spell of unemployment. The research examines in detail the ways in which work history information concerning young people can be used to untangle the causal processes which lie behind the observed relationship between youth employment and later economic status, and concludes that work history information provides enhanced understanding of the labour processes behind this work pattern (Ladkin 1999: 41). An understanding of labour processes could therefore help us to better understand the relationships between various social and economic factors and tourism careers.

Burchell (1992: 350) analyses labour market flows using work history data, and provides an alternative labour market approach within the field of work history data and labour market analysis. Here, there is a shift in emphasis away from the individual or the job as a means to examine work histories, to a focus on the job change as the unit of analysis. Support for this unit of analysis is derived from the notion that in an attempt to understand the structure and dynamic processes of the labour market, the actual jobs that people do are not as significant as the way in which individuals move between those jobs. This gives information on labour market mobility, and linkages between jobs.

The life and work history methodology applied to tourism and hospitality research

The first half of this chapter was concerned with exploring the historical developments of the life and work history methodology. The remainder of the chapter seeks to explore how the life and work history methodology can be applied within tourism and hospitality research. It must be stated at the outset that the application of the technique is largely speculative, as to date it is an underutilised approach in the tourism and hospitality field. The technique offers a range of opportunities for tourism research, and it is hoped that this chapter will encourage tourism researchers, whatever their epistemological or ontological stance, to consider using the technique.

Essentially, life and work histories are concerned with the collection of data that span a given period of time. Although the focus of the data collected is different – that is, life history focuses on the many aspects of an individual's life, and work history focuses on the work element of a person's life – they are both concerned with understanding the past. This understanding of the past could be used in a number of areas relevant to research in tourism and hospitality, and is used in Chapter 15 by Small to examine women's holiday experiences.

Butler (1993) has argued that there is a lack of research into change in tourism destinations following the development of tourism, and in particular a lack of baseline studies. One potential area for using the life history technique is to provide a historical narrative of holiday and/or travel experiences of individuals. As an industry, tourism has not remained static over time, and holiday and travel experiences are very different from what they were in the past. Detailed histories of past travel experiences could provide an insight into the ways in which travel and societies have changed over time. The method could be used to gain an insight, for example, into childhood memories of holidays, compared across cohorts of different age groups.

The same principle could also be applied to the history of tourism development in a particular destination or region. Taking detailed recollections of the nature and character of destinations in the past from older residents and charting how they have changed over time would provide an insight into the transformation of destinations with the arrival and subsequent development of tourism. This would give a perspective on the individuals and the society which has been altered with the arrival of tourist activities.

Detailed life histories could also be used to help us understand patterns of holiday choice and travel experiences. Taking this a stage further, patterns of holiday choice could be related to the different stages of the life cycle in order to test the assumptions and characteristics of travel demand. Associated with the idea of life and work histories being able to uncover patterns and processes, a further use for the methodology may be in the area of customer loyalty. The use of a variety of approaches to the study of loyalty research in tourism has been explored by Riley *et al.* (2001: 24), and perhaps the life history approach could unravel some of the complexities behind repeat travel purchases and loyalty. Indeed, the motivations behind travel decisions have been explored using the 'travel career ladder'. This concept was first fully described by Pearce (1988) and has been extensively used and discussed by other researchers, for example Holden (1999), Ryan (1998) and Kim *et al.* (1996). Based on Maslow's hierarchy of needs and conceptualisations of psychological maturation towards a goal of self-actualisation, the travel career ladder postulates that as tourists become more experienced, they increasingly seek satisfaction of higher needs (Pearce 1991: 46). In order to test the travel career ladder, travel decision-making histories are obtained from individuals.

To turning attention to work history, it is in this area where most empirical work has been undertaken in tourism and hospitality. Recently, work history data have been collected and analysed for use in conjunction with labour market analysis, in order to understand career patterns. This approach has been used extensively in exploring the careers of hotel general managers (Ladkin 2002: 380; Ladkin 1999: 38; Ladkin and Juwaheer 2000: 120; Ladkin and Riley 1996: 444) and employees working in the convention and exhibition industry (McCabe and Ladkin 2002: 159). In an ever-growing industry that provides a range of employment opportunities, an understanding of how individuals develop their careers has important implications for tourism labour markets and the demand and supply of labour for the industry. Although, to date, work history has focused on the careers of hotel managers and managers in the convention and exhibition industry, the technique is not exclusive to this occupation, and it could be applied to a whole range of professions within both the tourism and the hospitality industry.

Following on from the use of work history data to explore career patterns and processes, such data could also be used to trace the development of skills within any particular job or sector. The collection of detailed work histories focusing on skill development would enable an assessment of where important skills are learned, and at what stage in a person's career. This has implications for education and training for the tourism and hospitality industry.

Problems of the method

Despite there being a range of areas in which the life and work history methodology could be applied, there are a number of problems associated with the application of the technique. Essentially, the problems are the same as the criticisms that apply generally – for example, whether the technique is able to discover the truth (Faraday and Plummer 1979: 779), the problems associated with bias in recording events, and the fallibility of human memory when using the memory recall method. The epistemological issue is wider than the application of the life and work history methodology specifically in tourism, and the issue of bias again takes place in a wider context. However, the issue of memory recall does present specific problems for tourism and hospitality research. Unlike the collection of work history data, which requires respondents to recall specific jobs and career details, data on, for example, holiday patterns or skill development may be more difficult to recall. Evidence suggests that information important to 'self' assists the recall process (Robinson and Swanson 1990: 327). While jobs and careers are important to individuals in relation to their 'self', holidays and travel experiences and the learning of specific skills are less likely to be so. While holidays and travelling are memorable events over the short term, the actual dates and details of travel events are forgotten over time. As skills are built gradually over time, it may be difficult to recall when they were first acquired.

A further problem facing the use of the life and work history methodology lies in the nature of the data collected in terms of whether those data should be analysed in an individual or in an aggregate form. Taking the example of work history data, in effect what is collected is a range of detailed information relating to one person's career history. If you collect work history information from 200 individuals, the question arises as to how you analyse those data. While it may be interesting to have 200 separate cases, it might also be useful to consider all the cases in aggregate form. Essentially, if we are looking at patterns of jobs, patterns of skill development in specific occupations or patterns of holiday choices, the data need to be considered in aggregate form. The danger with this is that the detailed information related to an individual may be lost. If life and work history data are valuable because of the richness and complexity of the individual information obtained, is there any merit in looking at the data in aggregate form? The need to uncover patterns and processes suggests that there is a need for aggregate analysis also. However, Phillimore and Goodson argue there is a tendency in tourism research to focus mainly upon aggregate analysis and generalisation without paying sufficient attention to the local and individual. Life and work history offer the opportunity for both micro- and more macro-scale research.

Methods of data collection

As with any research approach, there are a variety of methods in which data can be collected. Given that the life and work history methodology can be used to generate both quantitative and qualitative data, there are a number of methods that can be employed for data collection purposes. Life and work histories involve the collection of historical information. There are essentially two main approaches to data collection: by the longitudinal method, or by memory recall. The merits of these two different approaches are discussed elsewhere (Dex 1991: 5). In longitudinal research exploring complex social issues through the collection of life history data, a range of research methods have been employed, including interviews, surveys by questionnaire, observation, and document research.

Using the recall method to collect life and work histories opens up a number of different opportunities to collect data. Collecting histories using memory recall relies on subjects being able to recall past effects. These memory recollections can be triggered using either interview or question-naire methods, and the most appropriate method is determined, as with any type of research, according to the nature and type of the information required.

In the tourism and hospitality field, empirical research has been under-taken in relation to the collection of work history data (Ladkin 2002: 380; Ladkin 1999: 38; Ladkin and Juwaheer 2000: 120; Ladkin and Riley 1996: 444; McCabe and Ladkin 2002: 159). For these different studies, the instrument of data collection was a postal questionnaire. Using the guiding principles identified by Gittens (1985: 92), the questionnaire was designed along the principles of work biography, and was organised in order to facilitate memory recall. This meant that the respondents were encouraged to think about their most recent jobs first, and to work back chronologically over time. The questions were all specific, rather than general, in order to prompt accurate memories. Although this may seem to be debatable, there is considerable evidence which supports this claim. For example, Norman and Bobrow (1979: 109); Gittens (1985: 92) and Cohen *et al.* (1993: 57) all argue that asking respondents to recall specific rather than general events improves the accuracy of memory recall. Struc-turing questions in this way means that it is the researcher who frames any memory by setting its parameters, and thus the respondent is not given the opportunity to consider the issues in a more spontaneous and naturalistic fashion. This will inevitably shape their response and restrict the range of possible answers. Reiser *et al.* (1985: 97) have, however, demonstrated that goal-based events are easier to recall, and Robinson and Swanson (1990: 327) explain the importance of relating events to the self in order to improve memory accuracy. The questionnaire sought information in relation to five main areas: career length, educational achievements, job levels and job functions, career mobility, and career ambition.

It is important to note that work history information can generate both quantitative and qualitative data. For example, asking subjects to recall the number of times they have changed jobs in their working lives gives an exact number of job moves as a measure of labour mobility. In addition, qualitative data can be generated if subjects are asked to explain the motivations and reasons for undertaking these job changes. It is the com-bination of both quantitative and qualitative data that makes work history analysis an important technique for developing an understanding of labour markets. In this case, quantitative data were generated relating to the structure of careers in terms of the number and type of jobs and moves, while qualitative data explained the patterns behind these patterns. The qualitative data were concerned with career mobility and initiatives, ambitions, and career choice and motivation.

Examples of qualitative research findings

In order to illustrate the type of qualitative data that can be collected using the work history methodology on career mobility and initiatives, career ambitions, and career choice and motivation, empirical findings are presented from two research studies. The first was a survey of hotel general managers in Australia. This study sought detailed work histories of the hotel managers, and was based on a sample of 180 hotel managers of hotels with more than 20 bedrooms. The research is published in its entirety elsewhere (Ladkin 2002: 383). The second study explored detailed career histories of people working in the conventions and exhibitions industry, using a cross-cultural comparison of the United Kingdom and Australia. The UK sample comprised 85 respondents, and the Australian sample consisted of 126 respondents. This work is in progress, although findings relating to career motivations and commitment have been published elsewhere (McCabe and Ladkin 2002: 171). While these two studies have been used to illustrate qualitative research findings, it should be noted that other areas would also be worthy of study – for example, the informal accommodation sector of family-run bed and breakfast establishments, and the careers of women working in this sector.

Beginning with career mobility and initiatives, and taking the sample of Australian hotel managers, the extent of the use of the labour market and the initiator of job changes were explored by asking respondents to recall their work histories. The research revealed that the hotel general managers have throughout their careers, held an average of 6.8 jobs, and they change jobs every two to three years. By asking respondents to state which of their job changes had taken place within the same company and which had involved a change of company, it was possible to gain a picture of the use of both the internal and the external labour market.

For the sample, the majority of the job moves involved use of the internal labour market (54.7 per cent). These are moves *within* companies. External moves took place 45.3 per cent of the time. Therefore, individuals make use of company opportunities for skill development to a slightly greater extent than they change companies to advance their careers. By asking respondents to explain these job changes, qualitative comments revealed that the respondents who had worked predominantly in chain hotels were following recognised career routes within the company, which partly explains this finding. Despite the dominance of internal moves, both the internal and the external markets are used for job changes. The respondents' careers typically involve the use of both the internal and the external labour market.

By asking respondents to record who is the initiator for the job moves, the data revealed that 40.2 per cent of the job moves were initiated by the company and 59.8 per cent by the individuals. Thus, it is predominantly the individuals who are responsible for the career moves, and not the industry. The respondents explained that they were prepared to search both the internal and the external labour markets when seeking job changes and promotions.

These findings on mobility demonstrate that work history analysis is a successful technique for gaining information on mobility within occupations. The collection of work history data relating to the number of jobs held and the nature and motivation for those job changes enables us through career analysis to develop a clear picture of mobility patterns within one particular occupation.

With regard to career ambition, and again using the example of the Australian hotel managers, the collection of work histories allowed for an exploration of the ambitions and strategies that hotel managers had used in seeking to advance their careers. In order to achieve this, sets of questions were developed concerning behaviour in relation to career advancement. Twenty-four questions were devised in relation to job search techniques, career planning and personal ambition. Respondents were asked to identify those personal strategies that they had used to advance their career, and to offer an explanation for the strategies they had used. A number of important themes emerged, including a willingness to relocate and to change jobs to gain experience, and having long-term career goals. Qualitative comments indicated that it was the thrill of working in a dynamic industry that acted as a main motivation to pursue the career, and it is challenging and exciting jobs that assist career development, rather than just going after the highest salary (Ladkin 2002: 386).

These findings on career ambition demonstrate that work history analysis can be used as an effective means of collecting motivational information. By using qualitative work history data, detailed and individual accounts of motivational factors that drive career decisions and development can be identified.

Finally, using the sample of Australian and UK conference and exhibition employees, work histories have been used to provide information on career choice and motivation. Respondents in both samples were asked to consider their present career and to recall what had motivated them to choose the industry as an employer, and what motivated them to remain in the industry. The reasons given for selecting the industry as a career option included the excitement of working in a 'people' industry, creative opportunities, and the image of the industry as being fun, challenging and dynamic. When asked to consider their career histories, the respondents indicated that the same reasons have kept them within the industry (McCabe and Ladkin 2002: 174).

For both these studies into work histories, the memory recall technique was the approach taken, and the method of data collection was a postal questionnaire. As has been stated previously, although traditionally life and work history information has been collected using the longitudinal approach and the interview method, because of time and cost constraints an alternative approach was tested. This technique proved to be successful, although it is likely that using face-to-face interviews would allow for more in-depth information to be gathered. Although the process of memory recall was unaffected by using a questionnaire, more detailed qualitative data could be achieved using the interview method.

Conclusion

Despite the long history and significant utilisation of the life and work history methodology in the social sciences, it is an underused technique in tourism and hospitality research. While steps have been taken to use the approach for career analysis, there is clearly scope for its use in other areas of tourism and hospitality research where detailed historical information is required. Using memory recall, it is possible to capture both quantitative and qualitative data at a relatively low cost. This, combined with the proven success of the technique, might encourage those researching in tourism and hospitality to consider using the life and work history methodology in new areas and research directions.

Summary

To summarise, there are a number of key points that have arisen from this exploration of the life and work history methodology. First, despite a long historical development and use in the social sciences, it has yet to receive much attention within tourism and hospitality research. Second, the life and work history methodology is capable of generating a wide range of both quantitative and qualitative data on many aspects of an individual's life. Third, life and work history data can be generated using the memory recall technique, which is a more cost-effective and less time-consuming way of gathering historical data. Fourth, life and work history data can be gathered using either the interview or the questionnaire technique. Fifth, work history data have been extensively used in research into the career paths of hotel managers, which makes the assumption that we can learn about current labour market conditions by studying the past. Finally, the life and work history methodology could be used in a variety of research areas in tourism and hospitality which are currently at the developmental stage.

Questions

1   Identify the differences in the historical development and current use of the life and work history methods.
2   Explain the main advantages and disadvantages of the life and work history method.
3   What are the issues surrounding the use of memory recall to collect historical data?
4   Identify areas where the life and work history methodology might be used in tourism and hospitality research.

References

Babbie, E. (2001) *The Practice of Social Research.* Belmont, CA: Wadsworth.

Baker, M. and Elias, P. (1991) Youth employment and work histories'. In S. Dex (ed.) *Life and Work History Analysis: Qualitative and Quantitative Development.* London: Routledge.

Burchell, B. (1992) 'Towards a social psychology of the labour market: or why we need to understand the labour market before we can understand unemployment', *Journal of Occupational and Organizational Psychology,* 65: 345–354.

Butler, R. (1993) Pre- and post-impact assessment of tourism development'. In D. Pearce and R. Butler (eds) *Tourism Research: Critiques and Challenges.* London: Routledge.

Cohen, G., Kiss, G. and Le Voi, M. (1993) *Memory: Current Issues,* 2nd edn. Buckingham: Open University Press.

Denzin, N. K. and Lincoln, Y. S. (1998) *The Landscape of Qualitative Research: Theories and Issues.* London: Sage.

Dex, S. (1984) 'Work history analysis, women and large data sets', *Sociological Review,* 32 (4): 631–637.

Dex, S. (ed.) (1991) *Life and Work History Analyses: Qualitative and Quantitative Developments.* London: Routledge.

Elder, G. (1978) 'Family history and the life course'. In T. K. Hareven (ed.) *Transitions: The Family and the Life Course in Historical Perspective.* London: Academic Press.

Elliott, B. (1990) 'Biography, family history and the analysis of social change'. In S. Kendrick, P. Straw and D. McCrone (eds) *Interpreting the Past, Understanding the Present.* London: Macmillan.

Faraday, A. and Plummer, K. (1979) 'Doing life histories', *Sociological Review,* 27 (4): 773–789.

Gittens, D. (1985) 'Oral history, reliability and recollection'. In L. Moss and H. Goldstein (eds) *The Recall Method in Social Surveys.* London: Heinemann.

Hareven, T. K. (1978) *Transitions: The Family and the Life Course in Historical Perspective.* London: Academic Press.

Holden, A. (1999) 'Understanding skiers' motivations using Pearce's "travel career" construct', *Annals of Tourism Research,* 26 (2): 435–438.

Howes, J. L. and Katz, A. N. (1992) 'Remote memory: recalling autobiographical and public events from across the lifespan', *Canadian Journal of Psychology,* 46 (1): 92–116.

Kim, Y. T., Pearce, P. L., Morrison, A. and O'Leary, J. T. (1996) 'Mature versus youth travellers: the Korean market', *Asia Pacific Journal of Tourism Research*, 1 (1): 102–112.

Ladkin, A. (1999) 'Life and work history analysis: the value of this research method for hospitality and tourism', *Tourism Management*, 20 (1): 37–45.

Ladkin, A. (2002) 'Career analysis: a case study of hotel general managers in Australia', *Tourism Management*, 23 (4): 379–388.

Ladkin, A. and Juwaheer, R. (2000) 'The careers of hotel managers in Mauritius', *International Journal of Contemporary Hospitality Management*, 12 (2): 119–125.

Ladkin, A. and Riley, M. (1996) 'Mobility and structure in the career paths of UK hotel managers: a bureaucratic model?', *Tourism Management*, 17 (6): 443–452.

Larsen, S. F. (1988) 'Remembering without experiencing: memory for reported events'. In U. Neisser and E. Winograd (eds) *Remembering Reconsidered: Ecological and Transitional Approaches to Memory*. Cambridge: Cambridge University Press.

McCabe, V. and Ladkin, A (2002) 'Career motivations and commitment in the conventions and exhibitions industry: exploring evidence from the UK and Australia'. Paper given at Australian Centre for Event Management Conference: Events and Place Making, 15–16 July 2002, University of Technology, Sydney.

Norman, D. A. and Bobrow, D. G. (1979) 'Descriptions: an intermediate stage in memory retrieval', *Cognitive Psychology*, 11: 107–123.

Pearce, P. L. (1988) *The Ulysses Factor: Evaluating Visitors in Tourism Settings*. New York, NY: Springer-Verlag.

Pearce, P. L. (1991) 'Analysing tourist attractions', *Journal of Tourism Studies*, 2 (1): 46–55.

Plummer, K. (1983) *Documents of Life: An Introduction to the Problems and Literature of a Humanistic Method*. London: George Allen and Unwin.

Reiser, B. J., Black, J. B. and Abelson, R. P. (1985) 'Knowledge structures in the organisation and retrieval of autobiographical memories', *Cognitive Psychology*, 17 (1): 89–137.

Riley, M. (1990) 'The role of age distributions in career path analysis', *Tourism Management*, 11 (1): 38–44.

Riley, M., Ladkin, A. and Szivas, E. (2002) *Tourism Employment: Analysis and Planning*. Clevedon: Channel View Publications.

Riley, M., Niininen, O., Szivas, E. and Willis, T. (2001) 'The case for process approaches in loyalty research in tourism', *International Journal of Tourism Research*, 3 (1): 23–32.

Robinson, J. A. and Swanson, K. L. (1990) 'Autobiographical memory: the next phase', *Applied Cognitive Psychology*, 4 (4): 321–335.

Ryan, C. (1998) 'The travel career ladder: an appraisal', *Annals of Tourism Research*, 25 (4): 936–957.

Smith, L. M. (1994) 'Biographical method'. In N. K. Denzin and Y. S. Lincoln (eds) *Handbook of Qualitative Research*. London: Sage.

Stanfield, J. H., (1987) 'Life history analysis and racial stratification research', *American Behavioral Scientist*, 30 (4): 429–440.

Walby, S. (1991) 'Labour markets and industrial structures in women's working lives'. In S. Dex (ed.) *Life and Work History Analyses: Qualitative and Quantitative Developments*. London: Routledge.

15 Memory-work

Jennie Small

Aims of the chapter

- To introduce the research method 'memory-work': the philosophy, the procedure and the application of the method.
- To examine the way the method was employed in a study of women's and girls' tourist experiences.
- To compare the writer's experience of the method with the experience of other users.
- To consider the emotional experience of using the method.

Introduction

A review of tourism research at the end of the twentieth century by Dann and Phillips (2000) has suggested methodological progress from the findings of a previous review by Dann *et al.* (1988). The earlier review had found that tourism researchers tended to neglect questions of theory and method, and their interrelationship. Dann and Phillips have indicated that in the intervening years, qualitative research has been undertaken which 'has managed to achieve a happy blend of theory and method' (2000: 260). Arguing for the application of qualitative research methods in tourism research, their model, which portrays the 'state of the art' of qualitative tourism research, nonetheless indicates that there remain gaps in the application of qualitative methods in tourism studies.

The relatively recent emergence in tourism studies of paradigms such as feminism (Aitchison 1996; Craik 1997; Davidson 1996; Deem 1996; Kinnaird and Hall 1994, 1996; Norris and Wall 1994; Richter 1994; Sinclair 1997; Swain 1995, Swain and Momsen 2002) requires tourism researchers to follow the path of researchers in other fields of study and reassess the usefulness of traditional research methodology. Questions need to be addressed concerning the best way to uncover lived experiences of different socio-cultural groups.

This chapter describes and critiques memory-work (Haug 1987), an innovative qualitative research method which has an acknowledged relationship with paradigm and theory. Memory-work was developed as a feminist research method and has been acknowledged by its users as a valuable method for researching women. Memory-work has some features in common with life history work, which is described in general terms by Ladkin in Chapter 14 of this book. However, unlike life history, it is the *collective* construction of memories that is important in memory-work, not the individual history. Memory-work attempts to avoid biography. The method has primarily been used by women researching women; only occasionally has it been used with or by men (Crawford *et al.* 1992), so as yet the usefulness of the method in researching men's experiences remains to be examined. The present chapter discusses the application of the method in a study of women and girls as tourists: their good and bad tourist/holiday experiences at different stages of the life span. In investigating women's and girls' understanding of holidays, the study attempts to identify patterns of experience. It is suggested that memory-work, as a research method in tourism studies, need not be confined to tourists but might also be employed in studies of hosts at the tourist destination region or those working in the tourism industry. The method would be particularly useful in this regard because it would help to plug gaps in knowledge (Butler 1993) about how destinations change over time as tourism develops and how the host population perceives those changes.

The philosophy of memory-work

Memory-work cannot be understood without knowledge of the paradigm feminist social constructionism, from which the method emanated. The premises which guide the method, interpretation and implication of the findings are as follows:

- Ontology is relativist. 'Reality' is a mental intangible construction. Unlike positivism, a social constructionist approach considers that there is no one truth or reality. There are multiple 'realities' which are socially constructed through language and other social processes. Constructions are also historically specific. Gender is a construction, and women's knowledge or reality will be different from that of men. Realities will differ between women and men, and of course within different groups of women and men.
- Epistemology cannot be distinguished from ontology since what there is to be known and how it is known is a social construction. The social constructionist approach considers that anyone can be a 'knower'. Feminists argue that women and their accounts have been ignored. The aim of feminist research is to acknowledge women as knowers.
- The methodological approach of a social constructionist paradigm assumes that in the research process, 'constructions can be elicited and

refined only through interaction *between and among* investigator and respondents' (Guba and Lincoln 1994: 111).

Appreciating the link between ontology, epistemology, methodology, theory and method, German feminists and socialists (Haug 1987, 1999) among others developed 'memory-work' as a research method. In the mid-1980s, while a visiting scholar in Sydney, Haug introduced the concept of memory-work to a group of Australian academics. Crawford *et al.* (1992) employed the method in a study of the social construction of emotion. In the process of their research they expanded and made more explicit the 'rules' or guidelines of the method.

As explained by Crawford *et al.*, memory-work is 'not merely a technique for data collection but includes analysing and theorizing the data, interpreting and re-interpreting them in the light of the overall theory' (1992: 38). The method is a social constructionist method as it focuses primarily 'on the process whereby individuals construct themselves into existing social relations' (Haug 1987: 33). As Crawford *et al.* (1992) explained, significant events which are remembered and the way they are constructed play an important part in the construction of self. Conversely, the construction of self at any moment plays an important part in how the event is constructed. Since the self is socially constructed through reflection, Haug (1987) used memories as her initial data, hence the name of the method. Memory-work has the benefit of enabling the researcher to tap into the past. Crawford *et al.* (1992) referred to this act of reflection as one self-engaging with one's memories, having a conversation with them and responding to them. The procedure of memory-work commences with the individual producing a memory, thus indicating the process of constructions.

The method proceeds to a collective examination of the memories in which theorisation of the memories and new meanings result. 'The two foci of memory-work capture something of the duality of self. The self talking with itself is phase 1 and responding to itself as others respond to it is phase 2' (Crawford *et al.* 1992: 40). Although the method is expressed as two separate phases, Crawford *et al.* stressed that 'intersubjectivity precedes subjectivity' (1992: 52).

Memory-work breaks down the barriers between the subject and object of research. Everyday experience is the basis of knowledge. *What* is known and *who* knows are one and the same. As explained by Crawford *et al.* (1992: 41), 'This collapsing of the subject and object of research, the "knower" and "known", constitutes or sets aside a space where the experiential can be placed in relation to the theoretical.' The researched become researchers. The academic researcher joins the collective and becomes the researched. The hierarchy of 'experimenter' and 'subject' is thus eliminated. Haug (1987) referred to the participants as 'co-researchers'.

At a superficial level, memory-work has characteristics similar to those of other research methods, such as critical incidents, narrative accounts (e.g. oral history) and focus groups. However, there are fundamental differences between memory-work and other such methods in the positioning of the researcher and researched and, subsequently, the likely outcome of the research. It is the avowedly inter-subjective nature of memory-work, as opposed to a subjective or objective claim for the status of knowledge produced, which differentiates memory-work from other research methods. A summary of the advantages and disadvantages of memory work is presented in Box 15.1.

The procedure

There are three distinct phases to memory-work. Crawford *et al.* (1992) outlined the steps in each phase:

Phase 1

A group is formed and each participant is asked to write a memory. In memory-work, the specified memory is considered a trigger. Memory-work requires that the memories are written according to the following set of rules:

1   *Write one to two pages about each memory.* The writing of the memory has a number of benefits. It provides a discipline for the group, the group remembers more through writing and it gives status to the everyday experiences of life, which is considered of particular importance for women.
2   *Write in the third person using a pseudonym.* The advantage of writing in the third person is that the participant can stand back and view the memory from the outside, which helps to avoid justification of the experience.
3   *Write in as much detail as possible, including even what might be considered trivial or inconsequential.* By asking for the trivial, it is hoped to avoid an evaluation by the participants of what was important or unimportant. Such an evaluation might well be socially defined.
4   *Describe the experience, don't try to interpret the experience.* Interpretation smoothes the rough edges and covers up the absences and inconsistencies.

Phase 2: The collective analysis of the memories

1   *The group meets and each participant reads their memory.* The discussion is taped.
2   *The participants discuss the written memories.* It is suggested that the group might look for similarities/commonalities in the memories, and also dissimilarities; identify clichés and contradictions in the memor-

---

Box 15.1 Advantages and limitations of memory-work

Advantages
- The relationship of theory and method is acknowledged.
- The method is appropriate for a social constructionist paradigm.
- The method reflects feminist principles of research.
- The method breaks down the barriers between the subject and object of research.
- The hierarchy of researcher and researched is eliminated.
- Inter-subjective knowledge is examined, as opposed to subjective or objective knowledge.
- *Writing* a memory has a number of benefits, especially for women.
- The method allows for the social construction of the memories to be realised and can generate the rich, complex, multi-layered themes of lived experience.
- Memory-work is liberating.
- Those who have used the method vouch for its worth.
- It has been successfully employed by researchers from a wide range of disciplines and research topics.
- The method has always been seen by the original users to have a creative, dynamic and flexible framework.

Limitations
- The method can be time-consuming.
- It can be difficult to retain all the participants over a number of meetings.
- The method may be too confronting for the study of sensitive topics.
- For researchers situated within positivist disciplines/academic departments, it can be problematic and anxiety-producing to adopt a non-traditional method such as memory-work.
- Participants may find it difficult to take ownership of the method and findings.
- The guidelines may not always be well understood, especially by children.
- There is no instruction for the analysis of the data.
- The original form of the method cannot be followed if the researcher cannot be positioned with the group, or if the researcher is a student who is required to take ownership of the research.

---

ies; and look for gaps, or what is not being written in the memories but which might be expected to be.

3 *The group rewrites the memories.* The group revises their original stories/memories in the light of the deeper understanding gained from the collective discussion. A new version is produced. While Haug's (1987) collective found that increased understanding from rewriting

the memories was incalculable, the experience of Crawford *et al.* (1992) was less convincing, leading them eventually to omit this step.

The essence of Phase 2 is the collective search for common understanding. The method allows for the social nature of the construction of the memories to be realised.

Phase 3: Further examination and theorisation of the collective analysis from Phase 2

In the earlier studies (Crawford *et al.* 1992; Haug 1987), the further examination and theorisation was normally the work of one of the participants with feedback sought from the collective. A draft of the discussion from Phase 2 would be circulated to the participants and the group would meet to discuss.

Applications of the method

A review of applications of the method (Onyx and Small 2001; Small and Onyx 2001) has indicated that memory-work is growing in popularity as a research method by those seeking a method that fits with a feminist, social constructionist paradigm. The method has been adopted by researchers from diverse fields of study: psychology, sociology, nursing, education, management, marketing as well as tourism studies (Small 2002, 2003) and leisure studies (Grant and Friend 1997; Friend *et al.* 2000; McCormack 1995, 1998). Memory-work has also been applied within a postmodern paradigm with a different understanding of the nature of the subject and consequently a different stance on agency (Davies *et al.* 2001). In Sydney in 2000, the number of researchers in Australia and New Zealand who were using the method proved sufficient for a Memory Work Research Conference (convened by Small and Onyx) to be held. The purpose of the conference was to discuss application and to critique the method. A number of critiques of the method have been published (Cadman *et al.* 2001; Farrar 2001; Ingleton 2001; Johnston 2001; Koutroulis 1993, 1996, 2001; Onyx and Small 2001; Small 1999, 2001, Small and Onyx 2001; Stephenson 2001).

Users of the method have turned to both Haug (1987) and Crawford *et al.* (1992) to enhance their understanding and employment of the method. However, while Haug (1987) and Crawford *et al.* (1992) used the method with fellow academics in non-degree research, most researchers have employed the method in higher degree (most notably, doctoral) study. While they have adopted the basic ideology of memory-work, various adaptations of the method have been necessary for those involved in student research.

The critique which follows is based on the writer's experience of the method in a study of women's and girls' good and bad holiday/tourist experiences at different ages/life stages. The focus is on the part tourist experiences play in the construction of self and the part the construction of self plays in the construction of tourist experiences. Memory-work was selected as the most appropriate research method as it appeared to closely fit the researcher's conceptual approach and the topic of study, which focused upon women's and girls' reflections of holidays over their life span.

The participants in the study were 20 groups of Australian, white, urban, middle-class women and girls residing in Sydney. There were four to six members per group. In most groups the women or girls were friends. Four age groups were studied: girls aged 12, young single women (in their early twenties), middle-aged women (in their forties), and older women (65 and older). The middle-aged and older women had children. Different age groups were studied in an attempt to gain an understanding of the social constructions of various age cohorts. To help the researcher understand further the construction of tourist experiences and the construction of self, the groups (with the exception of the 12-year-olds) were asked to write memories at specified younger ages. The specified memory was the focus for a particular meeting. The number of meetings was related to the number of aged memories being investigated (see Table 15.1). Two broad triggers were chosen: (a) a positive holiday experience, and (b) a negative (or not so good) holiday experience. The researcher was a member, a 'co-researcher', of one of the middle-aged groups. For the meetings of the other middle-aged groups and groups of other ages, the researcher was absent.

Evaluation of the method

The research design for this study, involving groups of different ages (and more than one of each age), necessitated modifications to the method as prescribed by Haug (1987) and Crawford *et al.* (1992). The doctoral nature of the study also conflicted with the original intention of non-hierarchical research in that the doctoral student is compelled, by the regulations surrounding the award of a PhD, to take sole ownership of the research

Table 15.1 Memory-work: research design

Participants	Memories at different ages			
	12 years	*Early twenties*	*Forties*	*Sixty-five and over*
Aged 12 ($n = 26$)	✓			
Aged early twenties ($n = 25$)	✓	✓		
Aged in forties ($n = 18$)	✓	✓	✓	
Aged 65+ ($n = 17$)	✓	✓	✓	✓

and its presentation. The following discussion examines the method's components in contributing to an understanding of women's and girls' tourist/holiday experiences.

Phase 1: Writing the memory

Despite some expressed doubts concerning their ability to recall or write a memory, nearly all women and girls arrived with a memory, and most of the memories were written. In total there was a potential for 396 written memories. When this total was adjusted to 371 (to take into account a few who had no memory or who, inadvertently, had not received the instructions for a written memory), it was calculated that 96 per cent of the memories were written. The group least likely to submit a written memory was the 40-year-old age group, which was perhaps the most time restricted of all the groups. Nonetheless, an impressive number of memories (86 per cent) were written by this group.

The benefits cited by Crawford *et al.* (1992) of writing a memory were evident in this research. The writing provided a discipline for all age groups. It created boundaries around a particular experience, which allowed the groups to focus on that experience. Participants reported that through writing they were able to recall far more of the detail of the holiday than if they had merely spoken the memory. The act of putting into print one's experiences elevated the everyday life of co-researchers because the printed voice appeared to have an authority that the spoken voice sometimes lacks.

Not only were the women and girls committed to the task through the act of writing, but some had dedicated a considerable amount of time and effort to producing an elaborate, literary account. A rough indication of the detail of the written memories is whether the memories were written in prose (sentences) or brief, bullet point format. On reviewing the memories, it was found that a high percentage were written in prose, with the older groups' memories (96 per cent) more likely to be written as such than the 12-year-olds' memories (86 per cent) or the 20-year-olds' (83 per cent). The instruction to write in the third person was more likely to be followed in the memories of the 40-year-olds (94 per cent) and the 20-year-olds (89 per cent) than the 65-year-olds (84 per cent). The memories of the 12-year-olds were least likely to be written in the third person (61 per cent). It might be that the girls had not read the instructions, did not understand what was meant by 'third person' or felt self-conscious doing so. The latter is a possible explanation, because even when the girls (and some of the women) had written as instructed, they often read their written memory using the first person.

Phase 2: The collective meets

The discussion generated many memories beyond the two written memories of each woman. Through the collective discussion of all age groups, shared meanings of the tourist experience were generated and themes were identified. The commonality in the experiences assisted the group dynamics.

Despite the fact that a research setting was not the normal environment for a discussion of holidays, the women and girls fell naturally into the situation and seemed to enjoy the sessions. This naturalness was evident in the group's apparent dismissal of many of the Phase 2 guidelines/ 'suggestions' for discussion. The guidelines had been presented as possible areas for discussion. An analysis of the collective discussions found that the groups were much more likely to identify similarities in their memories than dissimilarities, popular sayings, contradictions or silences. The age group most likely to address the guidelines was the 40-year-old age group. The age groups least likely to do so were the oldest (65-year-olds and older) and the youngest (12-year-olds). The dismissal of certain guidelines was possibly due to the participants' perception that they were a constraint on the natural flow of conversation, or perhaps they were difficult concepts to apply to the discussion. These concepts might not be as useful with 'everyday'/non-academic girls and women. In this study the researcher made no attempt to restrict the mode of response to that suggested in the guidelines. The point of comparison between age groups was the dynamic emergent themes. Of interest was the flow of thinking. An insistence on slavishly following the guidelines would have resulted in a loss of capacity for themes to emerge. The discussion appeared to reflect the normal social construction process concerning women's and girls' holidays, a typical conversation with one's friends, daughters, mothers or grandmothers. Despite rejection of the guidelines or distraction at times from the topic area (on the part of some of the girls), the transcriptions revealed a wealth of rich and complex meanings of the holiday experience for the participants.

The method attempts to break down the hierarchy between researcher and participant. However, there were difficulties in the present research in achieving the relationship of co-researcher. First, the research design involved different age groups. A social constructionist approach considers that knowledge of the world is socio-culturally and historically specific. Quite simply, the researcher cannot be a co-researcher of other age groups (according to the method) as she cannot position herself with the other age cohorts. Also, compared to the single group of participants that Haug (1987), Crawford *et al.* (1992) and others have engaged in memory-work, in this study there were a number of groups in which the researcher could position herself (she participated in only one). Thus, the notion of the researcher being a 'co-researcher' applied to only one of the groups. In order to ensure clarification of the groups' discussion and meanings when the researcher was absent, the collective was asked, within the discussion (and at the end of the meeting with the researcher), to summarise the themes. In the present study the collective did not rewrite/revise the memories as Haug's (1987) collective had done. Since the participants in the present study did not have a full and equal investment in the research, requesting them to rewrite their memories was seen as too great a demand. Also, the value of rewriting had been questioned by Crawford *et al.* (1992).

The second obstacle to being a co-researcher was the institutional requirement of student research that the student take ownership of the research. Equal ownership among participants was therefore not possible. Others have similarly commented on this impediment (Cadman *et al.* 2001; Ingleton 2001; Koutroulis 2001; Stephenson 2001). In Chapter 16, Belsky also reflects on the nature and value of participatory research.

When the researcher participated in a group, it was clear that the constraints included not only the formal rules of academe but also the way the women had constructed the researcher–researched roles over their lifetimes. The roles of each were so entrenched that it was difficult to equalise them. It was difficult for the women to see themselves as equal participants with the researcher. For the researcher, it was tempting to take on the role of facilitator rather than the role of co-researcher. There was a tension between letting the discussion take its natural course and seeking out the feminist concerns of the researcher; the researcher wanted the experience to be liberating for the women. By the end of the research, it appeared that the experience of the process of writing and sharing memories had been empowering despite members of the collective varying in their strength of commitment to a feminist analysis.

The researcher's experiences of memory-work were confirmed by other memory-workers also undertaking student research (Cadman *et al.* 2001). The Cadman *et al.* study involved 11 memory-workers (including the present researcher) 'doing memory-work on memory-work' to explore unresolved issues of power. The experience of others using the method highlighted a collective ambiguity in the role of researcher.

The collective described the confusion in attempting to disrupt the binaries of traditional research: speaking–silence, researcher–researched, objectivity–subjectivity, rational–emotional, male–female. The contradictions and ambiguities of these binaries 'appeared at times to be mediating against the researcher's intention to be, and to experience the method as, collaborative and participatory' (Cadman *et al.* 2001: 81).

While the present researcher was not able to achieve a co-researcher relationship with the participants, as had Haug (1987) and Crawford *et al.* (1992), it is still possible to identify in the present study, *when compared with other research methods*, a breakdown in the hierarchy between researcher and researched. Although it was the researcher who had set up the study, the participants experienced a fair degree of freedom in the outcome of Phases 1 and 2. The triggers for the written memories were broad (a good and a bad holiday experience), allowing the participants a freedom in selecting which memory to report. In Phase 2 the rules were kept to a minimum and the collective determined the outcome of the analysis, especially where the researcher was absent. The emphasis was on 'telling' rather than asking, and thus had the benefit of guarding against some of the dangers of manipulation and exploitation. The women and girls had control over what and how much they wanted to tell. Where the researcher participated as a member of a group, the researcher was positioned with women similar in social background and life experiences to herself. The researcher completed the same task as the other participants in Phase 1. In Phase 2 there was a shared understanding of holiday experiences among researcher and the other women in the collective. Self-disclosures of the group were similar; nuances, laughter, subtleties and hesitancies were understood by all. For a large part of the process, the researcher was positioned, if not together, at least closely with the women in Phase 2.

While acknowledging deviation from the 'prescribed' method, it is stressed that the value of any research method is reflected in the relevance and quality of the findings. In the final analysis, the test of the method is to be found in the usefulness of the data. It was clear in the present study that memory-work had generated many layers of meaning from which patterns of experience could be identified across all age groups. The participants could go beyond the 'rehearsed and sanitised' narratives which Deem (1996) associates with holidays. The two phases of the method – the individual talking to herself and the collective talking together – allowed for the contrast between women's public (sanitised) talk and private (unsanitised) talk. All groups went beyond the public discourse of the travel brochure in their discussion of good and bad experiences.

It has been possible to trace women's and girls' construction of the tourist experience. The tourist experience could be seen within the broader social context of the time in which the experience occurred and the time in which it was being remembered. The women imagined how their own experiences at a younger age compared to the experiences of those currently of that age. They imagined how their mothers would have understood holidays. They imagined how men would construct the holiday experience. The broadness of the trigger allowed themes which might not have been anticipated to surface.

Phase 3: Further examination and theorisation

Collective theorising according to Haug (1987) and Crawford *et al.* (1992) occurs in both Phase 2 and Phase 3. However, the institutional academic requirements regarding student ownership of the research meant that Phase 3 was the responsibility of the researcher, not the collective.

Independently of the academic status of the present research, the number of participants (86) and number of groups (20) of different ages would have made it operationally difficult and theoretically impossible for all participants to be equally engaged in the overall analysis in Phase 3, the writing up of the findings. Someone would have to take responsibility for the collation of the findings across all groups and the identification of themes and interpretation of them. Thus, whereas Phase 2 was how the collective made sense of the memories, Phase 3 was where the researcher made sense of them. It was the researcher who thematised the collective discussions and selected specific memories to illustrate the themes. In the words of Koutroulis, what the researcher can offer in Phase 3 is a 'meta analysis, not necessarily a better analysis' (1993: 90). A dilemma remains as to which memories to select for publication. The Cadman (2001) collective voiced the concern that researchers might judge some memories more appropriate for the public arena than others. When this happens, those whose memories are not made public are subjugated. The researcher could be accused of colluding with repressive discourses and thus repro-ducing the traditional social formation.

Questions arise as to how well the voices of the different age groups were re/presented. To ensure that the researcher was hearing the voices of her own collective, themes were identified by the researcher after each meeting of the collective (Phase 2) and presented to the collective at the next meeting for verification. Overall, there was strong consensus among the members that the identified themes represented the discussion. To ensure that the voices of the other collectives were heard, the researcher was present at the end of each meeting to hear a summary of their discus-sion.

The emotional experience of memory-work

As a feminist researcher, I was committed to the ethical values of memory-work. I was not the neutral researcher that traditional research would assert because I was physically and emotionally involved. The Cadman (2001) collective of memory-workers found that owing to the highly personal nature of memory-work, the experience was very emo-tional for all involved; research was an embodied experience (see Swain, Chapter 6, for a more detailed discussion of embodiment in the research process). The collective stressed that acknowledging emotion in the research process challenges the rational/irrational binary of positivist tradition.

As well as experiencing the positive features of memory-work, I, like other memory-workers, experienced anxiety. There was anxiety in disrupt-ing traditional research methodology, especially for those situated in posi-tivist academic departments and disciplines. There was concern as to how the method and findings would be received. This apprehension was combined with a tension in trying hard to 'be true to the method' while, at the same time, having to modify the method. Being 'true to the method' involved a very close relationship with the participants and concern for their well-being. Others in the Cadman (2001) collective felt similarly. Not only was there anxiety about the collective research process, but also there was concern for the physical comfort of the participants. Nurturance and 'hostessing' were a part of the process. Most of the memories referred to the provision of food and the question of hostessing competence. The Cadman (2001) collective recognised that these aspects of the research role went beyond our academic training as researchers to our social construction as women.

It became apparent from my own experience and that of other memory-workers that the levels of social formation in which women memory-workers were involved were multiple and highly complex.

This critique should not be seen as a weakness of the method. The research findings are the result of the method. Memory-work recognises that the holiday experiences which eventuate from the written and discussed memories are socially constructed, and the context of the research contributes to that construction. The complexity of that context must be recognised. Although I acknowledge some anxiety, this was born out of a commitment to the method rather than any doubt as to its relevance or appropriateness to the study of women's and girls' holiday experiences. Modifications were made to tailor the method to the groups. I concur with other memory-workers that 'memory-work, as a research methodology, generated great joy' (Cadman *et al.* 2001: 91). Comments by the participants (mostly the older two groups) confirmed that the research experience was an enriching and empowering experience.

Conclusion

It is the paradigm of the researcher that determines the appropriateness of memory-work for a research topic, but the questions being asked establish the usefulness of the method. It is not a positivist method, or a quantitative method. It cannot be used as an objective test of a predetermined hypothesis. The memory-worker is not trying to test objective validity but to seek out authentic voices. Memory-work provides a rich *source* of insights of concepts, as opposed to a *test* of insights. The method is concerned with emergent themes. The data which resulted from the discussion groups were the collective's understanding of good and bad holiday experiences. The method allowed for the construction of the tourist experience across different ages and at different life stages to emerge. Memory-work created a space for the girls' and women's multilayered voices to be heard. Memory-work allows the participants the opportunity to discuss the more personal lived holiday experiences within a collective of friends. Tourism cannot be disconnected from the broader social and temporal context. The construction of the tourist experience is related to the construction of everyday life experiences. Through theorising the tourist experience, the women and girls were also theorising their everyday non-touristic experiences. Memory-work enables the researcher to locate the tourist experience within the larger context.

The chapter has highlighted the constraints in being a 'true' co-researcher when the researcher is not positioned with the participating socio-cultural research group or when academe requires one person to take ownership of the research. In the present study the researcher–researched relationship did not conform to the method as initially outlined by Haug (1987) and Crawford *et al.* (1992). However, the researcher's feminist ideals, awareness of a researcher's responsibilities and commitment to representing all participants hopefully minimised the possibility of manipulation and exploitation of the participants. Although researchers have, from necessity, modified the method, memory-work was always seen by the original users as having a creative, dynamic and flexible framework. As Haug (1987: 71) said, 'the very heterogeneity of everyday life demands similarly heterogeneous methods if it is to be understood'. Future users might adapt the method to suit their own particular research topic and design. The method has been successfully employed with a wide range of research topics. While memory-work has been well suited to the study of sensitive topics, Farrar (2001) cautions that some topics may be too sensitive, painful, traumatic and personal to explore with a method such as memory-work. As Farrar says, 'perhaps there is a point beyond which we cannot intrude into memories of experiences that are too raw to allow for the deconstruction and scrutiny that memory-work involves' (2001: 12). In terms of participants, the method works well with women; it is yet to be seen how effective it is in hearing and theorising the voices of men. The question also arises whether the method is as effective with those who are not self-reflective.

For the researcher, memory-work was an emotional experience. As the Cadman collective found, feminist researchers could at times feel powerless employing the method 'within the dominant positivist discourses and patriarchal structures of academia' (2001: 76).

Despite the various challenges of the method, memory-workers have concluded that the method makes possible the examination of the multi-layered complexities of lived experience. Since much tourism research skims the top layer of lived experience, there is a need for a method which can reveal and theorise the multilayered complexities. The strength of the method has been demonstrated in a variety of disciplines and fields of study. Tourism researchers committed to a social constructionist paradigm should consider the inter-subjective, non-hierarchical, liberationist strengths of the method. The method could be employed in studies of tourists, hosts, or those employed in the tourism industry. Another area of study is tourism education. In different disciplines, memory-work has been employed to study the student assessment process, enhancement of student learning, emotion and learning, and experiences of teachers (Onyx and Small 2001). The potential of the method will be further explored when tourism researchers employ the method with men, and groups based on socio-cultural features such as ethnicity, age, socio-economic position, disability/health condition and sexuality. The possible topic areas in tourism for which the method is suited are innumerable and diverse. Memory work could be used with different groups in host communities to explore their experiences of tourism development over time, helping to fill at least some of the gaps in existing knowledge about how tourism's impacts are perceived both over time and between different stakeholders in host societies. Wherever tourism researchers decide to use the technique, they should consider using memory-work in joint research in which they can avoid the possibility of hierarchical researchers and thus be true 'co-researchers'.

Summary

- Tourism researchers working within a social constructionist paradigm and who support the feminist ethos of non-hierarchical research should consider memory-work as a research method.
- Experience of the method in a range of topic areas suggests the applicability of memory-work in different areas of tourism such as tourism education, tourism employment issues, community attitudes to tourism development, and tourist experiences.
- The many layers of meaning generated by the method allow a tourism researcher to trace the social construction of tourism experiences, whether they be those of tourist, host, employee or tourism student.
- Memory-work aims to be inter-subjective, non-hierarchical and liberationist. It is thus suited to the study of marginalised groups in tourism such as tourists with disabilities, senior tourists, women tourists, indigenous communities, hosts in less developed countries, and tourism employees performing menial roles in casual positions.
- There are three stages to the method: writing a memory, collective discussion of written memories, and further examination and theorisation of the memories. The act of writing and control over the theorisation of the memories can be empowering for marginalised tourism groups.
- There are obstacles to being a 'co-researcher' when the research involves: first, educational institution requirements that a student takes ownership of their research, or, second, participants with whom the researcher cannot be positioned. In a study of work relations in the tourism industry, the academic researcher could not be positioned with tourism industry participants. Neither could a researcher be positioned with a host community different from their own in a study of community attitudes towards tourism.

- There are unresolved issues of power in using the method, particularly for tourism researchers working within a traditional positivist culture.
- Research is an embodied practice. Participating in memory-work is a highly emotional experience. Tourism researchers should acknowledge the role of emotion but need to be aware that memory-work may be too emotional for highly sensitive subject matter such as sex tourism or 'dark' tourism.

Questions

1   What elements of the method differentiate it from other methods such as critical incidents and the use of focus groups?
2   In what situations might a researcher need to adapt the method?
3   In what way is memory-work an emotional experience? Should this be acknowledged in the research report?
4   How could memory-work be employed in other tourism studies?

References

Aitchison, C. (1996) 'Gendered tourist spaces and places: the masculinisation and militarisation of Scotland's heritage', *Leisure Studies Association Newsletter*, 45: 16–23.

Butler, R. (1993) 'Pre- and post-impact assessment of tourism development'. In D. Percse and R. Butler (eds) *Tourism Research: Critiques and Challenges*. London: Routledge.

Cadman, K., Friend, L., Gannon, S., Ingleton, C., Koutroulis, G., McCormack, C., Mitchell, P., Onyx, J., O'Regan, K., Rocco, S. and Small, J. (2001) 'Memory-workers doing memory-work on memory-work: exploring unresolved power'. In J. Small and J. Onyx (eds) *Memory-Work: A Critique*. Working Paper Series 20/01, School of Management, University of Technology, Sydney.

Craik, J. (1997) 'The culture of tourism'. In C. Rojek and J. Urry (eds) *Touring Cultures: Transformations of Travel and Theory*. London: Routledge.

Crawford, J., Kippax, S., Onyx, J., Gault, U. and Benton, P. (1992) *Emotion and Gender: Constructing Meaning from Memory*. London: Sage.

Dann, G. and Phillips, J. (2000) 'Qualitative tourism research in the late twentieth century and beyond'. In B. Faulkner, G. Moscardo and E. Laws (eds) *Tourism in the Twenty-first Century*. London: Continuum.

Dann, G., Nash, D. and Pearce, P. (1988) 'Methodology in tourism research', *Annals of Tourism Research*, 15 (1): 1–28.

Davidson, P. (1996) 'The holiday and work experiences of women with young children', *Leisure Studies*, 15: 89–103.

Davies, B., Dormer, S., Gannon, S., Laws, C., Lenz Taguchi, H., McCann, H. and Rocco, S. (2001) 'Becoming schoolgirls: the ambivalent project of subjectification', *Gender and Education*, 13 (2): 167–182.

Deem, R. (1996) 'Women, the city and holidays', *Leisure Studies*, 15: 105–119.

Farrar, P. (2001) 'Too painful to remember: memory-work as a method to explore sensitive research topics'. In J. Small and J. Onyx (eds), *Memory-Work: A*

Critique. Working Paper Series 20/01, School of Management, University of Technology, Sydney.

Friend, L. A., Grant, B. C. and Gunson, L. (2000) 'Memories...', *Australian Leisure Management*, 20 (April/June): 24–25.

Grant, B. C. and Friend, L. A. (1997) 'Analysing leisure experiences through "memory-work"'. In D. Rowe and P. Brown (eds) *Proceedings of the 1997 Australian and New Zealand Association for Leisure Studies Conference*, Newcastle, NSW: Australian and New Zealand Association for Leisure Studies and Department of Leisure and Tourism Studies, University of Newcastle.

Guba, E. and Lincoln, Y. (1994) 'Competing paradigms in qualitative research'. In N. K. Denzin and Y. S. Lincoln (eds) *Handbook of Qualitative Research*, Thousand Oaks, CA: Sage.

Haug, F. (1987) *Female Sexualization*. London: Verso.

Haug, F. (1999) *Female Sexualization*, 2nd edn. London: Verso.

Ingleton, C. (2001) 'Meaning-making: issues of analysis in memory-work'. In J. Small and J. Onyx (eds) *Memory-Work A Critique.* Working Paper Series 20/01, School of Management, University of Technology, Sydney.

Johnston, B. (2001) 'Memory-work: the power of the mundane'. In J. Small and J. Onyx (eds) *Memory-Work: A Critique.* Working Paper Series 20/01, School of Management, University of Technology, Sydney.

Kinnaird, V. and Hall, D. (1994) *Tourism: A Gender Analysis*. Chichester: Wiley.

Kinnaird, V. and Hall, D. (1996) 'Understanding tourism processes: a gender-aware framework', *Tourism Management*, 17 (2): 95–102.

Koutroulis, G. (1993) 'Memory-work: a critique', *Annual Review of Health Social Science: Methodological Issues in Health Research*, Centre for the Study of the Body and Society, Deakin University, Geelong.

Koutroulis, G. (1996) 'Memory-work: process, practice and pitfalls', in D. Colquhoun and A. Kellehear (eds) *Health Research in Practice*, vol. 2. London: Chapman and Hall.

Koutroulis, G. (2001) 'Talk about men and sidelining! The "other" text in a memory-work study about menstruation'. In J. Small and J. Onyx (eds) *Memory-Work: A Critique.* Working Paper Series 20/01, School of Management, University of Technology, Sydney.

McCormack, C. (1995) '"My heart is singing": women giving meaning to leisure', unpublished master's thesis, University of Canberra.

McCormack, C. (1998) 'Memories bridge the gap between theory and practice in women's leisure', *Annals of Leisure Research*, 1: 7–49.

Norris, J. and Wall, G. (1994) 'Gender and tourism'. In C. P. Cooper and A. Lockwood (eds) *Progress in Tourism, Recreation and Hospitality Management*, New York, NY: Wiley.

Onyx, J. and Small, J. (2001) 'Memory-work: the method', *Qualitative Inquiry*, 7 (6): 773–786.

Richter, L. (1994) 'Exploring the political role of gender in tourism research'. In W. F. Theobald (ed.) *Global Tourism The Next Decade*. Oxford: Butterworth-Heinemann.

Sinclair, M. T. (ed.) (1997) *Gender, Work and Tourism*. London: Routledge.

Small, J. (1999) 'Memory-work: a method for researching women's tourist experiences', *Tourism Management*, 20 (1): 25–35.

Small, J. (2001) 'Researching different age groups through memory-work'. In

J. Small and J. Onyx (eds) *Memory-Work: A Critique*. Working Paper Series 20/01, School of Management, University of Technology, Sydney.

Small, J. (2002) 'Good and bad holiday experiences: women's perspective'. In M. Swain and J. Momsen (eds) *Gender/Tourism/Fun(?)*. Elmsford, NY: Cognizant.

Small, J. (2003) 'Voices of older women tourists', *Tourism Recreation Research*, 28 (2): 31–39.

Small, J. and Onyx, J. (eds) (2001) *Memory-Work: A Critique*. Working Paper Series 20/01, School of Management, University of Technology, Sydney.

Stephenson, N. (2001) 'If parties are battles what are we? Practising collectivity in memory-work'. In J. Small and J. Onyx (eds) *Memory-Work: A Critique*. Working Paper Series 20/01, School of Management, University of Technology, Sydney.

Swain, M. (ed.) (1995) 'Gender in tourism' [Special issue], *Annals of Tourism Research*, 22 (2).

Swain, M. and Momsen, J. (eds) (2002) *Gender/Tourism/Fun(?)*. Elmsford, NY: Cognizant.

16 Contributions of qualitative research to understanding the politics of community ecotourism

Jill Belsky

Aims of the chapter

- To briefly review the literature on the politics of tourism research and practices.
- To discuss two qualitative methods – participant observation and in-depth interviewing – that I emphasised in my teaching about and research on community ecotourism in Gales Point, Belize.
- To discuss opportunities and challenges that these two methods afforded this project.
- To offer some suggestions for pushing the benefits of participant observation and in-depth interviewing further through critical reflection and, possibly, participatory research.

Introduction

Qualitative research has been instrumental for uncovering and elucidating the political dimensions and tensions of tourism. Among the many tools in the qualitative research toolbox, participant observation and in-depth interviewing can help to develop a holistic perspective on the context and political dynamics of politics. Though not without its own challenges, participant observation can enable opportunities for observing everyday tourism activities and for in-depth dialogue between researcher and subjects. These conditions, as opposed to formal settings and procedures that can work against dialogue and a fuller viewing of tourism in practice, can reveal interconnections and power dynamics associated with tourism practices, and help in the creation of new theories. Given the newness of alternative tourisms such as community ecotourism, these qualitative methods are particularly suitable for assisting in theory discovery and generation following Burawoy's (1991) extended case method.

To illustrate these points, I draw upon my multi-year research on community ecotourism in Gales Point, Belize (Central America). The substantive findings of this research have been published elsewhere (Belsky 1999, 2001, 2003; Outside Television 2000).

Though still experimental and fraught with challenges, participatory research (Finn 1994; Greenwood *et al.* 1993; Park 1997, 1999) may be well suited to responding to some of the limitations of participant observation and in-depth interviewing, especially for generating qualitative tourism research attuned to political dynamics and action that benefits participating subjects. This discussion also underlies the importance of interdisciplinary engagement of tourism researchers with other social scientists and practitioners.

Politics of tourism and tourism research methodology

With few exceptions, tourism researchers rarely speak directly about the values that influence their choice of topics and the research methods they employ. Most of the attention of tourism research is devoted to the practical business and marketing of tourism and its economic costs and benefits. Only a few studies pay serious attention to tourism's political dimensions. Ritchie and Goeldner's book *Travel, Tourism, and Hospitality Research* (1987: 6–12) is the only introductory text with a chapter on tourism's political dimensions. Some suggest that the politics of tourism are muted because tourism research has not been particularly thick with ethnographic detail on the particular people and places in which the politics are embedded (Roche 1992).

The lack of research on tourism as embedded in particular social as well as environmental settings is particularly problematic given the increasing emphasis of politicians and development planners around the world on tourism as a means of economic development, and on development based on the problematic model of Western modernisation. For example, Hall notes:

> The dominant ideology of leisure and tourism in Western capitalist societies, which is increasingly being exported throughout the world through the modernization dynamic, portrays leisure and tourism as essentially a private and individual choice. Such an ideology serves to legitimize the relationship between the culture industry and dominant ideology, and misses opportunities for examining how capitalism itself affects and is affected by tourism.
>
> (1994: 193)

This is despite the fact that there is a vast history on Western modernisation and its impacts across different groups, sectors and nations in which tourism researchers could potentially draw upon to assess the links between tourism and the 'political-economy' of development (McMichael 2000). Most tourism research continues to emphasise a market-oriented view of development policies and structural adjustments to the changing global economy. It tends 'to focus on notions of prescription, efficiency and economy rather than ideals of equality and social justice' (Hall 1994: 7). Nonetheless, some tourism researchers have studied how tourism unevenly affects different social groups within and across nations (Britton 1991). Ethnographic works by anthropologists are particularly noteworthy for paying attention to asymmetries in terms of power between hosts and guests, and analysing the impacts these have on each (Stronza 2001). Anthropologists and other tourism researchers to a limited degree have addressed political themes, including tourism and the global–local nexus (Cameron 1997; Milne 1998); class, commodification and tourism (Greenwood 1976; Patullo 1996); tourism discourse and tourism marketing (Urry 1990); and cultural politics (Chambers 1997; Greenwood 1976, 1989; Smith 1989).[1] Cheong and Miller (2000) challenge tourism researchers to move beyond discussing how tourists, and especially wealthy Northern tourists, negatively impact Southern hosts, to how 'there is power everywhere in tourism' (Cheong and Miller 2000: 372). They advise increased scrutiny of how the exercise of power operates at individual, institutional, industry and state levels, and of how local peoples creatively respond to tourism not as passive recipients, but as active social agents. Similarly, Stronza (2002) notes that even in the anthropology of tourism attendant to political dynamics, the analysis is devoid of local voices, and especially fails to examine how local peoples themselves perceive and act on their perceptions regarding the array of pros and cons associated with tourism. Indeed, my research in Belize addresses these points exactly. It describes how different residents take advantage of opportunities differentially available to them through the community ecotourism project in Gales Point, with some joining, others resisting and a few people actively protesting against activities that they interpret as working against their own personal or household interests (Belsky 1999, 2000, 2003).

The rise of alternative tourism suggests an important opportunity for analyses of tourism politics precisely because of its explicit attention to the interaction of social and environmental forces and values. Alternative tourism includes 'forms of tourism that are consistent with natural, social, and community values, and which allow both hosts and guests to enjoy positive and worthwhile interaction and shared experiences' (Eadington and Smith 1992: 3). Ecotourism is an increasingly popular example of an alternative tourism. An often cited definition of ecotourism is 'a form of tourism inspired primarily by the natural history of an area, including its indigenous cultures' (Ziffer 1989). The goal of ecotourism is for eco-tourists to appreciate natural and cultural resources while contributing revenue and attention to local conservation efforts. Importantly, eco-tourism researchers and others concerned about 'sustainable tourism' are paying close attention to the links between tourism and environmental sustainability, and to how culture further influences this relationship (McCool and Moisey 2001). Nonetheless, ecotourism has been criticised as environmentally destructive and 'business as usual'. There remains limited consideration to how alternative tourism is influenced by capitalism and how it affects and is itself affected by social differences such as by class, gender, race, or political affiliation at micro (individual, household, community) or macro (regional, national, global) scales (Belsky 1999). What is germane to the main point of this chapter is that some research methods are more suited than others for revealing these dynamics.

In one of the few books on the political dimensions of tourism, Hall (1994) in *Tourism and Politics: Policy, Power and Place* argues that the choice of tourism topics and research methods is not solely the result of a rational and objective decision-making process, as usually purported, but is highly political in itself. Hall (1994: 15) suggests three problems in the study of the political dimensions of tourism: a reluctance to acknowledge the values that underlie the research, how the researcher(s) themselves influence the research process, and lack of an agreed methodological or theoretical base from which to conduct such research. He says that another reason tourism researchers do not recognise the political nature of tourism studies or the methodological options available to them is the dominance of the positivistic tradition, the conventional science paradigm that sees scientific inquiry as objective, empirical, truth-seeking, apolitical and quantifiable. The positivist scientific approach places attention on 'hard' data, measurement and prescription. Rather than seeing that there are a variety of possibilities and legitimate approaches to knowledge, positivist researchers view their approach as 'science' – it can be no other way – rather than as reflecting particular choices and values. The difficulty of obtaining quantitative measures on questions of ideology, power and structure may limit researchers' attention to these subjects (Ritchie 1984). A lack of intellectual cross-fertilisation of ideas and debate among tourism researchers as well as with other critical social, political and geographic scholars also contributes to the prevalence of positivistic approaches (Burton 1982). Hall argues for interdisciplinary exchange and recognition that research – tourism research in particular – is not 'value-free'. He advises tourism researchers:

> to cast their net for appropriate research methodologies and approaches far wider than has been the case in the past, and to place greater emphasis on some of those central aspects of society, such as power and values, which influence the patterns and processes of tourism development.
>
> (Hall 1994: 16)

In 'casting a wider net' to tourism research, and especially to a more contextual and political understanding of tourism practice, what are the benefits and limitations of qualitative research? It seems appropriate here to acknowledge Sandra Harding's important distinction between *methodol-*

ogy and *method*. She writes, 'a research methodology is a theory and analysis of how research does or should proceed' whereas 'a research method is a technique for (or way of proceeding in) gathering evidence' (1987: 2–3). Methodology frames the questions being asked, determines the methods and types of evidence to be used, and shapes the analyses. As Burawoy (1991: 271) says, 'If technique is concerned with the instruments and strategies of data collection, then methodology is concerned with the reciprocal relationship between data and theory.' Qualitative methodology and its portfolio of available techniques would seem well suited to tourism research that is critical of a value-free epistemology and attuned to particular contexts as well as the political dimensions of tourism practice. Indeed, Denzin and Lincoln (2000: 3) define the key, enduring features of qualitative research as contextuality, interpretation and subjectivity:

> Qualitative research is a situated activity that locates the observer in the world ... qualitative research involves an interpretive, naturalistic approach to the world. This means that qualitative researchers study things in their natural settings, attempting to make sense of, or to interpret phenomena in terms of the meanings people bring to them.

A basic assumption of participant observation is that it is critical of the researcher as a neutral, objective interpreter and seeker of truth, emphasising instead the subjectivity of the research process and empathy between researcher and research 'subjects' (Berg 2001; Taylor and Bogdan 1998). Participant observation refers to the method of researchers making observations in the course of taking part in the activities of the people they study. Like other observational methods, it excels in the possibility of gaining an in-depth understanding of a situation in its natural or usual social context, and especially by providing a sense of what has been called an 'insider's view' of that situation and context. For example, participant observers have documented the benefit of this method for studying marginalised populations uncomfortable with interacting with professional researchers from a different socio-economic and cultural background. Women researchers have also noted the benefits of participant observation, as well as informal, in-depth interviewing, for researching other women. Women are often ill at ease talking to male researchers, especially if their concerns include critiques of men. Nonetheless, feminist and other scholars also continue to challenge participant observation on issues of scale, authority, significance and balance between researcher and research subjects' views (Clifford and Marcus 1985, Stacey 1991). Box 16.1, adapted from Adler and Clark (1992) and Neuman (2003), sets out the advantanges and disadvantages of participant observation. It is to these complex issues that I turn now. What follows is based on my community ecotourism research in Belize.

Box 16.1 Advantages and limitations of participant observation

Advantages
- It can be used to gain an in-depth understanding of a natural situation.
- It acknowledges the subjectivity of researchers.
- It permits interpretation through subjects' perspectives.
- It facilitates dialogue with research subjects.
- It maintains flexibility to shift the focus as events unfold.
- It contributes to theory discovery, generation, modification.
- It is combinable with other data collecting methods.

Limitations
- It can be time-consuming.
- It involves waiting around for 'events of interest' to occur.
- Distortions result when people know they are being observed.
- It is susceptible to researchers' bias.
- It has limited generalisability beyond the situation observed.
- The researchers lack control over the process.

Critical reflections on ecotourism research in Belize

Mezirow (1998: 185–186) explains reflection as a 'turning back' on experience, but suggests that it can mean a variety of things, including simple awareness of objects, events or states of being, and consideration of something to see its alternative possibilities. Social theorists and social 'justice' workers Finn and Jacobson (2003: 355) suggest:

> *Critical reflection* is a structured, analytical and emotional process that helps us examine the ways in which we make meaning of circumstances, events and situations. Critical reflection pushes us to interpret experience, question our taken-for-granted assumptions of how things ought to work and reframe our inquiry to open up new possibilities for thought and action. Posing critical questions is key to critical reflection.

Finn and Jacobson (2003) also suggest that critical reflection promotes continuous self-assessment, fosters connections and linkages between personal and social concerns, legitimises the challenging of dominant explanations and observations, opens up and strengthens spaces of possibility, and, perhaps most importantly, aids problem-posing. Out of this often uncomfortable condition, we make discoveries and work towards meaningful change. Critical reflection, many argue, also provides a way to counter the disadvantages of participant observation and other qualitative methods, as I discuss below.

I first went to Belize in 1976 as an undergraduate participant on a student study group to learn anthropological field methods for understanding applied development issues. Belize was still British Honduras (it did not receive its full independence until 1981) and there were few foreign travellers, at least in the remote coastal community of Placencia where I was 'immersed' for five months. The fishing economy and rural social structure were still very much thriving. When I revisited Placencia in 1996 I caught a glimpse of how transformation had occurred in one place. Whereas 20 years earlier one reached Placencia by air or boat, a road now connects the village to the mainland. The road is dotted with driveways with cars bearing foreign licence plates (mostly from California). The once coconut-lined beaches had been replaced by small lodges, restaurants, bars, diving supply shops and other tourist facilities. The fishing co-operative that was inaugurated during my stay in 1976 had closed that week, signalling quite dramatically the restructuring of Placencia from a natural resource-based economy to one largely focused on tourism, or ecotourism as its promoters contend. Expensive yachts replace the small wooden dories I remembered from my earlier time. In trying to learn how residents are making sense of these changes, I was able to reconnect with an elderly woman who, quite amazingly, remembered me from my earlier student days. Her comment exquisitely sums up at least one perspective on the changes. She told me, 'We have become a community without a soul.' Among other concerns, she told me about the poverty, prostitution and drug addiction that plagued the community and her own family. Her story remained with me as I moved through my academic studies of rural ecotourism in the nineties, as did the tremendous richness and insights afforded by the ethnographic methods I had learned so many years earlier.[2]

This research project began in 1992 when I returned to Belize as a co-instructor of a student field practicum on conservation and development issues. The field trip was one component of a larger, five-year linkage project between the University of Montana (UM) and University College of Belize (UCB) that developed exchange experiences for both American and Belizean students and faculty. The larger project was designed to be international, interdisciplinary, collaborative and action orientated in the form of fostering engaged scholarship and creating a new degree programme at UCB. Each year from 1992 to 1998, I and another faculty member from the University of Montana's School of Forestry brought 10 to 12 students to Belize to teach field research methods and examine conservation and development issues. Participants in the field trips included both Belizeans and Americans, and were diversified too in terms of social and natural science background and gender. Those attending UM were required to complete a semester-long background course, while Belizeans were provided with key texts from the seminar.

We first visited the village of Gales Point in 1992 to investigate the community ecotourism project that had just begun in this rural Creole community. Ours was one of the first groups to stay in the village as paying ecotourists. During this first year, I realised that Gales Point could provide an excellent site not only for teaching about field research methods, community ecotourism and its challenges, but for longitudinal research. I began to think of ways to combine pedagogical goals with conducting a long-term research project. For the next five years (1993–1998), I returned to Gales Point with student groups and we developed a case study, using largely, though not exclusively, qualitative methods.[3] The entire project combined research methods not necessarily to improve objectivity, but rather to improve opportunities for viewing the subject matter from different perspectives. Using more than one interpretive practice helps to 'make the world visible in a different way' (Denzin and Lincoln 2000).

Qualitative methods, and especially participant observation and in-depth interviewing, seemed particularly appropriate for this project for many reasons. First, given that community ecotourism was such a new practice, I wanted the students and myself to maximise our opportunities for new learning rather than hypothesis-testing. Second, given the so-called community organisation of the ecotourism effort, I saw these methods as particularly robust for seeing how ecotourism is embedded in rural community dynamics. Third, these methods are sensitive to addressing the cost and benefits of ecotourism through the perspectives of its participants, especially those for whom community ecotourism is theorised to offer alternatives to non-sustainable livelihood activities. Naturalistic methods such as participant observation and in-depth interviewing enabled the students and myself not only to participate as ecotourists ourselves, but to combine them with observation and informal conversations with ecotourist service providers as we engaged in ecotourist activities.

We focused our participant observation around the two major ecotourist activities in Gales Point: lodging and taking meals in the various homestays or bed and breakfasts and participating in guided nature tours and local cultural events. Each student as well as myself boarded with a local Gales Point family serving as a homestay. This arrangement was crucial not only for permitting ample observational opportunities, but for building rapport with the ecotourist hostess. Over the course of our stay, it enabled us to dialogue freely with our hosts as they conducted their business, and to pursue pertinent questions. In addition, we conducted in-depth interviews with local natural resource users from across the community (fishers, hunters, farmers, loggers and food gatherers); local ecotourism organisation leaders (i.e. concerned with bed and breakfast provision); leaders of the ecotourist associations and their umbrella organisation, Gales Point Progressive Cooperative; and with leaders from the Belizean government and affiliated non-governmental organisations instrumental in originating and co-administrating the Gales Point Ecotourism project. The students were given training in interviewing, participant observation and, especially, note-taking. With regard to the latter, we kept copious notes, carefully differentiating between direct observation, inference, analysis and our own personal feelings (Neuman 2003). We also met daily as a team to compare what we had learned, identify themes for further investigation and validation, and draw up lists of questions for guiding if not structuring interviews.

There were many advantages to conducting participant observation and in-depth field interviews. These included observing the daily activities and demands of ecotourist hosts and organisers, holding conversations in an informal, more naturalistic and less hierarchical manner than is possible in formal interviews, and, related to this, developing relationships and trust between researcher and research subjects. Rather than being the 'object' of research with little or no control over what is discussed, many ecotourist hosts raised important issues during interviews. Many of the bed and breakfast hostesses were particularly comfortable sharing their concerns regarding increased workload, trade-offs between tourism and historical natural resource extractive activities, and ensuing conflicts over leadership and policies upheld in the various ecotourism associations (Belsky 1999). The bed and breakfast hostesses as well as the nature and boat tour guides whom we paid for their services greatly appreciated the economic business our group provided, an important but rarely acknowledged benefit of participant observation. The bed and breakfast hostesses were particularly pleased about having a guest for multiple days and that every bed and breakfast association member received a guest. During interviews with bed and breakfast hostesses many complained about inequities in the allocation of and competition between bed and breakfast hostesses for guests (Belsky 1999). Lastly, the fact that most of the student participant observers were young and some were Belizean (though from different ethnic groups and from urban areas) afforded additional intersections to reduce difference and hierarchy between researcher and research subjects. I think these characteristics increased the willingness of many in the village to engage with us, not just answering questions but posing questions to us as well. Our ongoing interaction offset some of the inevitable discomfort rural Belizeans may have felt with us and especially me, a formally educated, non-Creole-speaking foreign professor.[4]

I suggest that another advantage of participant observation, most pertinent to the role of qualitative methods in tourism research, was an ability to gain an understanding of power dynamics and the politics of ecotourism from the perspective of different community members. These methods were particularly successful for facilitating communication with women, especially the least materially well off women from the least powerful families and factions in Gales Point. The views of women, and especially the poorest women, are usually not sought or given adequate attention when development projects are discussed or implemented (Smith 1999). Formal approaches such as requesting quantitative information can be intimidating and off-putting for male and female respondents alike, on both routine issues and those of a more sensitive and non-quantifiable nature such as politics. By definition, they do not permit 'respondents' to raise issues that the researcher may not see as pertinent, but which are major concerns to them. Indeed, the project's attention to ecotourism politics emerged during unstructured discussions with men and women in Gales Point.

Despite these advantages, there are important questions regarding qualitative research that need to be addressed. Many quantitative or positivist researchers remain concerned about conventional notions of validity (assumptions about causation devoid of researchers' bias) and reliability (whether our research instruments/methods capture what we think we are studying, repeatedly, with consistency and producing the same results) of the study. Indeed, editors of an academic journal to which I first submitted publication of the research raised questions about the reliability and validity of the data, given that they were collected and interpreted in addition to myself by different groups of students of different genders, nationalities, first languages and academic disciplines. I had attempted to deal with this in the following ways. First, following Lincoln and Guba (1985: 314–316), we increased the validity and reliability of our qualitative data by conducting 'member checks', or group discussion where raw data and personal interpretation were shared, analysed and checked with members of the entire research team. The idea here is that through discussion, differences in interpretation are worked out and a higher degree of reliability and rigour is reached. Though I am not claiming to have conducted rigorous 'member checks' in the manner described above, each evening the entire student group met and discussed the data and our interpretations. In fact, in some years I assigned the students to prepare a collaborative final paper that necessitated intensive discussion and negotiation. Both Jordan and Gibson in Chapter 13 and Small in Chapter 15 discuss the merits and drawbacks of collaborative working.

Second, we addressed reliability and validity through 'checks' with research subjects themselves. Every year I returned with a (new) group of students and I provided them with the material collected from the previous year. This included a set list of questions students were assigned to complete with their particular bed and breakfast hostesses. Each year, students reviewed reports from the previous year with the hostesses, checking to see whether the reports had captured how they felt and serving as a base for discussing changes and new issues.[5]

Lastly, two other traditional criticisms of participant observation are that it is incapable of generalisation and therefore not a true science and that it is inherently 'micro' and ahistorical and therefore not true sociology. Burawoy (1991) argues that 'the extended case method' can help to overcome these concerns. In this approach, the case study gives rise to generalisations through reconstructing theory based on comparative analysis. By working to explain the particulars of a single case, but also why there are differences across cases, it becomes possible to acknowledge the historically specific causality of a case, but to move to broader generalisations by checking how it informs or challenges some pre-existing theory that is then reconstructed. The significance of a case then relates to what it tells us about the world in which it is embedded.

In addition to Gales Point, we visited and revisited other rural communities involved in ecotourism and integrated conservation and development projects. These included the Community Baboon Sanctuary, the Crooked Tree Wildlife Sanctuary, the Maya Center and the Cockscomb Basin Wildlife (Jaguar) Sanctuary, and the Rio Bravo Conservation and Management Area. Where possible, we replicated exercises we conducted in Gales Point, including staying with and conducting participant observation with bed and breakfast hostesses, tour guides and random household surveys. As Burawoy (1991) suggests, these other cases provided a broader context for understanding the particulars of Gales Point and enabled me to develop some comparative observations and work towards challenging and reconstructing theories about community ecotourism. This was particularly helpful, since some scholars have argued that our understanding of community conservation models has been based on uncritical popular images and boosterism, rather than on what is actually happening on the ground in particular places (Brosius *et al.* 1998). The case study approach, which combines multiple qualitative methods attendant to the politics of practice and representation, seemed well suited to addressing to these concerns.

The Gales Point case study did not suffer from the common limitations of case studies to explicate the link between micro and macro levels of analysis and constituting the social situation in terms of the particular external forces that influence it. My approach to community ecotourism in Gales Point explicitly situated the case study within a nested-scaled approach across multiple spatial and temporal scales. I deliberately sought information on both place- and non-place-based political-economic-cultural factors and assessed their influence on attitudes and practices in Gales Point. This included collecting oral histories from residents as well as project planners as to the reasons for environmental change; studying the available literature; and holding interviews with key state and private actors on colonialism, nation-building, racism, debt and structural adjustment mandates, and programmes involved in the conservation of biological diversity. In contrast to critiques that tourism research typically fails to assess how macro forces influence micro events (such as community ecotourism in Gales Point), I suggest that this study was deliberate about making such connections.

From participant observation to participatory research

In this chapter I have suggested a number of benefits of qualitative methodology for revealing the political dimension of tourism research. Indeed, I think it helps to bring us closer to a critical perspective on tourism. By 'critical' I am referring not only to the standard political economic concerns of critical theory, but to critiques of Western (positivistic) science that it serves largely to legitimate the assumptions and projects of Western scholars and development workers rather than to accurately record and communicate the concerns of historically marginalised groups of people (Smith 1999). Critical indigenous researchers draw from Freire's (1970) work to further challenge a culture of silence among politically marginalised groups, arguing for a research methodology that involves listening to, affirmation of, reflecting on, and analysis of personal stories and experiences 'from the ground up'. They have raised the metaphor of 'voice' as a response to silence, secrets, lies, talking back, talking in a different voice, and contesting the voice of authority, and a preoccupation if you will with women and other marginalised peoples speaking from and about their own experiences (Maguire 2001).

Qualitative methodology including participant observation and intensive interviewing has helped move research in this direction. Nonetheless, qualitative researchers continue to struggle with the obtrusiveness of some of these methods, and dilemmas regarding how to listen to and report research subjects' concerns, while still reserving space for their own voice and analyses. What should researchers do when their analyses raise questions and concerns that are uncomfortable or even dangerous for research subjects, or when they come to different conclusions from those of some of the research subjects themselves? In trying to resolve this contradiction, Stacey suggests an important but incomplete way out. She advocates that the researcher be intensively self-aware and humble about the 'partiality of the ethnographic vision and [its] capacity to represent self and other' (1991: 117). This is a good start, but it begs the question regarding whose views ultimately matter most, and what good our visions and analysis, however 'accurate', bring to research subjects themselves.

As critical researchers develop policies and ethics over research and the knowledge that research produces, creating more accurate representations and better theories is extremely valuable but may not be sufficient for effecting change on the ground. Revealing the ways ex-colonial peoples have been understood by and marketed to the rest of the world via tourism images and ideology is an important area of critical tourism research (Roche 1992; Urry 1990; Belsky 2000). Explorations into the ideological nature of tourism entail close textual reading and critique that expand the (qualitative) methodological 'net' for tourism research. However, the 'postmodern' assumptions and procedures inherent in discourse analysis are likely to be resisted by positivistic tourism researchers, suggesting again Hall's critique that tourism research is political and value laden. Discourse analysis emphasises the value-laden and reflexive nature of research. In *Decolonizing Methodologies,* Smith reminds us that:

> Research is one of the ways in which the underlying code of imperialism and colonialism is both regulated and realized. It is regulated through the formal rules of individual scholarly disciplines and scientific paradigms, and the institutions that support them (including the state). It is realized in the myriad of representations and ideological constructions of the Other in scholarly and 'popular' works, and the principles that help to select and recontextualize those constructions in such ways as the media, official histories, and school curricula.
>
> (1999: 7)

Others critique the 'deconstruction' method on other grounds, especially that it is long on critique and short on constructive alternatives. Indeed, Smith argues that in a 'decolonizing framework', unravelling meanings and their deployment for particular agendas is part of a much larger project:

> Taking apart a story, revealing underlying texts, and giving voice to things that are often known intuitively does not help people to improve their current conditions. It provides words, perhaps an insight that explains certain experiences – but it does not prevent someone from dying.
>
> (1999: 3)

The latter point suggests the importance of tourism research that not only documents subjects' concerns, but also results in action.

Participatory action research combines concerns that research is both 'participatory' and 'action orientated'. *Participatory research* differentiates itself from other forms of action research by the fact that it emerges from the felt needs of a community, while *action research* refers largely to research with a practical outcome. Neither of these types of research is motivated by academic concerns or those of an outside expert's views of what ails a particular group of people. Rather, they are driven by concerns and, in the case of action research, practical outcomes as defined by the people themselves, who take control of the research topic, process and product. Participatory research entails 'people learning with and from each other about themselves and, secondarily, about the social conditions affecting them' (Park 1999: 3). Under the right circumstances the investigative activities can lead to a cycle of community actions, further reflection, knowledge generation and additional collective action. The ideal participatory action research process and products are emancipatory and empowering. The fact that the 'research subjects' provide their own interpretation of a situation has made it attractive to anthropologists concerned about 'representation', while its emphasis on participants' defining and implementing social change on the ground makes it attractive to social workers with a critical perspective. Are these also relevant concerns and methods for (qualitative) tourism research?

In reflecting on my community ecotourism research in Belize, one of its strengths was that it included the perspective of participants often on the sidelines of such projects. As a result, the research produced important theoretical insights about community, power, gender, environmental conservation and discourse, and their intersections. However, I acknowledge that the research process and products still largely reflected my concerns as the researcher and afforded few tangible benefits to the research subjects themselves. I did not design the research in Gales Point as participatory action research, and it did not emerge as it either. I say this because I don't think that our groups' conversations with Gales Point residents, or my own published accounts, contributed to the energising of participants (i.e. through better knowing reality or gaining knowledge to transform it) or to collective action. Yet these are the central goals of participatory action research in most definitions of it (Finn 1994; Finn and Jacobson 2003; Park 1999). In fact, I think it fair to say that the research contributed little to the empowerment of resident peoples. Its major contribution, I submit, centred on improving the understandings of Western researchers ourselves regarding the complexities involved in the practice and politics of (Western-inspired) community-based ecotourism.

But in reflecting further on this effort, a question I pose to myself now is how might the research be organised differently to engage critical thinking as well as move towards participatory action research? What trade-offs might this entail? A participatory action approach would focus on researching how participants understand the process, but move forward to include how they could transform tourism practice. One theme that might have emerged for developing a participatory action project was the desire of bed and breakfast providers to develop separate cabanas or rooms to host guests. Clearly, many wanted to build on the market and expertise developed in the community ecotourism effort to move towards a more individualistic mode of ecotourism development. A more participatory action approach would begin with these topics as the focus of inquiry and proceed from there. Questions for research emanate from the research subjects themselves rather than from the researcher, from the literature or from theoretical debates. A crucial role remains for the researcher as a facilitator of discussion, reflection and action along the way, but it leaves the values to be acted upon up to participants themselves. Supporters of this and other types of what is being called 'indigenous research' see participatory action research as particularly relevant in post-colonial nations, where Western research has been viewed as neo-colonial and even imperialistic (Smith 1999). In the case of Belize (an independent country only since 1981), participatory action research might lead to identification of topics and strategies to understand if not directly challenge the domination of its tourism industries by foreigners (increasingly Americans). But it is also fraught with challenges.

Taking up the example in Gales Point illustrates the range of complexities and challenges faced in participatory action research. Though desirable to many residents, working on behalf of understanding and developing a more individualistic approach to tourism, if not ecotourism, might be beneficial to a few individuals and households, and could undermine the ecological commitment hoped for through ecotourism. It raises questions regarding how to decide among the interests of competing 'participants' in considering how best to base a participatory action research effort, and, though this is not noted as important, what happens if they conflict with those of the researcher? In Gales Point there is no unified community interest among resident 'participants' (Belsky 1999). Even in the relatively small, rural community of Gales Point it would be extremely challenging to identify issues that would be relevant to a broad and inclusive group. Rather, perspectives and possible action agendas would depend on particular individuals and groups within the 'community' of place, or within the 'community' of interests (i.e. local and foreign conservation organisations, state and other regional actors). Neither group of potential participants is homogeneous, nor would they be in agreement with the other regarding key topics and preferred actions. An example is that attention to ecological sustainability, though the primary concern of biodiversity conservationists involved in the Gales Point ecotourism project, was not the key concern among residents struggling with livelihood security, though it became important to residents benefiting from ecotourism. As a result, there were considerable tensions and conflicts over the practice and meaning of 'biodiversity' both among residents and with those of outside organisers, and most likely there would be questions regarding some of these views as against my own.

In thinking about the applicability of participatory action research to tourism, it is likely that (conventional) tourism researchers would raise both practical and theoretical issues with both the methodology and its particular techniques. If qualitative research may be viewed as overly subjective and unscientific, it is likely that participant action research would be criticised as even more so, and derided as advocacy rather than research. Nonetheless, supporters of participatory action research would credit it as a valid knowledge-building process, and no less subjective and advocacy based – only in this case, advocacy on behalf of particular groups of peoples and social justice values.

Conclusion

In conclusion, this chapter has reviewed the notion that tourism research has insufficiently recognised politics in tourism practice and the practice of tourism research. By drawing on one case study it highlighted both opportunities and constraints of qualitative methodology, especially participant observation and in-depth interviewing, for understanding the political and contextual dimensions of tourism. The chapter also discussed ways this research addressed standard critiques of qualitative methods and the case study method itself through 'member checks' and through 'extending' it to additional cases, drawing on analyses of extra-local and historical social forces, and using these data to challenge and reconstruct theory.

In the latter part of the chapter I tried to push the discussion further by raising the promise and perils of participatory action research, and its possible application to tourism research. This methodology squarely recognises the political and value-laden process of research and seeks to reorient research, largely though not exclusively qualitative research, to work towards promoting the values of self-determination, equity and social justice. As my case study in Belize suggests, putting these goals into practice is fraught with complexities, among others over whose voices and concerns are highlighted and legitimised, even in one small rural community. I have also noted that conventional tourism researchers are likely to view participatory action research as even more value-laden, political and problematic than standard qualitative methods. Nonetheless, my intent here is to stimulate discussion among tourism researchers of different theoretical and epistemological positions, as well as with researchers from other academic disciplines. Indeed, political studies of tourism can be conducted through a variety of methodologies, techniques and values, each capturing a different and valid dimension of the phenomena. I hope that this chapter will help nourish such a wide-ranging dialogue among advocates of different criteria, methodologies and methods in tourism research.

Questions

1   What do you see as the advantages of participant observation and in-depth interviewing for understanding alternative tourisms such as community ecotourism?
2   How does politics affect not only *what* we study about tourism, but *how* we go about studying it?
3   What do you see as the benefits and costs of participatory action research generally, and as applied to tourism research?

Notes

1   In this brief literature review I have no doubt omitted other important works. Nonetheless, I think the generalisation remains valid.
2   For a discussion of the theoretical frameworks that informed the research approach in Gales Point, notably political ecology, feminism and critical materialism, see Belsky (1999).

3 The student group also assisted me in conducting a randomised household survey in 1994 using both open- and close-ended questions. The field practicum included ecological studies using sampling methods along random transects and vegetative plots. I would like to acknowledge my co-instructor, Stephen Siebert, who was also the innovator and principal investigator of the entire UM–UCB linkage project.

4 Though all researchers and research subjects spoke English, most of the Belizean students held conversations in their native Creole language. Speaking in the local dialect also increased the naturalness of the setting and the ease of subjects. To facilitate translation as well as enhance interdisciplinary learning, whenever possible American and Belizean students worked in teams, paired to balance nationality, gender and scientific discipline.

5 To maximise the naturalistic setting, sense of informality and comfort of the bed and breakfast hostesses to raise critiques, I chose not to tape-record any interview or conversation. Instead, the students and I kept extensive field notes.

References

Adler, E. S. and Clark, R. (1992) *How It's Done: An Invitation to Social Research.* Belmont, CA: Wadsworth Press.

Belsky, J. M. (1999) 'Misrepresenting communities: the politics of community-based rural ecotourism in Gales Point Manatee, Belize', *Rural Sociology*, 64 (4): 641–666.

Belsky, J. M. (2000) 'The meaning of the manatee: community-based ecotourism discourse and practice in Gales Point, Belize'. In C. Zerner (ed.) *Plants, People and Justice: Conservation and Resource Extraction in Tropical Developing Countries.* New York, NY: Columbia University Press.

Belsky, J. M. (2003) 'Unmasking the "local": gender, community and the politics of rural ecotourism in Gales Point Manatee, Belize. In S. R. Brechin, P. R. Wilschusen, C. L. Fortwangler and P. C. West (eds) *Contested Nature: Promoting International Biodiversity with Social Justice in the Twenty-first Century.* Albany, NY: SUNY Press.

Berg, B. L. (2001) *Qualitative Research Methods for the Social Sciences.* London: Allyn and Bacon.

Britton, S. G. (1991) 'Tourism, capital and place: towards a critical geography of tourism, environment and planning', *Society and Space*, 9 (4): 451–478.

Brosius, J. P., Zerner, C. and Tsing, A. L. (1998) 'Representing communities: histories, and politics of community-based resource management', *Society and Natural Resources*, 11: 157–168.

Burawoy, M. (1991) 'The extended case method'. In M. Burawoy, A. Burton, A. A. Ferguson, K. J. Fox, J. Gamson, N. Gartrell, L. Hurst, C. Kurzman, L. Salzinger, J. Schiffman and S. Ui (eds) (1991) *Ethnography Unbound: Power and Resistance in the Modern Metropolis.* Berkeley, CA: University of California Press.

Burton, T. L. (1982) 'A framework for leisure policy research', *Leisure Studies*, 1: 323–335.

Cameron, C. M. (1997) 'Dilemmas of the crossover experience: tourism work in Bethlehem, Pennsylvania'. In E. Chambers (ed.) *Tourism and Culture: An Applied Perspective.* Albany, NY: SUNY Press.

Chambers, E. (ed.) (1997) *Tourism and Culture: An Applied Perspective*. Albany, NY: SUNY Press.

Cheong, SoMin and Miller, M. (2000) 'Power and tourism: a Foucauldian observation', *Annals of Tourism Research*, 27 (2): 371–390.

Clifford, J. and Marcus, G. (1985) *Writing Culture: The Poetics and Politics of Ethnography*. Berkeley and Los Angeles, CA: University of California Press.

Denzin, N. K. and Lincoln, Y. S. (2000) 'The discipline and practice of qualitative research'. In N. K. Denzin and Y. S. Lincoln (eds) *Handbook of Qualitative Research*, 2nd edn. Thousand Oaks, CA: Sage.

Eadington, W. R. and Smith, V. L. (1992) *Tourism Alternatives: Potentials and Problems in the Development of Tourism*. Philadelphia, PA: University of Pennsylvania Press.

Finn, J. L. (1994) 'The promise of participatory research', *Journal of Progressive Human Services*, 5 (2): 25–42.

Finn, J. L. and Jacobson, M. (2003) *Just Practice: A Social Justice Approach to Social Work*. Peosta, IA: Eddie Bowers.

Freire, P. (1970). *Pedagogy of the Oppressed*. New York, NY: Seabury Press.

Greenwood, D. J. (1976) 'Tourism as an agent of change: a Spanish Basque case', *Annals of Tourism Research*, 3 (3): 128–142.

Greenwood, D. J. (1989) 'Culture by the pound: an anthropological perspective on tourism as cultural commoditization'. In V. Smith (ed.) *Hosts and Guests: The Anthropology of Tourism*, 2nd edn. Philadelphia, PA: University of Pennsylvania Press.

Greenwood, D. J., Whyte, W. F. and Harkavy, I. (1993) 'Participatory action research as a process and as a goal', *Human Relations*, 46 (2): 175–192.

Hall, C. M. (1994) *Tourism and Politics: Policy, Power, and Place*. New York, NY: Wiley.

Harding, S. (1987) *Feminism and Methodology*. Bloomington, IN: Indiana University.

Lincoln, Y. and Guba, E. (1985) *Naturalistic Inquiry*. Thousand Oaks, CA: Sage.

McCool, S. F. and Moisey, R. N. (2001) *Tourism, Recreation and Sustainability: Linking Culture and the Environment*. Oxford: CABI Publishing.

McMichael, P. (2000) *Development and Social Change: A Global Perspective*. Thousand Oaks, CA: Pine Forge Press.

Maguire, P. (2001) 'Uneven ground: feminisms and action research'. In P. Reason and H. Bradbury (eds) *Handbook of Action Research: Participatory Inquire and Practice*. London: Sage.

Mezirow, J. (1998) 'On critical reflection', *Adult Education Quarterly*, 48 (3): 185–198.

Milne, S. (1998) 'Tourism and sustainable development: exploring the global–local nexus'. In C. Hall and A. Lew (eds) *Sustainable Tourism: A Geographical Perspective*. New York, NY: Longman.

Neuman, W. L. (2003) *Social Research Methods: Qualitative and Quantitative Approaches*, 5th edn. Boston, MA: Allyn and Bacon.

Outside Television (2000) *Eco-Sanctuary Belize*. Santa Monica Video, Inc. (dub-date 30 November 2000).

Park, P. (1997) 'Participatory research, democracy and community', *Practicing Anthropology*, 19 (3): 8–13.

Park, P. (1999) 'People, knowledge and change in participatory research', *Management Learning*, 30 (2): 141–157.

Patullo, P. (1996) *Last Resorts: The Cost of Tourism in the Caribbean*. New York, NY: Monthly Review Press.

Ritchie, J. R. (1984) 'Assessing the impact of hallmark events: conceptual and research issues', *Journal of Travel Research*, 23 (1): 2–11.

Ritchie, J. R. and Goeldner, C. (1987) *Travel, Tourism, and Hospitality Research: A Handbook for Managers and Researchers*. New York, NY: Wiley.

Roche, M. (1992) 'Mega-events and micro-modernization: on the sociology of the new urban tourism', *British Journal of Sociology*, 43 (4): 563–600.

Smith, L. T. (1999) *Decolonizing Methodologies: Research and Indigenous Peoples*. London: Zed Books.

Smith, V. (1989) *Hosts and Guests: The Antropology of Tourism*, 2nd edn. Oxford: Blackwell.

Stacey, J. (1991). 'Can there be a feminist ethnography?'. In S. B. Gluck and D. Patai (eds) *Women's Words: The Feminist Practice of Oral History*. New York, NY: Routledge.

Stronza, A. (2001) 'Anthropology of tourism: forging new ground for ecotourism and other alternatives', *Annual Reviews of Anthropology*, 30 (1): 261–283.

Taylor, S. J. and Bogdan, R. (1998) *Introduction to Qualitative Research Methods: A Guidebook and Resource*, 3rd edn. New York, NY: Wiley.

Urry, J. (1990) *The Tourist Gaze: Leisure and Travel in Contemporary Societies*. London: Sage.

Ziffer, K. (1989) *Ecotourism: The Uneasy Alliance*. Washington, DC: Conservation International.

17 Shared benefits

Longitudinal research in eastern Indonesia

Stroma Cole

Aims of the chapter

- To discuss how rapport, trust and power relations change over time and affect the nature of the data gathered in the field.
- To demonstrate how focus groups can be adapted to a non-Western setting and be a valuable anthropological method.
- To demonstrate how tourism research can be undertaken in a way that provides benefits for both researchers and the researched.

Introduction

Extensive fieldwork is seen as the hallmark of anthropology and usually involves undertaking participant observation for a year or more. However, in the field of tourism many anthropological studies often represent a snapshot (Wilson 1993) and are justly criticised for missing the diachronic nature of change brought about by tourism (Nash 1996). The study reported in this chapter was carried out over a ten-year period, 1989–1999, during which my position as researcher evolved and changed. The research was carried out in three distinct phases: a phase when I was a practitioner-cum-researcher, a phase of participant rural appraisal and a phase of long ethnographic fieldwork. The first two phases are very important to set the context and for understanding the researcher–researchee relationships of the third phase, which is the main focus of the chapter. This chapter discusses how the research process evolved as the relationships and understanding of the research setting changed over time. It examines the nature of relationships between researcher and respondent and how this affects the data-gathering process and impacts upon the ways in which research methods were applied, the quality of the data collected and the eventual uses the data can be put to.

The essentially action-orientated approach employed in the final stage of the research relied upon a high level of trust and confidence, sharing of knowledge and experience, and personal involvement. Although, to establish a good rapport, I favoured non-hierarchical relationships between respondent and researcher, this was difficult to achieve because as an educated, white, Western researcher in a post-colonial setting, I was ascribed status. Both self-disclosure and reciprocity were used to minimise the hierarchical pitfalls (Reinharz 1992). As relationships developed over time and became enduring, there was a tendency to find a reciprocal balance so that both the researcher and respondent felt comfortable with the relationship. Unlike in the case of a once- or twice-in-the-field study, when relationships are continued for a number of years and the researcher wishes to be welcomed on a continuous basis, there is a need for the researcher and the respondents to develop a comfortable mutuality. Following a discussion of how I gained access, established all-important rapport and trust, and the power dimensions of the respondent–researcher relationship, the chapter discusses how I adapted focus group methodology for the mutual benefit of researcher and researched.

The setting

The research took place in two villages, Wogo and Bena, in Ngada regency of Flores, Nusa Tenggara Timor, Indonesia. The Catholic villagers are largely swidden peasants eking out a hand-to-mouth existence on poor soils. The rugged mountainous area began to be visited by 'drifters' in the 1980s and has seen increasing numbers of tourists ever since. The most popular village, Bena, received 9,000 tourists in 1997 (Regency Department of Education and Culture 1998). The area is one of the poorest in Indonesia, and tourism is considered the area's best option for economic development (Umbu Peku Djawang 1991).

Traditional hamlets provide a complex of attractions: clan 'totems', megaliths and traditional houses. Four of the Ngadha villages, including those where the research was carried out, have been given 'prime tourist attraction' status (*obyek wisata unggulan*: *objek* = object, *wisata* = tourist, *unggulan* = superior) by the government. Ngadha houses, beyond being important tourist attractions, are a central organising principle in Ngadha society. All members of Ngadha society belong to a named house and a clan. Houses belonging to a number of clans are arranged in two parallel lines or around the sides of a rectangle, providing the tourist with a 'feeling of being enclosed in antiquity' (Cole 1997b).

Phase 1: Research as a tour operator

The first time I visited Wogo was as a tour operator accompanied by my Indonesian husband. My initial acceptance was eased by my being the wife of an Indonesian. Not only was my status, as a married woman, not a threat to local women (see Angrosino 1986 on the threat an unattached male anthropologist poses), but I also had the empathy of a shared nation of residence. Further, as I spoke Indonesian, I was respected (Indonesian-speaking Westerners were, and remain, rare). Although the villagers' first language is Sara Ngadha, the vast majority of them had a good command of Indonesian. While it would have been desirable to conduct the research in the villagers' first language, Indonesian was accepted as the language of mutual communication.

The research conducted during the early years was intended to aid the company's success. As a tour leader, leading cultural tours marketed as 'led by anthropologists', my business depended on providing detailed cultural information for my clients. Furthermore, as an anthropologist I clearly had an intellectual interest in exploring the cultural background of the village. Furthermore, the success of the business depended upon bringing further tours to the village. To this end, we needed to comply with the villagers' requests that largely focused upon avoiding behaviour that would cause offence, such as public displays of affection and the wearing of skimpy clothing.

Many villagers had positive experiences of tourism as a direct result of my tour company's actions: for example, bringing tourists, significant income and piped water to the village (Cole 1997a). During these visits I established rapport with the villagers. This was critical when I embarked upon my research, given that the relationship between a field-worker and respondents is considered crucial to the research process (Bernard 1988). Angrosino states that 'the process of building acceptable local persona is a trial and error affair' (1986: 66). By the time I came to do my long fieldwork, I had found a persona that was both comfortable to me and non-threatening to my hosts. I was *au fait* with the cultural norms, dress, behaviour and patterns of communication in the village. Not only did the success of the tours taken to Wogo influence the selection of the research site for my PhD, but also the status I gained from running successful tours affected the collection and analysis of the data, and clearly impacted on research relationships.

Phase 2: Rapid rural appraisal

The second type of research I conducted in 1990 was a 'fast and dirty' (Hampton 1997: 370) participatory rural appraisal (PRA) for an academic paper (Cole 1997b). The purpose of PRA is to generate knowledge and information in a relatively short time (Koning and Martin 1996). The background, rapport and language established during the previous visits meant I was able to access participants with relative ease and gain their consent to conduct 30 questionnaire-based interviews in ten days. The purpose of these interviews was to explore villagers' ideas and views about tourism. Without the trust built up during the first phase of the research it would not have been possible to conduct research of this intensity.

When I arrived in Wogo to undertake this rapid research I met my first challenge in undertaking research in a village where I was well known. I wanted to conduct research, to obtain some 'hard data' in a limited amount of time. Most of the villagers wanted to 'catch up'. I struggled to steer conversations on to tourism and collect the data I wanted. However, the trip proved invaluable to reconnect with the villagers. The continuity of relationships and friendship bonds that were cemented meant I was readily accepted when I returned to carry out further research.

My second challenge was to overcome 'conformist responses'. The feeling that everyone was providing a set of 'the answers' resulted from two well-documented aspects of Indonesian culture. First, Indonesians are well known for 'telling you whatever they think you want to hear' (Draine and Hall 1996: 79); and second, the hierarchical nature of Indonesian authority means that the opinion of someone in higher authority cannot be questioned (Reisinger and Turner 1997). I discuss how I overcame this problem by using focus groups later in the chapter.

One-to-one interviews were extremely difficult to arrange. Houses consist of three spaces: a dark inner sacred room, a middle room and a terrace. Villagers who were at home but not busy would be on their terrace, chewing betel, mending mats, shelling corn, etc., and were therefore easily noticeable and approachable. However, as I approached, sat down and began chatting, I was easily spotted, and a potential one-to-one interview with any villager became a gathering. If I had my notebook out, someone would be trying to read over my shoulder, and others would gently but surely become part of the interview, which would turn into a general discussion. When an elder male had provided his opinion in front of other villagers, this opinion was repeated as *the* opinion that they assumed I wanted.

However, opinions could be gleaned, stories with important subscripts were told, and the nature of questions put to me often revealed important data. It was because the villagers knew me as a successful tour operator, and thus trusted my knowledge of tourism, that they quizzed me. For example, the question 'Do you think tourists would want to stay in a guest-house if it was built outside the village?' alerted me to the fact that villagers were already discussing accommodating tourists.

I learned that one-to-one spontaneous, indoor, fireside chats were a more successful technique than attempting to carry out questionnaire-based interviews. They were far easier to manage and the information gathered was more insightful. As Berno (1996) discusses, the cultural context of research is given little consideration in the tourism literature. Her analysis of tourism in the Cook Islands suggests that owing to 'a lack of familiarity with structured questionnaires and social science research in general, subjects tended to acquiesce, and a positive response bias was evident on questions that dealt with satisfaction with tourism and tourists' (1996: 384). Fear of giving 'the wrong answer' to an authority on tourism was apparent in my research. However, quiet one-to-one conversations disclosed information on topics that were not openly discussed at other times. Some topics were not related to tourism but were important contextual detail such as the phenomenon of divorce in the predominantly Catholic Indonesian society. Some information was essential for understanding how views about tourism were changing; for example, I was informed how donations from tourists were being appropriated and not shared or used for community projects as they had been in previous years when I took tours to the village.

Although a researcher who was less familiar could have carried out the PRA, it is unlikely they could have obtained the same quality of data in the limited time. The time I lost in catching up, a hardened, Indonesian-speaking anthropologist would have taken to settle. Any researcher would have to face the problem of conformist responses. However, because of familiarity, I had ample opportunity for fireside chats, which revealed more honest and insightful data. Furthermore, the success of my tour operations meant that I was trusted for my knowledge in the area of the research and could provide examples for the villagers to draw upon in discussions.

Phase 3: Long fieldwork

My position

When I arrived for my long fieldwork (eight months) in 1998 I was more prepared than an anthropologist entering the field for the first time would have been. I knew the necessary 'impression management' (Hammersley and Atkinson 1995). I already had a high degree of awareness about self-presentation. Although culture shock 'is the stock-in-trade of social and cultural anthropology' (Hammersley and Atkinson 1995: 102), I did not suffer from a culture clash; I knew about the toilet facilities, the monotonous diet and lack of privacy. I therefore worked on my coping strategies: going to town once a week for a relatively private night in a guesthouse, to get a bath and have a different meal.

My age and gender were important aspects of my role and influenced the nature of the data that I obtained. Although these factors were no doubt important in the earlier phases of my research, it was not until this third stage that it became more apparent. In Ngadha it is the men who deal with outsiders. I dealt with men because I was an outsider, but as a woman I could also gain the female perspective, in a way a male would have found impossible. This ambiguous yet advantageous position has been identified by other female anthropologists (Hammersley and Atkinson 1995). As I noted in the first phase, as a married woman I was not a threat; because I was married I was not considered as in competition with other women.

As I grew older, my role changed. When I first started visiting Wogo, before my daughter was born, I was excluded from some women's matters, although I only realised this afterwards. I then (mistakenly) repeated local cultural patterns by excluding unmarried, mature women who were not mothers from some aspects of my own research. However, when I returned for my long fieldwork I was a mother; I had become a 'real' woman. Furthermore, my daughter's friendships gave me fruitful new openings that would not have otherwise existed. Her friends' parents easily became respondents as we got to know one another. Casual chats that began from our shared experiences in child rearing readily moved onto other tourism-related issues and opened up new topics for discussion.

One of the problems of returning to the field is that villagers assume that the researcher remembers what they have been told on previous visits, or that they know about certain aspects of their culture. The researcher is assumed to understand things, and they are therefore difficult to learn about (Foster *et al.* 1979). Many aspects of the culture, so different from my own, I needed explaining more than once. I could use my daughter as an excuse for eliciting explanations and further clarification on subjects the villagers thought were obvious and I should know about.

Participant observation involves prolonged immersion in the life of a community, group or organisation in order to discern people's habits and thoughts. The researcher attempts to become immersed in the everyday life of the community being researched to obtain an insider's perspective. The method has both pros and cons (see Box 17.1).

The role of a researcher who carries out participant observation, referred to as an ethnographer, has been seen as developmental, moving through a series of phases as the research progresses (Burgess 1984; Bernard 1988). During the eight months of my long-term research, three distinct phases could be discerned:

- Getting to know you again: although I was very familiar, there was a re-settling-in phase characterised by being treated like a guest: being served first at mealtimes, for example.

Box 17.1 Advantages and limitation of participant observation (2)

Advantages
- It enables the researcher to gain an insiders' perspective.
- It gives the researcher direct experience.
- It provides insights and depth of data.

Limitations
- Collection and analysis of data are time-consuming.
- The researcher's presence may affect the data.
- The researcher needs to fit into the research environment.
- The method is subjective.

- Acceptance: characterised by being treated as one of the family, as a labour source to be tapped, and as someone who could be used to achieve political ends. During this phase it is likely that the head of household would be served first, but in the family setting, order was unimportant. In Bena I ate from a shared gourd with female members of the family.
- Imminent departure: characterised by special treatment and an effort to help me out. This phase was somewhat interrupted by Christmas, New Year, *Reba*, and my family visiting from England, at which point I reverted to guest status. (*Reba* is the annual gathering of clan and harvest festival held in all Ngadha villages.

These three phases were apparent during the long research. However, over the total research time, different phases are discernible with different key informants. Over time people change, their circumstances change and so does their relationship with the researcher. I will describe the development of relationships with three of my key informants, in three key areas, that have important implications for ethnographic research: access, rapport and trust, and power. Table 17.1 summarises how my roles and my relationships with these three key informants have changed.

Access

Our initial access to Wogo was negotiated with the Department of Education and Culture, and with a villager, Pak Anis, who worked there. Pak Anis confirmed bookings for our groups of tourists and later wrote to me to say the villagers would welcome our return and were enthusiastic about

Table 17.1 Changing roles, activities and relationships over time

	1989–1993	*1996*	*1998–1999*
My role	Tour operator	PRA[a]	PhD student
Purpose of research	Business success	Academic paper	Data for PhD
Activities	Constant question-and-answer sessions Shepherding tourists	30 questionnaire based interviews and important fire-side chats	Participant observation, interviews and focus groups
Relationships	Started formal, later adopted	Hurriedly 'catching up'; friendship bonds were cemented	Accepted, helped and used

Note
a PRA = participatory rural appraisal.

the research. Pak Anis was a gatekeeper, an actor with control and access to potential participants and avenues of opportunity.

Rapport and trust

Knowing foreigners brings pride; having them to stay in one's house is even more highly regarded. As Pak Anis had acted as gatekeeper, we were expected to take a room in his house, which we did. Initially we were still treated as guests, but quickly I learned that Pak Anis saw himself not only as my mentor but also as my 'father'. As a foreigner I could expect honour and as a researcher support. As his daughter, I was expected to show Pak Anis deference and obedience. I felt as though I was on a see-saw trying to find a balance between deference and respect, honour and obedience. My categorical sisters were constantly bossed about, and I felt my behaviour was directed. However, it was when I was requested to undertake domestic duties for the family – for example, I was left in charge of fire-lighting and getting the rice and beans under way – that I felt like a real daughter, and I knew true rapport was established.

Nene Yuli was my first host when I took tourists to Wogo. (*Nene* means grandmother and grandfather in Bahasa Ngadha. It means grandmother in Indonesian.) She was a sharp lady who put up with no nonsense. Our rapport was established quickly, and over the years we have grown fonder of each other. Her age demanded my respect, my foreignness hers. A bond developed based on a deep and loving trust. As she has got older we have taken fewer long, demanding walks and spent more time chatting by the fire. These conversations with this honest and forthright woman generated very insightful data. She told me she was the last woman in the village to have had her teeth filed, an initiation into adulthood ceremony that had been banned by the Church. The abandonment of this ceremony, combined with youth observing uninhibited Western tourists, was the reason given by some villagers for the rising rate of teenage pregnancies. Nene Yuli would give me little pieces of jigsaw that helped me fit the complete picture of the villagers' culture together. Her openness to discussing issues rarely covered by other women was a function of our lengthy, enduring bond developed over the years. Furthermore, Nene Yuli's position as 'clan-house head' had important bearings on other relationships. (Ngadha clans have a number of hierarchically ordered houses, each overseen by a woman. Nene Yuli has that role for one clan's primary house.) Other members of the house and clan would help me because Nene Yuli told them to.

Having such a deep bond with respondents has pitfalls. As Nene Yuli has aged, her increasing frailty has been a worry. The emotional energy expended was at times burdensome and detracted from my research.

Power

Power relations also change over time. As I have suggested, my relations with Pak Anis see-sawed between deference and respect as I balanced my role of daughter and researcher. The power relations between us also see-sawed. As a gatekeeper, Pak Anis introduced me to some people before others (perhaps at the expense of others). Although he never prevented me from interviewing members of the village, he made some meetings much easier than others. He would advise me, and talk in depth on some subjects while avoiding others. I would sometimes get exasperated, and (by going through his wife) stress that an investigation was essential for my research. Although eventually I could persuade Pak Anis of my needs, I became aware that over-reliance on any key informant was in some ways detrimental to the research process, but in others offered opportunities that might otherwise not have been open to me. By the end of my long research we had developed a mutual respect, and the relative power of our positions appeared more balanced.

Another example of how power relationships change over time is evidenced in my relationship with Sipri. He was 15 when I first visited Wogo. Using his excellent English, Sipri helped with our tours by translating between the villagers and our clients. Subsequently he left for university on Java and I didn't see him for many years. Towards the end of my long fieldwork he returned from Java. We had an interesting discussion about our kin relationship. Sipri was keen to point out that I still had a lot to learn about Ngadha, and he promised to help me. While I was writing up my PhD, Sipri worked for the United Nations High Commission for Refugees and had access to email. This was especially important in the data analysis and interpretation. Not only was I able to check the spelling and meaning of Ngadha terms, but also I could bounce ideas off an insider. Sipri provided me with 'respondent validation' of my work. I could check my interpretation with him, and at times he would refer thorny cultural issues to village elders. 'Respondent validation' was especially important as I was often interpreting notes taken in two foreign languages. Although my age and foreignness commanded a certain respect from Sipri, he was aware of my need of him. Fortunately, my enduring bond with his aunt, Nene Yuli, meant that Sipri kept his promise to help me, and has enabled me to stay in touch with my family-cum-respondents.

The various roles that ethnographers adopt have been analysed, and a distinction has been made between 'complete participant', 'participant as observer', 'observer as participant', and 'complete observer' (Gold 1958 and Junker 1960 in May 1997). For by far the greater part of my research I could not be a true 'complete participant' because the villagers were aware of my research. (The exception was research with tourists on guided tours, where I concealed my research until the end of the tour.) However, in Wogo I did have a problem of feeling too at home, which, as Hammersley and Atkinson (1995) suggest, leads to over-rapport. When life became too routine, I found that I was participating fully but not observing new things. My researcher role became eclipsed and I found the days passing without my accumulating any new data. As a strategy (partly subconscious) to remain marginal, I moved to another village: Bena. This necessitated a return to the role of participant observer.

An appropriate methodology

As I had learned in my attempt to carry out PRA, conducting interviews with individuals was problematic. When I returned to carry out my long-term research I was keen to try alternative research methods. Most of the literature on focus groups is culturally bound to their use in Westernised developed countries, except a few examples such as Khan and Manderson (1992). However, I felt that with some adaptation, focus groups were an appropriate method to try in a rural location in eastern Indonesia for the following reasons:

- Focus groups felt socially appropriate because they were not dissimilar to *arisan*[14] gatherings. *Arisan* is a 'regular social gathering whose members contribute to and take turns at winning an aggregate sum of money' (Echols and Shadily 1989). It is thus a communal savings system found all over the Indonesian archipelago and is widely used in the villages. It was while I was attending an *arisan*, where many villagers were participating in a discussion in a social setting, sharing a communal meal, that the idea of focus groups came to me. When I suggested to Pak Anis that I could provide a meal, palm toddy and the topic of conversation, he was sure that the idea would be well received and an effective way to gather data.
- Morgan (1997) suggests that groups should be homogeneous and made up of strangers. My groups were homogeneous but not strangers. It has been suggested that acquaintances rely on taken-for-granted assumptions that the researcher is trying to find out (Agar and Macdonald 1995). However, as Kitzinger points out, interaction between clusters of people who already know each other approximates to 'naturally occurring' data, and it is useful to work with such groups because 'they provide one of the social contexts within which ideas are formed and decisions made' (1994: 105).

I was interested to find out the opinions from different sectors of the village population, and the villagers were also keen on this aspect of my work. As Stewart and Shamdasan (1990) point out, there is a difference in group dynamics and outcomes between single-sex and mixed-sex groups. I considered it important to separate women and men because the women in Ngadha rarely openly contradict the men's opinions. Likewise, the young people would be unlikely to contradict the views of their elders. The same interview guide was employed so that comparisons could be made across the different groups.

- Although some villagers were keen to offer their opinions, others were shy. One reason for trying focus groups was that shy people might find the interviews easier in a group situation even if their participation was limited to agreement with some of the opinions expressed by other members of a group. Indonesians have a concept of *malu*, which is a culturally sanctioned expression lying somewhere between shyness, shame and embarrassment. Indonesian villagers easily feel embarrassed or belittled by authority and are much more relaxed in a group than on their own. Because of the 'safety in numbers', focus groups enable 'less inhibited members of the group [to] "break the ice" for shyer participants ... but also being with other people who share similar experiences encourages participants to express, clarify and even develop particular perspectives' (Kitzinger 1994: 111–112). This is particularly important in the Indonesian village situation, where strength in numbers made it easier for members of the group to express themselves.
- Accessing individuals was also difficult. Villagers in Wogo are rarely found alone except when working on agricultural tasks. When carrying out informal interviews while participating in such tasks, I felt I was hindering their work.
- Conflict existed between being adopted and being a researcher. I wanted to reciprocate the villagers' time in some way, but paying participants was out of the question as I was considered as family. Providing a special meal for people was a culturally appropriate way to reciprocate for time given up participating in my research, but would have been unreasonably burdensome in shopping and cooking time if it had been carried out for individual households. By having focus groups I could provide a special meal for a number of people at the same time.

Adapting focus groups to the local setting

In each village, Bena and Wogo, three groups were held: one for women, one for men, and a mixed group of young people aged 15–20. The main criterion of selection was willingness to take part. However, an effort was made to ensure that all groups were representative of the village population, in terms of age and clan membership.

Focus groups have been widely used in the United States and have become more commonly used in Europe over the past few decades. Their use in the less developed world has been exceedingly limited. The method had thus to be adapted to a village in eastern Indonesia. The conduct of these focus groups varied considerably from recommendations made in the literature.

Morgan (1997) and Stewart and Shamdasani (1990) suggest that 'the common rule of thumb is to over-recruit by 20%', intuitively, the villagers also suggested over-recruiting in case of dropouts. This underestimated the enthusiasm of the villagers. I aimed for groups of six to ten, but nine to fourteen participants attended. Those who were not involved were disappointed. For example, unmarried women in their twenties were not invited to either the women's group or the young people's group. My categorical sister, who invited participants by word of mouth, was unsure which group they should belong in as they fell between the villagers' social categories. Furthermore, the groups were already full to capacity. This case reminds us to be mindful to deconstruct informants' social categories before accepting their offers of help. The women who missed out on the focus groups were upset they had been excluded from a social event, but because they were neighbours, I was able to interview them later and include their opinion in my study. As focus groups were only one of my research methods, the multi-method approach meant that I was able to adapt, and remedy the embarrassment of not including this small group of women.

Further differences that result from the cultural setting are also important. Morgan claims that 'the most important element for the site is a table for participants' (1997: 55); as facilities and custom dictated, we sat on the floor. Both Morgan (1997) and Kitzinger (1994) suggest beginning the discussion with writing something down as a useful way to start focus groups and to encourage participation. This would have been humiliating for some villagers and entirely culturally inappropriate, as not everyone would have been able to write. Finally, and most importantly, Morgan (1997) considers ensuring the quality of the recorded data to be crucial to the focus group method, and that conversation which is not recorded represents a loss of data. To ensure a good recording, aid facilitation and provide the opportunity for all group members to voice their opinions, it is suggested that six to ten participants is the right number. With more participants, people often break into small groups, and talking over one another results in a loss of data. None of the researchers describes how to deal with crying babies, barking dogs and squealing pigs as the major sources of noise interruption! Although I found recording the group's discussions useful, I also had to take notes. The tapes, or what could be gleaned from them, were transcribed the following day in all cases. Because of the level of background noise and the number of interruptions, word-for-word transcriptions could not be produced, which meant having to make detailed notes when the information was still fresh. Furthermore, my participants were all readily accessible, so we could clarify bits of conversations that were not clearly recorded. One of the successes of the focus groups was that issues were raised that could then be followed up at other opportunities.

The effect of long-established relationships on focus group dynamics

The make-up of each group differed across a number of different dimensions, including age difference between researcher and respondents, 'stranger value' (Simmel 1950 in Burgess 1990) versus long-established relationships, and the gender of the researcher and researchees. The three factors in combination affected the dynamics of each of the groups. I had long-established relationships with the villagers in Wogo, but not with those in Bena.

Being familiar was normally an advantage, but having long-established relationships made moderation difficult, especially as a woman dealing with a men's group. The first group was the men's group in Wogo. This was the first group I conducted; my lack of experience at this stage was perhaps a factor in why I found this group the hardest to moderate. Further, my having known these men over a period of time meant that there was very limited researcher 'stranger value'. I found it very hard to organise men who considered themselves my 'father' and 'uncle'. Pak Anis, in particular, took over moderation and steered questions about tourism to their issues. I had, as his 'daughter', to respect his authority in this setting. While this was frustrating at the time, the discussion was very useful in finding out issues that were important to this group, which provided a useful basis for further research.

Again, little 'stranger value' existed between researcher and respondents in the young people's group in Wogo, as these youngsters had known me since they were pre-school age. My age commanded enough respect for sufficient moderation, but there was limited need to steer conversations as they spoke freely but without going off the topic. In contrast, with the youth group in Bena I had the greatest 'stranger value'. The participants were very self-conscious and reserved, and not very forthcoming. This, combined with an age and authority gap, made it difficult to get anybody to say anything, and impossible to have a discussion. Perhaps the group did not even fulfil the criteria of a focus group, as Morgan states: 'the hall mark of a focus group is their explicit use of group interaction to produce data and insights' (1997: 2). Being older commanded respect with the young people, but when combined with being a stranger, the authority gap was too great for the group to be successful.

The difference between being familiar and having long-established relationships had little bearing on the running of the women's groups. Both the women's groups produced lively group discussions on my topic areas. The groups were very social and relaxed, and I felt very much at ease. Although I knew the women in Wogo much better than those in Bena, I was familiar to both groups, and this fact, combined with being a woman among women, meant that little difference could be discerned in the enthusiasm between these groups.

Mutual benefits resulting from the focus groups

In this final sub-section I will discuss how balancing the applied and interpretive aspects of my research proved more of a challenge than I had
expected. I wanted the research process to be reciprocal, but balancing the
giving and taking of knowledge proved increasingly challenging during my
stay. As I became more accepted by the communities, I felt that they
increasingly made use of me. The process whereby my interviews and
focus groups would be turned into the villagers interviewing me, and I
would be the giver rather than the taker of knowledge, was particularly
apparent in the Bena men's group. For every topic they wanted to know
what I thought, wanted to probe into my knowledge, ask my advice, and
glean information from my experiences elsewhere in Indonesia. It was
during this group session that I realised how important these focus groups
were for the villagers, not just as social events and meals but as a way for
them to access information. Data-gathering became a two-way educational
experience. As a result, interviews and focus groups took longer than
planned, and I learned that I had to allow extra time for sharing knowledge.

The need for community participation in tourism planning is a common
theme in the literature on socially sustainable tourism (Hunter and Green
1995; Tourism Concern 1992; Hitchcock 1993; Simmons 1994). Abram
argues that 'the difficulty for ordinary people in accessing technical discourses is often identified as a major barrier to full participation' (1998: 6).
Participation beyond rhetoric cannot be achieved without elucidation
(Cole 1999). The empowerment end of the participation ladder (see Pretty
1995) cannot be achieved until those who are participating have the knowledge behind the discourse. Knowledge of the tourism system and
decision-making process is essential if the villagers are to participate in the
planning and management of tourism.

Although Krueger would claim that 'focus groups are not intended to
teach, to inform or to tell' (1994: 223), the focus groups held in the villages
became the locus for the transfer of knowledge about tourism, as discussions often became a case of the villagers probing my knowledge and
experience. The focus groups allowed for the demystification of a 'technical discourse'. For example, the Department of Tourism had built home-
stays in Bena. The villagers wanted to know 'What are home-stays for?
How should they be equipped and managed?' In Indonesia, many aspects
of tourism are discussed in English, or English tourism expressions are
adopted into other languages without translation. *Turis* (tourist), *foto*
(photograph) and transport have become everyday Indonesian vocabu-
lary. 'Gateways', 'home-stays', and 'backpackers' occur in Department of
Tourism literature produced in Indonesian. (It is important to note that
the villagers also considered that the use of Bahasa Indonesia at meetings
put those who were able to articulate well in the language of authority at a

distinct advantage (see Cole 1999).) In this case the Department of Tourism had used the Australian-English term 'home-stay' and not translated it into a language meaningful to the villagers. This acted as a barrier to villagers' participation.

The focus groups allowed for a discussion of various topics, such as the culturally insensitive behaviour of tourists, and possible solutions to the problems. The villagers were able to learn from me how similar problems are addressed in other regions with a longer history of tourism or greater numbers of tourists, and discuss whether these solutions would be appropriate in their own setting. When considering tourists' inadequate clothing, for example, we discussed the possible use of codes of conduct in hotels, clothes at the village entrance that could be lent or hired to tourists and whether their use should be compulsory or voluntary. Some general issues to consider when using focus groups in non-Western settings are summarised in Box 17.2.

Box 17.2 Using focus groups in non-Western settings

- Focus group literature provides guidelines of how to conduct groups in Western (developed country) settings. In order to be used in a non-Western setting they need to be adapted.
- The dynamics of the different groups is a function of a number of factors. The difference in age, gender and stranger value between the researcher and respondents needs to be considered, as these will affect the rapport and authority the researcher has.
- When combined with participant observation and other fieldwork methods, focus groups can be used to uncover issues that can be discussed later on in the research. The timing of the use of focus groups within the overall research process therefore needs to be given consideration.
- If a population is subdivided, try to ensure that all groups within the population are represented. Researchers need to be mindful of the social categories of field assistants.
- Focus groups take considerable time and energy to set up and organise. Planning the focus groups should not begin until after rapport and trust have been established.
- Focus groups can be an excellent forum for respondents to learn from researchers. They can provide an excellent opportunity for research to inform and empower local people. Extra time needs to be allowed by the researcher for this.

Summary

Long-term research, because of the enduring ties that develop, is more likely to be action or advocacy orientated (Foster *et al.* 1979). The over-whelming sense of responsibility that I developed from having long, enduring relationships meant that respondents' needs and desires became important to me. Over the period of my research it has changed from being research *on* the people (for my company's benefit) to being research *for* the people. I have not been able to ignore the villagers' desire to benefit from my research.

There are distinct advantages for the researcher returning to the same field site over a number of years: re-entry is easier, culture shock is min-imised and full engagement occurs only hours after arrival. Furthermore, moving to and from the study site over a period of years allowed for periods of reflection after periods of fieldwork. Although it is not normally possible in conventional field studies, I was able to return for social visits and collect missing information. Through email I have been able to main-tain contact and continue to clarify points relevant to my research.

As relationships developed, so did trust and respect. This was a two-way process between researcher and respondents, and resulted in a depth of data unlikely to arise from a once-in-the-field visit or even several visits over a few years. Rich insights and a depth of understanding have developed with increased trust. The longitudinal nature of the study revealed how some aspects of the villagers' feelings changed over time while others have remained fundamentally the same. This has enabled an in-depth understanding of how tourism affects the villagers' lives.

Over time, power relationships also change. By entering and leaving the field a number of times I have stepped on and off a superior–subordinate see-saw. Establishing a comfortable mutuality where key informants and researcher regard each other as different but equal has been a vital aspect of this research. It has allowed the villagers to share an open and honest evaluation of their experiences of tourism. However, long-term relation-ships can be limiting as well as helpful. In some respects, my freedom was restricted and aspects of my research directed. Balancing the roles of family membership and researcher was a constant challenge. A once-in-the-field anthropologist could have paid in lieu for domestic chores, an option not open to me.

Focus groups were used because they were socially appropriate, to compare views on tourism between different groups and to reciprocate the villagers' time given up for my research. However, they varied consider-ably from those used in social science research in Western developed countries. The use of focus groups in non-Western settings requires the reconsidering of some of the accepted assumptions in the sociological liter-ature such as the setting, the use of tables and chairs, the importance of the recording, and suggestion that participants write things down. Focus groups are an effective methodology to add to the multi-method mix in the anthropologist's tool kit. However, the quality of the data collected in the focus group, and indeed potentially the direction of inquiry, is likely to be affected by the rapport, authority and respect the researcher has with respondents. Group dynamics are also likely to be affected by a number of different factors such as age difference, gender and amount of 'stranger value'.

Focus groups can be useful to raise issues that can then be discussed in later phases of fieldwork. They can provide a forum for respondents to gain some insight into the researcher's perspective, offering the potential for both the researcher and the researched to learn from the experience. The groups allowed information from outside the community to be transferred into it, which, as Connell (1997) suggests, is necessary for meaningful participation in development. The groups assisted the community to articulate, and to seek solutions to their problems. They gave the villagers the opportunity to discuss with and learn from a practitioner. The knowledge exchanged may help to empower the villagers to manage tourism more effectively. The researcher needs to be mindful that this knowledge-sharing takes time, and extra time needs to be allowed in order for focus groups to be the locus for knowledge exchange.

Questions

1   What are the pros and cons of long-term research?
2   The chapter discusses how the researcher–respondent relationships developed over the years of the research. In what ways was this specific to the setting, and what generalisations could be made about the development of research relationships over a period of ten years?
3   Identify the differences between the use focus groups in Western and non-Western settings. Three factors are identified in the chapter that affected the dynamics of the different focus groups. What were they? And what other factors might be important in a different setting?

References

Abram, S. (1998) 'Introduction'. In S. Abram. and J. Waldren (eds) *Anthropological Perspectives on Local Development*. London: Routledge.

Agar, M. and Macdonald, J. (1995) Focus groups and ethnology, *Human Organization*, 54: 78–86.

Angrosino, M. (1986) 'Son and lover: the anthropologist as non-threatening male'. In T. Whitehead and M. Ellen (eds) *Self, Sex and Gender in Cross-cultural Fieldwork*. Chicago, IL: University of Illinois Press.

Bernard, R. (1988) *Research Methods in Cultural Anthropology*. London: Sage.

Berno, T. (1996) 'Cross cultural research methods: content or context? A Cook Islands example'. In R. Butler and T. Hinch (eds) *Tourism and Indigenous Peoples*. London: International Thompson Business Press.

Burgess, R. (1984) *In the Field: An Introduction to Field Research.* London: Routledge.

Burgess, R. (1990) *In the Field: An introduction to Field Research*, 2nd edn. London: Routledge.

Cole, S. (1997a) 'Anthropologists, local communities and sustainable tourism development'. In M. Stabler (ed.) *Tourism and Sustainability.* Oxford: CABI Publishing.

Cole, S. (1997b) 'Cultural heritage tourism: the villagers' perspective'. A case study from Ngada, Flores'. In W. Nuryanti, W (ed.) *Tourism and Heritage Management.* Yogyakarta: Gadjah Mada University Press.

Cole, S. (1999) 'Education for participation: the villagers' perspective. Case study from Ngada, Flores, Indonesia'. In K. Bras, H. Dahles, M. Gunawan and G. Richards (eds) *Entrepreneurship and Education in Tourism.* ATLAS Asia conference proceedings, Bandung, Indonesia.

Connell D. (1997) 'Participatory development: an approach sensitive to class and gender', *Development in Practice*, 7 (3): 249–259.

Draine, C. and Hall, B. (1996) *Culture Shock: Indonesia.* London: Kuperard.

Echols, J. and Shadily, H. (1989) *Kamus Indonesia Inggris*, 3rd edn. Jakarta: Gramedia.

Foster, G., Scudder, T., Colson, E. and Kemper, R. (1979) 'Conclusion: the long-term study in perspective'. In G. Foster, T. Scudder, E. Colson and R. Kemper (eds) *Long-Term Field Research in Social Anthropology.* London: Academic Press.

Hammersley, M. and Atkinson, P. (1995) *Ethnography Principles in Practice*, 2nd edn. London: Routledge.

Hampton, M. (1997) 'Unpacking the rucksack: a new analysis of backpacker tourism in South East Asia'. In W. Nuryanti (ed.) *Tourism and Heritage Management.* Yogyakarta: Gadjah Mada University Press.

Hitchcock, M. (1993) 'Tourism in South East Asia: introduction'. In M. Hitchcock, V. King and M. Parnwell (eds) *Tourism in South East Asia.* London: Routledge.

Hunter, C. and Green, H. (1995) *Tourism and the Environment: A Sustainable Relationship?* London: Routledge.

Khan, M. E. and Manderson, E. (1992) 'Focus groups and rapid rural assessment procedures', *Food and Nutrition Bulletin*, (United Nations University), 14: 119–127.

Kitzinger, J. (1994) 'The methodology of focus groups: the importance of interaction between research participants', *Sociology of Health and Illness*, 16 (1): 103–121.

Koning, K. and Martin, M. (1996) *Participatory Research in Health.* London: Zed Books.

Krueger, R. A. (1994) *Focus groups*, 2nd edn. London: Sage.

May, T. (1997) *Social Research Issues, Methods and Processes.* Buckingham: Open University Press.

Morgan, D. (1997) *Focus Groups as Qualitative Research.* London: Sage.

Nash, D. (1996) *Anthropology of Tourism.* London: Pergamon Press.

Pretty, J. (1995) 'The many interpretations of participation', *In Focus*, 16: 4–5.

Reinharz, S. (1992) *Feminist Methods in Social Research*, New York, NY: Oxford University Press.

Reisinger, Y. and Turner, L. (1997) 'Cross-cultural differences in tourism: Indonesian tourists in Australia', *Tourism Management*, 18 (3): 139–147.

Simmons, D. (1994) 'Community participation in tourism planning', *Tourism Management*, 15 (2): 98–108.

Stewart, D. and Shamdasan, P. (1990) *Focus Groups: Theory and Practice.* London: Sage.

Tourism Concern (1992) *Beyond the Green Horizon.* WWF UK.

Umbu Peku Djawang (1991) 'The role of tourism in NTT development'. In C. Barlow, A. Bellis and K. Andrews (eds) *Nusa Tenggara Timor: The Challenge of Development. Political and Social Change.* Monograph 12, ANU University, Canberra.

Wilson, D. (1993) 'Time and tides in the anthropology of tourism'. In M. Hitchcock, V. King and M. Parnwell (eds) *Tourism in South East Asia.* London: Routledge.

18 Translators, trust and truth

Cross-cultural issues in sustainable tourism research

Guy Jobbins

Aims of the chapter

- To highlight some issues relevant to working in cross-cultural environments.
- To outline methods of managing translators in interview situations.
- To demonstrate problems of evaluating rapport in interviews.
- To demonstrate the utility of reflexivity in evaluating the quality of responses.

Introduction

This chapter is based on my involvement in the MECO project, a multidisciplinary and multinational research programme which considered the integrated sustainable management of Mediterranean sandy beaches. My role on the project was to analyse social, economic and institutional aspects of human uses of beaches and nearby coastlands and waters in Morocco and Tunisia. Of particular interest to the project was the establishment of guidelines for tourism development and management at the study sites, with reference to existing resource uses and practises. This meant that I was looking at the dynamics between a range of sectors including tourism, conservation and agriculture, and between the resource use and management regimes governing each sector. The political sensitivities of this work, touching on governmental decision-making processes, public accountability and, sometimes, illegality, meant that accurate information was difficult to obtain. This chapter reflects on my experiences in establishing dialogue with a wide range of stakeholders, and the problems of using translators and negotiating meaning across barriers of language and culture.

For someone like myself, with a background in ecosystem sciences, conducting qualitative research is an unnerving experience. Quantitative data offer a feeling of certainty, rightly or not, that I do not find in qualitative data. When first learning about qualitative research for the MECO project, I was naively horrified by the potential problems in ascribing truth-value to other people's statements (Whyte 1982). My beliefs about the nature of knowledge could have been described as somewhat positivistic, and my experience of research to date had been based upon the methods of the natural sciences. I was even more alarmed by the recognition that the problems of social science research are compounded in cross-cultural situations where researchers can be denied access to accurate information by suspicious subjects or by the lack of common experience shared by the researcher and the researched. When working in a foreign country one is placed into a context in which one's usual framework for interpreting the world may no longer be appropriate.

In the field of international development, methods pioneered by Robert Chambers such as participatory rural appraisal (PRA) have encouraged researchers to integrate themselves into communities (Chambers 2002). PRA was developed as a deliberate reaction to 'flying visit' research in which specialists based studies on brief missions, an approach thought to result in biased conclusions owing to a lack of understanding of local knowledge and perspectives by outside 'experts'. Through integration with communities over long periods of time it is argued that researchers can better deal with problems such as trust, their understanding of local perspectives on issues, and negotiating local micro-politics and power-relations. Additionally, there is a focus on transferring research and building capacity in the local community. While the experience of PRA has been instructive, many projects still operate on short timescales and draw upon 'expert' outsiders. As an example of this kind of research, the MECO project was designed around short field trips by researchers from outside the local communities. There is still a need to develop approaches that can account for differences in meaning and problems caused by ignorance of culturally based signals in communication, while not requiring the intensive outlay of time that ethnographic research or PRA requires.

Sustainable tourism

The concept of sustainable tourism was core to the MECO Project. While this chapter is not intended to be a comprehensive review of the debates about what sustainable tourism is or should be (see Butler 1999 for such a review), it is useful to situate our approach within the literature in order to explain the context of some practical methodological issues we faced.

The different definitions and operationalisations of the sustainable tourism concept reveal its contested nature. This seems to be at least partially the result of the confluence of two different concepts of sustainability: the sustainability of destinations in economic terms, and the Brundtland Report's linking of environmental, societal and economic goods through the concept of sustainable development (World Commission on Environment and Development 1987; Bramwell and Lane 1993). Critics have argued that the resulting confusion over the nature of sustainability with respect to sustainable tourism has allowed different and fundamentally opposed definitions to claim legitimacy (Wheeler 1993; Wall 1996). Some claim that insufficient recognition of the value-dependent nature of defining sustainable tourism has created a misleading illusion of a single unified concept, feeding the abuse of and confusion over the term (Butler 1999).

Coccossis (1996) identifies four broad interpretations of sustainable tourism, which can be arranged into a typology that progresses from 'soft' or 'weak' to 'hard' or 'strong' sustainability. At the 'weak' end of the sustainability scale is an approach based on a sectorally orientated concern with the economic sustainability of the touristic product. Second in the typology are approaches that protect certain aspects of environmental quality as a factor of economic sustainability. Next are interpretations that focus on tourism activities that are ecologically sustainable, in the sense that they are complementary to, and minimally impacting upon, the natural environment. Finally, at the 'strongly' sustainable end of the scale are approaches that situate tourism within strategies for sustainable development, where sustainability is assessed in terms of the whole ecological and human system.

Most discourse on sustainable tourism is concentrated on single-sector approaches specifically addressing the sustainability of tourism (Butler 1996). However, the socio-economic component of the MECO Project took a more holistic concept of sustainability, questioning tourism's sustainability in the sense of interactions between tourism and other aspects and sectors of the wider coastal system, thus allying with the 'strong' extreme of Coccossis's interpretations (Caffyn and Jobbins 2003). Our analytical framework was informed by concepts of Integrated Coastal Management which advocate the integrated management of all resources, uses and systems within the coastal zone (e.g. Post and Lundin 1996; Sorensen 1997; Kay and Alder 1999). This meant that rather than just analysing potentials for the development of sustainable and appropriate tourism products and the mitigation of their negative impacts, we were doing so with respect to a bottom line of system-wide sustainability. Thus, in those case studies where tourism development already existed, we determined to investigate interactions between tourism and other resource-using activities such as fishing, conservation and farming in order to suggest means of reducing inter-sectoral conflicts and identifying inter-sectoral opportunities. Where tourism development was proposed, we attempted to analyse potential impacts on other resource-using activities and assess whether tourism development could contribute to or detract from local and regional sustainability.

Bramwell *et al.* (1996) identify seven dimensions – adding cultural, managerial, governmental and political sustainability to the Brundtland Report's 'Big Three' of environmental, economic and social sustainability – that might comprise a systematic assessment of sustainability. The inclusion of these additional dimensions resonates with concerns about power relations in tourism development, specifically debates about participation in planning and management (Jamal and Getz 1995; Marien and Pizam 1997; Richards and Hall 2000). Similar concerns exist in the Integrated Coastal Management literature, especially with respect to fishing and biodiversity conservation (e.g. Christie and White 1997; Christie *et al.* 2000). We decided to extend our analysis of systemic sustainability to socio-political institutions and power relationships governing the interactions between stakeholders from state organisations, non-governmental organisations, local resource-using communities and the tourism industry (Caffyn and Jobbins in press).

This decision had several consequences, the first of which was that we were faced with a range of exciting research questions. A second was that several of these research questions either directly or indirectly approached issues of political sensitivity that subjects were reluctant to discuss. The ways in which reflexivity helped us cope with this problem are dealt with later in the chapter. Before moving on to that, however, it is worth stating that our analytical framework had also gone beyond a concept of sustainability as accepted by the Tunisian and Moroccan organisations that we needed to develop working relationships with, and even some of the MECO Project partners questioned the value of our framework. For many people, sustainability and environmental management are issues that lie in the province of what Kooiman (1999, 2000) terms first-order governing: auditing and solving societal problems through the acquisition and analysis of data, and the generation and application of tools and strategies. However, we were proposing to include an analysis of second-order governing, which occurs at the level of the frameworks and institutions within which first-order governing takes place. Such analysis may appear irrelevant to those who believe the institutional framework and decision-making system to be value-free, scientific, rational and apolitical. However, my research demonstrated the second order of governing in these case studies to be highly relevant, as it accounted for political and scientific bias in decision-making. This could be interpreted as threatening by those whose power and status the decision-making framework supported.

I mention this here in order to contextualise the fieldwork referred to in this chapter. However, it also serves to illustrate the fundamental and obvious point that reflexivity in tourism research does not begin in field interview situations. Reflexivity is equally important when articulating conceptual approaches, and especially when doing so in conjunction with project partners trained in different disciplines and cultures. Although our partners may not have wholeheartedly embraced our framework for analysing sustainability, understanding why they did not was a significant advantage when sharing results and ideas.

Having outlined the conceptual and theoretical background of the research, in the rest of this chapter I examine how working reflexively helped me cope with three practical problems faced in interviews: working with translators, negotiating trust, and evaluating the truth value of responses.

Translators

Working with translators is a haphazard affair, and their potential influence on the research interview cannot be overstated. Over the course of the MECO project I worked with ten different translators. A limited budget meant that only one of them was a professional translator, while most of the rest were project personnel or students employed locally. Though their abilities in English and interpreting varied wildly, it was their other characteristics that more greatly affected the success of an interview.

Obviously, when one is conducting an interview through a translator, they are the medium of interaction and dialogue, relaying questions and answers between researcher and subject. The tone of voice, body language, choice of words and manner of the translator become the vehicle for dialogue. This issue is particularly problematic from the point of view of developing rapport. In one of the first interviews I conducted in Tunisia, I carefully modulated my voice and prepared my most trustworthy smile when asking a delicate question of one official, but the translator turned to him and relayed the question as an aggressive demand. Over the course of the interview, the subject became increasingly irritated with the translator's manner, and finally brought the interview to a close. I had lengthy conversations with the translator about the need for sensitivity, tact and moderation, but he was never able to convey any of these qualities in interviews, and shortly I had to find a replacement.

Each translator had different strengths. One could quickly form bonds with farmers, small businessmen and youth, but his demeanour was completely inappropriate for interviewing officials. Another was too staid and intellectual to bond well with youths and farmers, but was excellent when talking with officials, university professors, business leaders, and the like. At one site I was able to choose between these two on a daily basis according to my itinerary. Elsewhere I rarely had that choice, usually working with a single translator for several weeks.

Working in such close proximity with another can be extremely stressful, especially on a busy schedule and when plans go awry. Just as important as the relationship between the translator and subject is the relationship between translator and researcher. This adds another level of complexity to the research interview, as one of the researcher's main priorities becomes maintaining a positive working relationship with the translator. This dividing of attention between translator and subject can be disastrous, because the translator can then become the focus of the subject's attention, rather than the researcher. On some occasions I found it difficult to reassert my authority as lengthy dialogues developed between the subject and translator. The obvious problem in such situations was that I was unable to record everything that was being reported by the subject, as the translator was not pausing to translate. Often these dialogues would result from the translator attempting to understand a concept such that it could be translated, but they left me unsure whether the subject's original idea had not been lost in the discussion. All that could be done in such situations was to try to go over the conversation with the translator and impress upon them the importance of allowing the subject's own ideas to be recorded.

On the other hand, insufficient monitoring of the researcher–translator relationship can lead to miscommunication and irritability. Translation is an exacting task, made more difficult by inexperience – doubly so when the translator has an inexpert grasp of one of the languages in use. The essential problem of translation can be understood as one of transferring concepts and meaning from one language to another (Smith 1996). This is problematic, owing to the interconnection of culture, language and meaning, and requires the mapping of ideas in one culture and language on to the categories of meaning from another (Sturge 1997). One only has to think of occurrences of miscommunication between two native language speakers to appreciate how problems increase in translation. On long days, having conducted perhaps three or four interviews of between one and three hours each, translators would be exhausted. Their ability to interpret effectively would be much lower, and their frustration, and mine, would increase.

My response to this problem was to spend as much social time as possible with translators. The translators were the one group of people I spent enough time with to establish true rapport and understanding. Good personal relationships developed outside of interview contexts greatly enhanced the quality of our relationships in them, thus allowing me to devote more attention to subjects.

Translators also became my most trusted sources of data quality control. In familiar environments we generally have some context for judging the likelihood of a particular claim, based on previous experience and background knowledge. In new environments one can misinterpret or be misled by basic statements because of a lack of familiarity with the situation. When one is conducting research in foreign country and culture, this ignorance can threaten to overwhelm one, and translators' responses to subjects' claims can be a useful indicator of veracity. This is also problematic, because of course one must account for any agendas and prejudices held by the translator. Again, this is why it is important to have a deeper knowledge of the translator, a powerful reason for developing a social relationship with them.

The amount of power the translator holds in an interview is considerable, and it was important to make sure that they were 'on my side'. When working with host partners, such as government agencies, you do not necessarily have the control of an employer, and the translator may well have other loyalties. One translator I worked with, a government employee, apparently censored both questions and responses in discussion with local resource users. When I returned to the site with a different translator, further conversation with several subjects indicated some manipulation and misrepresentation during previous interviews. Several also told me they had not liked the first translator because of class differences; they felt the translator had been condescending and aloof, and they had not felt comfortable about talking honestly. This was a surprise to me, because that translator had apparently communicated well with these subjects. I later made an effort to 'turn' other translators I worked with, such that they identified with the aims of my research and wanted it to succeed rather than seeing themselves as chaperones who might attempt to interpose an 'official' perspective during the process of interviewing.

Translators were invaluable in explanation of background contexts and cultural concepts. It became apparent that even had my language skills been sufficient for conducting interviews, substantial meaning would have been lost owing to my cultural ignorance. I found this to be the case in Malta, where I didn't work with translators. Sometimes it was only later that subjects' allusions would be clarified by conversation with my Maltese friends, such as references to Malta's colonial history or political current affairs. I also found it strange not working with a translator, after my experiences in Morocco and Tunisia. In those countries I developed an interviewing technique that allowed me to monitor the interview at a 'second order', particularly during long interchanges between translator and subject. I had the time to monitor body language, consider previous responses, and select the direction of questioning more effectively. Without the translator I found myself without the leisure to monitor the interview from that level, being constantly involved in the first-order process of asking questions and listening to responses.

Trust

As an outsider researcher, one is often confronted with trust issues. In the developed world, trust is a research issue usually connected with the protection of subjects' identities and the accurate representation of their views (Fontana and Frey 2000). In the developing world, concerns about trust are connected more often with suspicion about researchers as outsiders, and the potential political consequences of giving information (Bulmer 1993).

In Morocco and Tunisia many of the subjects appeared concerned about these issues. However, there were notable differences in the willingness of different stakeholders to trust me. Non-governmental stakeholders were generally easier to reassure than officials. In fact, many citizens wanted their opinions to be recorded, in the belief that the research would lead to tangible benefits, despite my disclaimers. Even on occasions when I arrived in a government car and with a state employee to translate, my outsider status appeared the basis of trust rather than being held against me. One villager explained, 'They speak to you more than they would a government man – you are independent.'

I found that the best way of setting people at ease was to spend several days in the area before conducting any interviews. Walking around the village making observations, visiting cafés and making a few friends would establish my presence at least at a minimal level. Soon people would approach me and ask what I was doing, which was an excellent way of beginning an interview. Knowing a few words of the local language, especially greetings, and being prepared to spend substantial time talking about things other than the research were important in establishing rapport. (In Chapter 17, Cole also discusses the importance of building rapport with her research participants in two remote Indonesian villages.)

There were cases where private citizens had their own reasons for being initially unwilling to talk to me. Notably this was the case with people who felt they might be criticised, such as loan sharks or some hoteliers. Time and persistence were needed to convince them that I was unbiased and simply interested in hearing what they had to say. Sometimes they agreed to talk to me after hearing that their 'enemies' had already spoken to me.

Governmental officials were generally more wary, often appearing defensive. This unease has many potential sources. Some ministries have past experiences of being criticised by external researchers, making them disinclined to open themselves to further punishment. Others believe that their work is sensitive in nature, and the older generation especially may have been conditioned by the nationalist ideologies of early independence and have a general suspicion of foreigners. Moroccan and Tunisian civil servants are also bound by a code of professional ethics that emphasises secrecy. Finally, the management structures that many officials operate within may disincentivise personal initiative. In other words, the subject may be worried about saying the wrong thing and find it easier to say nothing.

Where possible, I tried to talk to officials outside of their offices, where I found them more prepared to talk freely. Suggesting that perhaps the subject was busy but that I could meet them at a café after work was sometimes a useful ploy. However, this approach was more often unsuccessful, and most officials refused to meet me outside of their workplace. During interviews I usually monitored body language to gauge how relaxed and trusting the subject was. Obviously one can be deceived by body language, but I found it a useful guide to aid my line of questioning. I noticed that the body language and demeanour of officials became more relaxed away from their offices, and in less formal interview situations subjects appeared willing to discuss more sensitive issues, even if off the record.

Several officials insisted on bringing in colleagues to the interview, ostensibly so that they could contribute information. However, it appeared that there was also an element of the subject trying to prove to his colleagues that he had nothing to hide and was not betraying his organisation in speaking with me. This resulted in far from idealised interview situations in which I had to build trust not only with the subject but also with others in the room, and acknowledge that the presence of others might inhibit communication with the subject.

I did not identify any one set of solutions for winning trust with all subjects, and do not think that it should be taken in black and white terms, that either the subjects did or did not trust me. Rather, it is better to recognise that the majority of subjects were constantly monitoring the content and context of the dialogue, and making trust decisions on that basis. Often a subject would make an oblique remark about something that he could not talk openly about. It was clear on these occasions that the subject was monitoring my reaction. When I became sensitised to such veiled comments and was able to acknowledge them discreetly, subjects would often feel confident in proceeding with a line of conversation that was supposedly 'hypothetical'. Similarly, during interviews I myself found it necessary to constantly monitor the subjects' body language, tone of voice and the presence of evasive language in order to evaluate the degree to which I trusted each response.

Truth

The difficulties of interviewing subjects having been acknowledged, the obvious question was the extent to which I had confidence in the information collected. Recognising the difficulties of establishing that confidence was a positive step, as it meant that little was taken for granted. During interviews, significant points would be reapproached from different directions several times in order to expose inconsistencies or reveal additional layers of meaning.

Sometimes this would result in conflicting responses, the source of which was not always clear. In Tunisia, trying to understand the basics of household economics, I had the following conversation with a woman who herded goats:

Guy: How many goats do you own? (Translator relays question and response of herder.)
Woman: Six.
G: And how many of those are female? (Translator relays question and response of herder.)
W: Eighteen.

My initial response to this was to assume innumeracy on the part of the woman, which is something I had identified with other subjects (although not to the degree that they were unable to count). However, she had been poking fun at me earlier and it was entirely possible that she was continuing to do so. We approached the issue again, even trying to trace the lineage of the goats, yet she insisted that eighteen of her six goats were female. Matters got worse when I counted her flock and counted only twelve goats, eight of which were male. My translator at that time was a very competent biologist but not a very experienced translator, and it us took some time to realise that part of the problem might be the use of gender-specific nouns in Arabic. However, that did not solve the whole problem. Through persistent questioning we eventually revealed a complex set of property rights over goats, in which some of the animals in her flock were not hers, some of 'her' goats were part-owned, and others did not exist at all but were owed to her by other people. Thus my initial assumption was totally wrong, and far from being ignorant or poking fun, the woman was actually giving me an important insight into the interhousehold economics of the village, which were sophisticated enough to support the existence of virtual goats.

Following that experience, each time a subject offered a seemingly nonsensical response I would ask myself, 'Is this a Virtual Goat?' Sometimes the confusion might stem from poor translation, or poor phrasing of the question. More often the response would indicate different meaning attached to concepts and words. It is not surprising that as an outsider one can initially misinterpret information and leap to assumptions. The important step was to test those assumptions immediately by asking further questions. Structured approaches to interviewing would have often overlooked such insights, and flexibility in follow-up questioning was vital in negotiating meaning.

The problem with this approach was that while obvious instances of responses jarring with my expectations would be noticed, there might be cases where apparently sensible responses obscured radically different meanings. I found it necessary to consider everything as a potential 'Virtual Goat'. My response to this in interviews was to introduce each topic in general terms, and to move to specific questions after an initial discussion. The focus of such opening discussions was basic concepts, which I found to be the most frequent sources of misunderstanding. The terms *ecotourism*, *decentralisation*, *grass-roots*, *bottom-up decision-making* and *participation*, among others, were all revealed to have a meaning attached to them that differed from what one might expect.

In order to feel confident about the information I was collecting, a minimum of three questions were asked about important points where I felt that misrepresentation by the subject or confusion might be occurring. Coupled with a move from general to specific questions, this enabled triangulation of responses to reveal inconsistencies and differences in understanding. I would not claim that it revealed or resolved all such problems, and sometimes I would be left only with my notes on the interview transcript urging caution in the use of particular pieces of information. However, I would claim that the approach significantly improved my detection rate of translator error, misleading responses by subjects, differences in meaning, and erroneous assumptions on my part.

The same principle needed to be applied in trying to establish many basic facts where stakeholders offered different accounts of historical events or causes for them. Triangulating the accounts of different subjects would sometimes reveal a consensus about times, places and people, but more often it would not. For example, in the villages dates would often be vague, and something might have happened 'between ten and fifteen years ago'. Documented accounts from official sources would often indicate these dates to be wrong by many years. Triangulation of such different reports needed to be embedded within a stakeholder analysis accounting for the different agendas and political positions of the informers and their peer networks. Taken together, a subject's responses not only revealed what they knew, or wished to represent, about a particular issue, but, when deconstructed, revealed additional information to contextualise and aid understanding of their positions with respect to one another. It quickly became clear that no single response could be accepted as 'true'. Instead, subjects would construct representations of the world that conformed to their needs, goals and beliefs. Significant differences most often occurred along divisions of occupation, social standing, age and political allegiance. As an outsider trying to construct my own understanding, I had to accept all responses as tentative hypotheses yet remain sceptical about each one, and to keep asking questions of myself and others.

Conclusions

I hope that this short chapter has conveyed some of the problems encountered in conducting cross-cultural research, particularly the difficulty in obtaining access to information, and uncertainty in evaluating it. The approach I developed explicitly acknowledged this uncertainty and attempted to reduce it. It is clear that this approach requires a great deal of time, both for translation and for the negotiation of meaning within interviews. This necessitates a fairly limited number of topics to be discussed if any substantial progress is to be made in an interview lasting just one hour. Most of the interviews I conducted lasted between two and three hours, some longer. However, this is a balance between the much longer timescales needed for intensive research studies such as PRA and less reflexive approaches that, as I hope is shown here, would leave substantial meaning uncovered.

The key point that I learned was the necessity of constantly monitoring the interview from a second order, especially the demeanour of interviewees and my own understanding of what was being communicated. This entailed asking not only 'what does he mean by that?' but also 'what do I think he means by that, and why is that a valid assumption?' and 'why would the translator put it that way?'. The recognition that there was no basis for assuming anything about respondents' information was always at the forefront of my mind. Subsequently I have usefully applied this recognition when working in English and in cultures more similar to my own.

Summary

- When one is working in cross-cultural and interdisciplinary research environments, reflexivity is crucial to negotiating meaning.
- When one is working with translators it is important to monitor and account for their impact on the research interview, rather than assume they are neutral, undynamic conveyors of information between researcher and subject.
- Subjects may have valid reasons not to trust the outsider researcher, and a flexible, reflexive approach to dealings with them can at least help with assessing their level of trust.
- When working in unfamiliar environments one can be tempted to latch on to what appears familiar, yet reflection may lead one to find what at first appeared familiar even stranger than things that initially appeared different.

Questions

1 What basis does any subject have for trusting an interviewer? Would the level of trust remain constant no matter what questions were asked?
2 How can one distinguish between two empirically false responses, one of which the subject knows to be false and one he believes to be true?
3 In an interaction between a translator and a subject, how can one distinguish between body language that is culturally appropriate and accepted, and body language that is hindering the interview's development?

References

Bramwell, B. and Lane, B. (1993) 'Sustainable tourism: an evolving global approach', *Journal of Sustainable Tourism*, 1 (1): 1–5.
Bramwell, B., Henry, I., Jackson, G. and van der Straaten, J. (1996) 'A framework for understanding sustainable tourism management'. In B. Bramwell, I. Henry, G. Jackson, A. Prat, G. Richards and J. van der Straaten (eds) *Sustainable Tourism Management: Principles and Practices*. Tilburg: Tilburg University Press.
Bulmer, M. (1993) 'Interviewing and field organisation'. In M. Bulmer and D. Warwick (eds) *Social Research in Developing Countries*. Chichester: Wiley.
Butler, R. (1999) 'Sustainable tourism: a state of the art review', *Tourism Geographies*, 1 (1): 7–25.

Caffyn, A. and Jobbins, G. (in press) 'Governance capacity and stakeholder inter-action in the development and management of coastal tourism: examples from Morocco and Tunisia', *Journal of Sustainable Tourism*, 11 (2): 224–245.

Chambers, R. (2002) *Participatory Workshops*. London: Earthscan.

Christie, P. and White, A. T. (1997) 'Trends in development of coastal area man-agement in tropical countries: from central to community orientation', *Coastal Management*, 25: 155–181.

Christie, P., Bradford, D., Garth, R., Gonzalez, B., Hostetler, M., Morales, O., Rigby, R., Simmons, B., Tinkam, E., Vega, G., Vernooy, R. and White, N. (2000) *Taking Care of What We Have: Participatory Natural Resource Manage-ment on the Caribbean Coast of Nicaragua*. Managua: University of Central America; Ottawa: International Development Research Center.

Coccossis, H. (1996) 'Tourism and sustainability: perspectives and implications'. In G. H. Priestly, J. Edwards and H. Coccossis (eds) *Sustainable Tourism? Euro-pean Experiences*. Wallingford: CAB International.

Fontana, A. and Frey, J. H. (1998) *Interviewing: The Art of Science*. In N. K. Denzin and Y. S. Lincoln (eds) *Collecting and Interpreting Qualitative Materials*. Thousand Oaks, CA: Sage.

Jamal, T. B. and Getz, D. (1995) 'Collaboration theory and community tourism planning', *Annals of Tourism Research*, 22 (1): 186–204.

Kay, R. and Alder, J. (1999) *Coastal Management and Planning*. London: E. and F. N. Spon.

Kooiman, J. (1999) 'Experiences and opportunities: a governance analysis of Europe's fisheries''. In J. Kooiman, M. Van Vliet and S. Jentoft (eds) *Creative Governance: Opportunities for Fisheries in Europe*. Aldershot: Ashgate.

Kooiman, J. (2000) 'Societal governance: levels, modes and orders of social–political interaction'. In J. Pierre (ed.) *The Governance Debate: Authority, Steer-ing and Democracy*. Oxford: Oxford University Press.

Marien, C. and Pizam, A. (1997) 'Implementing sustainable tourism development through citizen participation in the planning process'. In S. Wahab and J. Pigram (eds) *Tourism Development and Growth, the Dhallenge of Sustainability*. London: Routledge.

Post, J. C. and Lundin, C. G. (1996) *Guidelines for Integrated Coastal Zone Man-agement*. Environmentally Sustainable Studies and Monograph Series 9. Wash-ington, DC: World Bank.

Richards, G. and Hall, D. (eds) (2000) *Tourism and Sustainable Community Devel-opment*. London: Routledge.

Smith, F. M. (1996) 'Problematizing language: limitations and possibilities in "foreign language" research', *Area*, 28 (2): 160–166.

Sorensen, J. (1997) 'National and international efforts at integrated coastal management: definitions, achievements and lessons', *Coastal Management*, 25: 3–41.

Sturge, K. (1997) 'Translational strategies in ethnography', *Translator*, 3 (1): 21–38.

Wall, G. (1996) 'Is ecotourism sustainable?', *Environmental Management*, 2 (3–4): 207–216.

Wheeler, B. (1993) 'Sustaining the ego', *Journal of Sustainable Tourism*, 1 (2): 121–129.

Whyte, W. (1982) 'Interviewing in field research'. In R. Burgess (ed.) *Field Research: A Sourcebook and Field Manual*. London: George Allen and Unwin.

World Commission on Environment and Development (1987) *Our Common Future* (the Brundtland Report). Oxford: Oxford University Press.

Index

Page numbers in *italics* denote figures or tables